BIO-NANOTECHNOLOGY
Concepts and Applications

BIO-NANOTECHNOLOGY
Concepts and Applications

Madhuri Sharon

✳

Maheshwar Sharon

✳

Sunil Pandey

✳

Goldie Oza

N.S. Nair Research Centre for
Nanotechnology & Bio-Nanotechnology
Ambernath (W), Maharashtra, India

Taylor & Francis
Taylor & Francis Group
Boca Raton   London   New York

CRC is an imprint of the Taylor & Francis Group,
an informa business

Ane Books Pvt. Ltd.

Bio-Nanotechnology
Madhuri Sharon, Maheshwar Sharon, Sunil Pandey and Goldie Oza

© Authors

First Published in 2012

Ane Books Pvt. Ltd.

4821 Parwana Bhawan, 1st Floor
24 Ansari Road, Darya Ganj, New Delhi -110 002, India
Tel: +91 (011) 2327 6843-44
Fax: +91 (011) 2327 6863
e-mail: kapoor@anebooks.com, anebooks@vsnl.net
Website: www.anebooks.com

For

CRC Press
Taylor & Francis Group
6000 Broken Sound Parkway, NW, Suite 300
Boca Raton, FL 33487 U.S.A.
Tel : 561 998 2541
Fax : 561 997 7249 or 561 998 2559
Web : www.taylorandfrancis.com

For distribution in rest of the world other than the Indian sub-continent

ISBN : 978 1 43985 2149

British Library Cataloguing in Publication Data
A catalogue record for this book is available from the British Library

Printed at: Gopson Papers Ltd., Noida

Foreword

Ever since the well-known talk entitled *"There's Plenty of Room at the Bottom"* by Richard Feynman in the American Physical Society Meeting at Caltech in 1959, nanotechnology and nanoscience have indeed realized then-impossible preparation of nanometer-scale materials and high-resolution imaging of such nanomaterials as well as high-precision nano-manipulation of nanomaterials composed of atoms and molecules. However, I often feel that it is extremely difficult to obtain general knowledge  on the forefront of rapidly growing areas of nanotechnology even though lots of books on nanotech have already been published during the past decade.

Under these circumstances, it is particularly nice to find that Professors Maheshwar Sharon and Madhuri Sharon have brought a nicely-written monograph on Bio-Nanotechnology to us, where many fascinating research topics and fundamentals in a cross-disciplinary area of nanotechnology and bioscience are aptly presented and discussed. I think that the authors are quite successful in 'fusing' otherwise diverging research topics of the rapidly emerging area. The present monograph is a good introduction to undergraduate/graduate students and researchers in such interdisciplinary research fields as nanotechnology, materials science and bio-nanotechnology.

I really hope that most of the readers of the current monograph may in future contribute to the further development of the fascinating world of bio-nanotechnology.

Hisanori Shinohara
Professor of Chemistry
University of Nagoya, Japan

Preface

"Smaller than the smallest and the most vast"—this concept of minuteness and vastness is all-pervading. Biological cells are very small but the nanomachineries like DNA, RNA, Proteins, Carbohydrates and Lipids, of which cells are constituted, are the smallest ones and are vastly pervading in all Biosystems existing in this world. This book focuses on all fundamental Bio-nanomachines including their applications towards the service to humankind. This book is a distillation of our thoughts and experiences gathered while doing our ground-breaking research in the field. This wondrous and fascinating gift is a humble attempt towards sharpening the knowledge bank of under-graduates, post-graduates and all researchers who are pertaining to pursue their research in the field of Nanobiotechnology. It is applicable for all individuals who wish to embark upon their career in this area. Improvements may include circuit lines less than 100 nanometers in distance and nanoelectronic devices replacing existing electronic devices. This book presents well documented and researched, information about how nanoscale technologies may enhance our lives and our health and significantly change the world we live in.

The revolution started when Richard Feynman mentioned in his speech that a day will come when all the books of library will find its space on the tip of a needle. This was a dream which has started to turn into a reality. Each day is poured with research papers in the area of Nanotechnology and Nanobiotechnology. This field has touched the zenith of medicine, engineering, optoelectronics, microbiology, environment and a never-ending list. In today's world if you are oblivious of this field, you are like a crippled person in the competitive scientific environment.

Organization and Coverage

Chapter 1 is aimed at presenting what is Nanotechnology and explains that nanotechnologies fall between the usual daily macrophysics and the quantum mechanics and that is why it is so mysterious. Moreover, superior electrical, chemical, mechanical or optical properties of nanostructures have also been touched upon.

Chapter 2 encompasses the domain of Bio-nanotechnology. Biological system realized the importance of efficiently functioning at nanoscale much before the present day nanotechnologists did. Various biological nanomachines and custom-made molecules by them are described in this chapter. For example, ATP synthase functioning as nanoturbine, flagellar motion in bacteria.

Chapter 3 Nanotech's vision is to assemble complex machines and circuits atom by atom; by bottom up or top down approach. This chapter covers most of the likely method of choice for building nanostructures.

Chapter 4 covers fundamentals and applications of noble metal nanoparticles and includes all the biosystems involved in fabrication of Nanoparticles like fungi, actinomycetes, plants, algae and bacteria, which are the versatile and dynamic Nanoengineers of nature. An account of *modus operandi* of their synthesis is dealt with utmost sharpness.

Chapter 5 is an up-to-date review of how DNA, RNA and proteins are so stable in a milieu fully crowded with other smaller organic and inorganic molecules, how molecular crowding endorses self-assembly. This chapter gives a bird's eye-view of electron transfer and transport in DNA and proteins and their applications in the field of sensors, electronics etc.

Chapters 6, 7 and 8 offer an in-depth look at the unique properties and potential applications of carbon nanomaterials (CNM). Beginning with a description of various CNM types, *Carbon Nanoforms and Applications* addresses the need to develop a new classification of carbon. After discussing the fundamental physics, it covers techniques for CNM synthesis and characterization. This authoritative resource then provides comprehensive information on the physicochemical and biosystems applications of CNMs.

Chapter 9 covers fundamental philosophy in manoeuvering artificial intelligent machines capable of following electronic instruction feed by us. Birth of nanorobots using natural spare parts and their organization into a functional machine is main attribute of the chapter. Focus is given to various attempts and methodologies of fueling nanomachines for navigation.

Chapter 10 and 11 deal with the marriage of nanotechnology and medicine that has produced fruitful offspring. This chapter is an all-embracing introduction to use of nanoscale architecture to fabricate nanocargoes carrying drugs to the target. Use of magnetic hyperthermia induced by iron oxide nanoparticles is main attraction of the chapter, particularly to combat the dreadful disease like cancer. Confrontation between the anomalies with nano-surgeons can cure anatomical complications such as removal of the thrombus causing cardiac arrest, is also one of the attributes of the chapter. Don't waste much time in only reading the highlights, rather enter into the depth of the chapters and enjoy!

Chapter 12 covers role of nanotechnology in environmental remediation, maintenance as well as enhancement of environmental quality. Special emphasis is given to revolutionary ideas like photocatalysis exerted by carbon nanotube and other metal nanoparticles like TiO_2 in presence of varying wavelength of light. The concern of the chapter is also to strengthen green technologies in environmental protection.

Chapter 13 talks about the initiatives in utilizing nanotechnology in food, agriculture and cosmetics. Use of metal nanoparticles like silver in cosmetics and in antimicrobial sprays highlight the new era of nanoparticles based chemicals used in our daily life.

Chapter 14 is the concluding chapter on synthetic nanoimplants being developed for its specific applications in biosystem and for other uses.

Authors

Detailed Contents

Nanoparticles: Physicochemical Methods; Electrodeposition; By Insulating Holes; Nanolithographic Self-assembly: *Soft Lithography*; *Photolithography or Optical Lithography*; *Magnetolithography (ML)*; *Nanoimprint Lithography (NIL)*; *Dip Pen Nanolithography or Atomic Force Microscopic Lithography*; *Enzymatic Nanolithography*; **Summary**

1

Basics of Nanotechnology

> I want to build a billion tiny factories, models of each other, which are manufacturing simultaneously. ... The principles of physics, as far as I can see, do not speak against the possibility of maneuvering things atom by atom. It is not an attempt to violate any law; it is something, in principle, that can be done; but in practice, it has not been done because we are too big.
>
> —**Richard Feynman (Nobel Prize winner in physics)**

1.1 An Introduction to Nano World

As per the *Vedas*, the origin of the world was from Nothingness. The *Yajur Veda* believes that before the creation of Universe there was only *Yatah* (meaning speed) and *Juh* (*i.e.,* space), hence the name *Yajur;* It was the collision of space and speed with a big *Naad* (according to the *Vedas* having a sound *Oum* or may be comparable to *Big-bang*) that created *Ghan-Vidyut* (+ ve charge) and *Rin-Vidyut* (– ve charge); which eventually ended in creation of matter and finally *Universe.* Hence, a concept of atom-wise fabrication of Universe was envisioned and visualized by ancient knowledge seekers. The importance of very small and yet very powerful with novel properties has been known to mankind from time immemorial. Today we are talking, reading and researching about various aspects of the very small (not visible to naked eye) matter under the concept of *Nanotechnology*. In future, we may be drifting to another technology of smaller material like Giga technology and so on…but the nature will still remain evasive with atomic and sub-atomic existences and functioning. In our quest to decipher, the natural operations at atomic and sub-atomic levels, which gave birth to Quantum Sciences and where boundaries of Physics, Chemistry and Biology do not exist we may be entering into a new vista of science and nature.

Anybody who has interest in Nanoscience and Nanotechnology is not unaware of the name of a great physicist and Nobel laureate *Richard Feynman;* who first introduced concept of Nanotechnology in 1959, in a talk entitled *"There's plenty of Room at the Bottom"*. Though the depth of his vision got the acceptance two decades later in 1980, when K. Eric Drexler accentuated the potential of molecular nanotechnology–MNT in his book "Engines of Creation". Today every Nanotechnologist has realized Feynman's dream and Nanotechnology has become a revolutionary field which has made a huge impact on several areas of material sciences and has applications in every field. Immense research is being carried out and progress has been made in the field of Nanoscience and Nanotechnology; and the revolution still continues. Who can understand it better than Biotechnologists, who unknowingly were studying and unfolding the Nanoscience and Nanotechnological functions being followed by living system for a long time?

Today we all accept nanotechnology as *"The design, characterization, production, and application of structures, devices, and systems by controlled manipulation of size and shape at the nanometer scale (atomic, molecular, and macromolecular scale) that produces structures, devices, and systems with at least one novel/superior characteristic or property."*

This definition includes a very important aspect of Nanotechnology, *i.e.*, nanosized particles have novel characteristics or property. We will discuss it later in this chapter.

With inputs from all the branches of sciences and technologies, Nanotechnology has become a highly diverse and multidisciplinary field, ranging from novel extension of conventional device physics, to completely new approaches based upon molecular self-assembly, to developing new materials with nanoscale dimensions, even to speculate on whether we can directly control matter on the atomic scale. Nanotechnology is being seen as a science having potential to create many new materials and devices with wide ranging applications, such as in medicine, electronics and energy production. On the other hand, nanotechnology raises many of the same issues as with any introduction of new technology that concerns about the toxicity, environmental impact of nanomaterials and their potential effects on global economics.

Let us first have a look at the basics of Nanoscience. As mentioned above, the idea of nanotechnology was first introduced by Richard Feynman in 1959. Whereas, the popular term *"Nanotechnology"* was coined by Norio Taniguchi, a Japanese Professor of Tokyo Science University, in 1974, who tried to engineer materials at nanometer level. However, the potential of Feynman's idea was popularized in mid-80s by K. Eric Drexler.

1.1.1 Nano

"Taniguchi" must have taken word Nano from a *Greek* word which means extremely small. Nanoparticles are neither visible to the naked eye (which cannot see things smaller than 0.2 mm) nor by light microscopes under which things smaller than 0.2 μm cannot be seen. The delay in realizing the potential of nanoparticles could be partly attributed to absence of machine like scanning tunneling microscope; to see them. Moreover, further advances in the tools that now allow atoms and molecules to be examined and probed with great precision have enabled the expansion and development of Nanoscience and Nanotechnologies.

1.1.2 The Unit Nanometer

Unit nanometer is 10^{-9} m. A nanometer (nm) is one thousand millionth of a meter. The acceptance of a nanoscale size for Nanotechnological purposes is accepted to be from 100 nm down to the size of atoms, because it is at this scale that the properties of materials can be very different from those at a larger scale. For comparison, typical carbon-carbon bond lengths or the spacing between carbon atoms in a molecule are in the range 0.12–0.15 nm, and a DNA double helix has a diameter around 2 nm. It must be mentioned here that Chemists have been making polymers, which are large molecules made up of nanoscale subunits, for many decades and technologists have been used to create the tiny features on computer chips for the past 20 years. However, advances in the tools that now allow atoms and molecules to be examined and probed with great precision have enabled the expansion and development of nanoscience and nanotechnologies.

1.1.3 Nanoscience

Nanoscience is the study of phenomena and manipulation of materials at atomic, molecular and macromolecular scales, their physical and chemical properties exhibited at atomic and molecular levels of dimensions ranging from a few nanometers to less than 100 nm; because at this scale properties differ significantly from those at a larger or bulk scale. This is the scale of large molecules; molecular chains (like plastics), proteins (from biology), nanocrystals (Nanocrystalline diamond) and new large molecules like fullerenes and nanotubes. Nanoscience is essentially the Physics and Chemistry of objects at the nanoscale.

1.1.4 Nanotechnology

Nanotechnology involves synthesis design, characterization, production and application of structures, devices and systems by controlling shape and size at the nanometer scale as well as application of nanoparticles. Since many physical, mechanical, electrical, thermal, catalytic and optical properties and phenomenon becomes altered as the size of the material decreases to nanolevel as compared to macroscopic or bulk sizes; Nanotechnologists utilize these properties for application in various fields.

Nanoparticles are fundamental units of Nanotechnology, and are used as raw materials for constructing Nanostructures, Nanomaterials, Nanomachines and Nanodevices. Nanoparticles may be defined as *"A particle having one or more dimensions of 100 nm or less"*. Novel properties that differentiate Nanoparticles from the bulk material typically develop at a critical length scale of "under 100 nm".

As a consequence to the understanding of novel properties of matter at nanoscale; efforts of scientists have created structures like quantum dots, nanoclusters, nanocrystals, nanowires, dendrimers, nanotubes and fullerenes for various uses. And no wonder that *"The next big thing is really small"* became an accepted quote. So much so that nanotechnology is now being referred to as a *general-purpose technology*.

1.2 Types of Nanomaterials

There are several man-made nanoparticles trying to compete with natural nanoparticles. Some of the synthesized nanoparticles are:

1.2.1 Quantum Dots

Quantum dots were discovered in 1980s by Louis Brus of Bell Labs. The term quantum dots was coined by Mark Reed. They are few nm in diameter, roughly spherical (some quantum dots have rod like structures), fluorescent, crystalline particles of semiconductors whose excitons are confined in all the three spatial dimensions. Apart from quantum dots, there are quantum wires and quantum wells also. Quantum wells confine electrons or holes in two dimensions allowing free propagation in the third. Whereas, quantum wires confine electrons or holes in one dimension and allow free propagation in two dimensions.

Their potential application in diverse fields can be attributed to the property of quantum confinement. The wavelength of the emitted light can be tuned by using particles of different diameters. The diameter of the semiconductor Nanocrystals is smaller than the equilibrium separation distance of the electron-hole pair (*excitons*) formed in the bulk material upon promotion of an electron from the valence band to the conduction band. The smaller the diameter of the Nanocrystals, the more confined the excitons are, resulting in a larger gap between the valence and conduction band. By changing the composition of the Nanocrystals and their diameter, the wavelength of emitted light can be varied. Quantum dots are making a huge impact in biology and light emitting diode technology due to the ability to tune its size *e.g.,* larger quantum dots have a greater spectrum shift towards red as compared to smaller dots.

For producing quantum dots, excitons have to be confined in semiconductors. This is done by different methods *viz.* by advanced epitaxial techniques (to grow nano dots), by chemical methods (to produce colloidal quantum dots) or by ion implantation (to produce Nanocrystals) and by lithographic technique (to make nanodevices). Moreover, Lee, *et al.*, (2002) have reported synthesis of quantum dots using genetically engineered M13 Bacteriophage virus.

1.2.2 Nanocrystals

The definition of Nanocrystal was given by Fahlman BD (2007) as a single crystalline material having at least one dimension of ≤ 100 nm. Whereas, any nanomaterial having less than 1000 nm or 1 μm diameter is referred to as Nanoparticles and not as Nanocrystals. When Nanocrystals are less than 10 nm they are referred to as quantum dots and more than Nanoparticles due to their novel or due to unusual crystal shapes and lattice order, exhibit higher chemical reactivity. Fluorescent Nanocrystals can be used as marker in biology.

Nanocrystals are helpful in studying the behavior of bulk sample of the similar material as they have no grain boundaries and other defects. Zeolite Nanocrystals are being used as a filter to turn crude oil into diesel in Exxon Mobil Oil refinery in Louisiana , USA. It is a more cost effective method than conventional one.

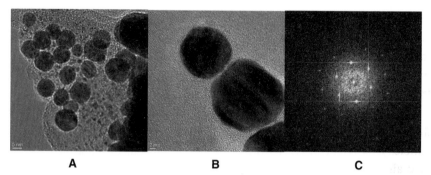

A   **B**   **C**

Fig. 1.1 (A) TEM images of spherical shaped gold nanocrystals (B) HRTEM of
two gold nanocrystals and (C) its SAED pattern.

1.2.3 Nanoparticles

This term is given to rather big particles *i.e.,* between 100–1000 nm sizes. Hence, it may or may not exhibit nanosize related properties that differ from the bulk material of the same sample. There is an accepted term **Nanocluster** also which refers to nanomaterials that have at least one dimension between 1–10 nm and has narrow size distribution.

1.2.4 Metallic Nanoparticles

Metallic nanoparticles possess a distinctive optical property—Surface Plasmon Resonance. Surface plasmon resonance is due to the collective oscillation of free electrons. Gold and silver nanoparticles are the most extensively studied metal nanocrystals and surface plasmon resonance for these occur in the vicinity of 400 and 500nm, respectively and is a function of nanocrystal size and composition, the molecules attached to the surface of the metallic nanoparticles and the surrounding medium. Michael Faraday in the mid 1800s for the first time prepared colloidal gold nanoparticles. He showed that by mere variation in the size of gold particles resulted in variety of different colours.

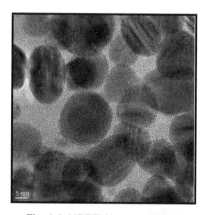

Fig. 1.2 HRTEM image of Silver nanoparticle.

1.2.5 Dendrimers

Dendrimers are spherical polymeric molecules where series of chemical shells are built on a small core molecule (each shell is called a generation). It is usually made from a core and alternating layers of 2 monomers: acrylic acid and di-amine. Its molecular structure has the form of a tree with many branches. Dendrimers are being considered as nanodevices for delivery of therapeutics. Dendritic nanoscale chelating agents are being synthesized for molecular modeling and environmental applications.

Dendrimers provide the necessary interface between chemistry and biology, possessing the unique traits to act as safe and effective drug-delivery vehicles as well as highly sophisticated diagnostic imaging agents. Dendrimers have already been commercialized in products designed for HIV prevention, anthrax detection, cardiac-marker diagnostics and gene transfection.

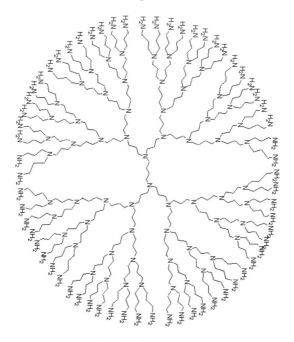

Fig. 1.3 A dendrimer.

1.2.6 Buckyballs

Buckyballs are spherical fullerenes (C_{60} is most stable and symmetrical and resembles a soccer ball). It has been named after architect R. Buckminster Fuller. In 1996, Nobel Prize in Chemistry awarded to Sir Harold William Kroto and Prof. Smally for their discovery. Endohedral fullerenes exhibit novel optical properties.

Fig. 1.4 A Buckyball composed of 60 carbon atoms.

1.2.7 Nanotubes

Carbon nanotubes (CNT) are elongated fullerenes that resemble graphene sheets wrapped into cylinders. Length to width ratio of CNT is very high (few nm in diameter and up to 1 mm in length). They are tubular structures having very small lumen and are composed of walls of single layer of atoms. Diameter of the Nanotubes decides its properties *e.g.,* superconducting properties. It has very high tensile strength; they are physically stable and chemically reactive with free radicals. Hence, their derivatives can be formed which can be more hydrophilic than fullerenes. Even new organic molecules can be generated using them. There have been successful attempts to place other atoms inside them *e.g.,* by doping with alkali metals.

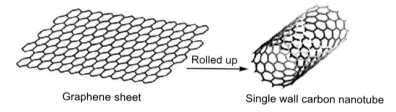

Graphene sheet　　　　Rolled up　　　　Single wall carbon nanotube

Fig. 1.5 Schematic diagram of a graphene sheet rolled to form carbon nanotube.

Table 1.1 Nanostructures, nanosize and example materials

Nanostructures	Nanosize	Example Materials
Quantum dots	1–10 nm Radius	Metals and Magnetic materials
Nanoclusters, Nanocrystals	1–10 nm Radius	Insulators, Semiconductors,
Nanowires	1–100 nm Diameter	Oxides, Sulphides, Nitrides, Metals, Semiconductors
Nanotubes	1–100 nm Diameter	Carbon including fullerene layered chalcogenites

1.3 Properties of Nanomaterials

The question arises why properties of nanomaterials are different from the bulk materials?

In one sentence it can be answered, because physical and chemical properties are size dependent.

Regardless of the composition, every substance when reduced to size of less than 100 nm exhibits new properties. The properties of bulk materials change at the nanolevel due to quantum effect. The optical, electrical, mechanical, magnetic and chemical properties can be systematically manipulated by adjusting the size, composition and shape of the nanoscale materials. Hence, the size of the particles or the scale of its features is the most important quality of nanoparticles.

"Nanomaterials (*e.g.*, CNT or nano metals) and composites of bulk and nanomateiral conjugates displays altered Physico-chemical properties. Carbon nano tubes are known to be 100 times stronger than bulk carbon and even steel. Bulk property of a material is manifestation of average of all the quantum forces. Whereas when the size is reduced to very small, at a particular point this averaging does not work".

Let's look at why properties of Nanomaterials are different from the bulk material.

1.3.1 Physical and Chemical Properties are Size Dependent at the Nanoscale

1.3.1.1 *Surface to Volume Ratio (SVR)*

Most of chemical reactions take place on the surface. Therefore, available surface for reaction is very important because it makes material chemically more reactive and affects their strength or electrical properties. Even some inert material at bulk level becomes reactive at nanolevel. Nanomaterials have a relatively larger surface area as compared to the same mass of material produced in a larger form. Thus, the concept of SVR is enhanced when the size is very small.

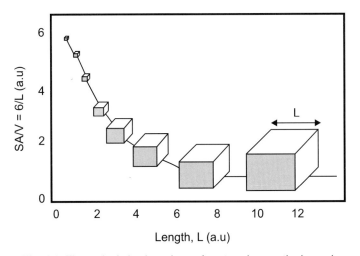

Fig. 1.6 Figure depicting how the surface to volume ratio depends upon the size.

1.3.1.2 *Smaller Particles have Higher Absorption Rate*

Making use of the fact that smaller particles have higher SVR, hence higher number of sites for hydrogen bonding; Carbon nanoparticles are being tried as a storage system for hydrogen. Hydrogen being a clean fuel, it is being considered to be used in future vehicles. A light and strong storage system like carbon nanomaterials (CNT and CNF) is a future prospect.

1.3.1.3 *Catalytic Properties*

The huge surface area to volume ratio of nanoparticles increase its chemical activities and therefore allow nanomaterials to become efficient catalyst. Bulk gold is catalytically inert. But when gold nanoparticles are bonded to an aluminum oxide support they become good catalyst and are used for oxidation of carbon monoxide even at a low temperature of −70°C.

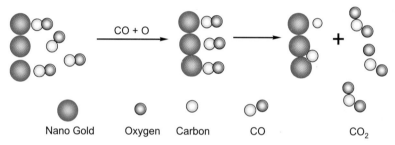

| | Nano Gold | Oxygen | Carbon | CO | CO_2 |

Fig. 1.7 Schematic diagram showing nanosized gold particles
that act as a catalyst and transform carbon monoxide
(CO) to carbon dioxide (CO_2).

1.3.1.4 *Photocatalytic Properties*

Nanoparticles of specific sizes or internal diameter (of CNT which acts as semiconductor) have been found to have higher photocatalytic rate. In photocatalytic process, when carbon nanoparticle of a semiconducting nature comes in contact with another material, it forms a depletion region (like a p:n junction). When its interface is illuminated with a light of photon energy greater than the band gap of the materials, formation of electron/hole pairs takes place in similar fashion as one gets when a p:n junction is illuminated. These photo-generated electron/hole pairs are highly reactive and can initiate oxidation / reduction process with any organic material which comes in their physical contact.

Moreover, it has been found (Chang, *et al.*, 2004) that the photocatalytic degradation rate of methylene blue depends on the size of titanium dioxide nanoparticles. Nowadays titanium dioxide is being used for treatment of organics in waste water.

1.3.1.5 *Optical Properties*

Unique optical property of Nanomaterials may also be due to *quantum size effects,* which arises primarily because of confinement of electrons within particles of dimension smaller than the bulk electron delocalization length. This effect is more pronounced for semiconductor nanoparticles, where the band gap increases with a decreasing size. The same quantum size effect is also shown by metal nanoparticles, when the particle size >2nm.

Different sizes of colloidal gold nanoparticles exhibit different colours. In ancient time, artists used this property of gold in using it as durable paints especially on the glasses. Mie in 1908 observed that gold nanoparticles are very good at scattering and absorbing light that is why they have been used to make stained glasses of different colours that adorn the windows of churches and cathedrals. He explained that the colour exhibited by metallic nanoparticles is due to the coherent excitation of all the "free" electrons within the conduction band, leading to an in phase oscillation and is know as surface *plasmon resonance.* Thus, the colour of metallic nanoparticles may change with their size due to surface plasmon resonance.

Although generally nanoparticles are considered an invention of modern science, they actually have a very long history. The light absorbing and scattering property of gold nanoparticles were used by artisans as far back as the 9th century in Mesopotamia for generating a glittering effect on the surface of pots. And nowadays this property of gold is being used to deliver them to cancer cells and make them shine. Gold NP is also being used for testing mercury in environment.

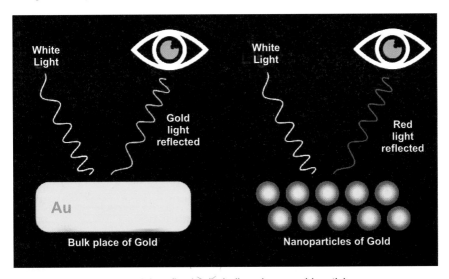

Fig. 1.8 Light reflection by bulk and nano gold particles.

1.3.1.6 *Magnetic Properties*

Magnetic properties of nanometals are distinctly different from that of bulk material. With reduction in size of the Ferromagnetic particles, it becomes unstable below a certain size, because there is an increase in the surface energy, which provides sufficient energy for domains to spontaneously switch polarization directions and become paramagnetic. But this transformed paramagnetism behaves differently from the conventional paramagnetism and is referred to as super-paramagnetism. In other words, ferromagnetism of bulk materials disappears and gets transferred to super-paramagnetism in the nanoscale due to the high surface energy. Quantum effects begin to dominate the behaviour of matter at the nanoscale–particularly at the lower end and affects the magnetic behaviour of materials.

1.3.1.7 *Mechanical Properties*

The mechanical properties of nanomaterials increase with decrease in size. Most of the studies have been focused on the mechanical properties of one-dimensional structure such as nanowire. The enhanced mechanical strength of nanocopper or any nanowire or nanorod is ascribed to the high internal perfection of the nanowires. Imperfections, such as dislocations, micro-twins, impurities that occur in bulk material; gets eliminated at the nanoscale dimension. At nanolevel, elimination of such imperfections is possible. No wonder we are already reading a lot about CNM impregnated polymers which are being used for bulletproof fabric, airplane and car body, even a sky city *i.e.,* wherever light weight, hard and non-corrosive material is needed.

1.3.1.8 *Electrical Properties*

Nanoparticles are good conductor of electricity, which make the electrical devices faster with low power consumption and with reduced imperfections.

1.3.1.9 *Thermal Properties*

Metal and semiconductor nanoparticles are found to have significantly lower melting point or phase transition temperature as compared to their bulk counterparts. The lowering of the melting points is observed when the particle size is, <100 nm and is attributed to increase in surface energy with a reduction of size. The decrease in the phase transition temperature can be ascribed to the changes in the ratio of surface energy to volume energy as a function of size.

Melting Point (MP) at macro level does not depend on size, but at nanoscale, M.P. is lower for smaller particles. This is because at macro-scale majority of atoms are almost inside the object and has very small effect on the percentage of atoms on the surface, therefore melting point which is responsible for thermal behaviour does not depend on size. Whereas at nano-scale, atoms are split between the inside and the surface of the object thus has big effect on the percentage of atoms on the surface and the melting point gets lowered for nanoparticles.

1.3.2 Some Physical Forces do not Apply at the Nanoscale

1.3.2.1 *Gravitational Forces*

At nanoscale, gravitational forces become negligible and molecular forces dominate for example:

(*a*) Mass of nanoparticles is so small that gravity forces are much smaller than intermolecular forces. That is why nanoparticles like CNT tend to bundle.

(*b*) Due to negligible gravitational forces, gold and silver nanoparticles remain in colloidal form and do not settle to the bottom. This is because intermolecular attractions, such as Van der Waals forces and electrostatic forces dominate over gravitational forces at nanoscale.

(*c*) Protein folding in biosystem is another example of nanoparticles avoiding physical forces. Protein folding occurs due to intermolecular forces between atoms and molecules and is not affected by gravity.

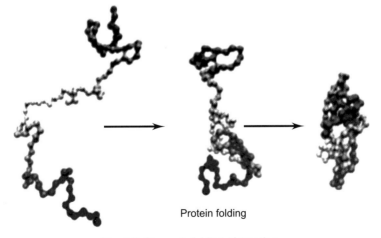

Protein folding

Fig. 1.9 Stages in folding of proteins.

1.3.2.2 *Friction*

At nanoscale, surface tension dominates over friction and force of inertia. Friction becomes negligible and surface tension dominates. This leads to 'NO-Slip' condition. The "NO-Slip" condition means that when a fluid flows through or passes on a solid surface it sticks to that surface; and there is no movement of fluid when it touches the surface of solid. But this "No-Slip" condition does not apply to surfaces having nanoscale features. When fluid or water passes through this surface, the surface tension keeps liquid out of the gaps allowing water to slip while flowing.

1.3.3 Some Physical Laws that do not Apply to the Nanoparticles

1.3.3.1 *Ohm's Law*

Ohm's law proposes that current is directly proportional to applied voltage to move in a single file. Resistance then varies with voltage and becomes large at low voltage. However, Ohm's Law do not apply to very thin (<40 nm) typical semiconducting nanowires because they are so narrow that electrons can not get around each other and there are examples that Ohm's law does not apply to—Ballistic (*i.e.,* independent of the length of nanotube) conductors.

1.3.3.2 *Fourier Law*

At the nanoscale, the continuum Fourier law for heat conduction in nano-particles can be significantly impeded below the predicted value by Fourier theory. Due to scattering, phonon (phonon = quantum of vibrational energy) transport is greatly impeded in a thin 1 D nanostructure having diameter less than the phonon mean free path (Yang, 2005).

1.3.4 Small Changes in Nanostructures Lead to Big Changes in Chemical and Physical Properties

Playing with the arrangement of atoms of a nanomaterial has yielded many interesting properties that can be exploited for various uses for example:

(*a*) Zig-zag arrangement of carbon atoms in a nanotube imparts semiconducting property.

(*b*) Whereas, arm-chair arrangement of carbon atoms in a nanotube displays metallic nature.

(*c*) Chiral CNT displays a tiny gap due to curvature effects, but displays metallic behaviour at room temperature.

(*d*) Cross-linking polymers make it stiffer.

(*e*) Synthesis of nanocomposites also alters and improves the physical and chemical properties.

1.4 Summary

Nanomaterials are getting popularised because of their special properties and possibility of their unique applications. It has been puzzling as to why the size of the material should influence the properties. Same materials of larger size exhibit certain property, but when its size is reduced to less than 100 nm that property gets altered. These are some of the interesting questions which have been tried to answer in this chapter. Efforts are also made to explain the various forms of nanomaterials which have been discovered so far.

Fundamentals of Bio-Nanotechnology

> There are at present fundamental problems in theoretical physics awaiting solution *e.g.*, the relativistic formulation of quantum mechanics and the nature of atomic nuclei (to be followed by more difficult ones such as the problems of life), the solutions of which problems will presumably require a more drastic revision of fundamental concepts than any that have gone before.
>
> **—Paul Adrien Maurice Dirac (Physicist)**

2.1 Introduction

The more we are finding out, the finer details of functioning of living system and some synthesized living molecules like synthetic DNA or self-replicating polymers; the more we are getting baffled about the difference between living and non-living forms. The dawn of fusion of nanotechnology and biotechnology, now called Bio-Nanotechnology, is likely to crack this confusion because the hypnotic versatility of Bio-Nanotechnology is expected to break through the "carbon barrier" of life. Hence, let us look at this subset of Nanotechnology. The interesting part is that not only Nanoscience has contributed to development of modern Bio-Nanotechnology, but understanding of functioning of Biosystem has also greatly contributed to the advancement of Nanotechnology.

2.2 Bio-Nanotechnology

NNI's definition of Bio-Nanotechnology:

> *"Bio-Nanotechnology is the understanding and control at the matter at dimensions of roughly 1–100 nm, where properties of matter differ*

fundamentally from those of individual atoms or molecules or bulk material. Encompassing nanoscale science and engineering and technology; nanotechnology involves imaging, measuring, modelling and manipulating matter at this length scale, to understand and create materials, devices and systems to exploit these phenomena for novel applications".

Mankind has learned chiseling materials for making weapons, agricultural instruments from time immemorial. This was perhaps, the first step towards miniaturization. During the search of secrets of immortality and the metamorphosis of gold from lead, alchemists sowed the seeds of controlled atomic manipulation and also their characterization. And that's how impressive birth of chemistry took place. During this giant leap towards atomicity, scientists learned over many decades, that macroscale science become deceptive when we enter into the realms of atoms. The shifts from *classical to quantum science* impart major obstacles in understanding the properties of nanometric objects and hence their design. At nanometric scale, inertia and gravity become futile giving versatile tactics for movement and interaction.

The major thrust of Nanoscience and nanotechnology is to fabricate nanometric objects capable of doing plethora of activities ranging from catalysis to modern nanorobots. K. Eric Drexler, in his illustrious manuscript, projected methods for construction of molecules by fabricating atoms into desired molecular construct. Starting with simple uncooked material, he envisioned assembling objects in an assembly line manner by directly gluing individual atoms. This persuasive idea helped to mimic the process called "Mechanosynthesis". The central theme of dubbing Mechanosynthesis is designing of nanomanipulator that assemble atoms as per given instructions. It's just a "Two week revolution" to construct a functional nanostructure once these nanomanipulators are made (Drexler, 1992). These assemblers would instantaneously allocate construction of large-scale factories, equipped with assemblers for building macroscale objects. Regrettably, mechanosynthesis is not yet in practice in nanotechnology.

The most overriding exploitation of nanotechnology can be seen in biological systems which are made of the miniature factories called cells. Functional molecules are cooked by these cells in order to carry out the life processes, from conception to death. In order to understand the functional efficiency of marvelous biological machines, toiling within the cell cytoplasm, knowledge of material science and molecular biology is compulsory. Intense research in Biology and nanotechnology has unlocked many rooms for fabricating nanoscale devices or nanorobots particularly to inject vital molecules in the body, the area called drug delivery. Drug delivery involves understanding the interaction between ligand and cell receptors at molecular level. This helps us in designing efficient drug delivery vehicles capable of injecting drugs via specific cell membrane protein or receptors. Let's now go for a very short voyage in the biological dynasty to understand complexities of cells.

A tiny biological cell, by exploiting the elementary principles of physics and chemistry, synthesizes enormous biological molecules. These molecules are assembled into functional biological work horses (biological machines) and their modus operandi can only be understood by applying principles of quantum sciences (and sometimes both quantum and statistical mechanics). We can now confidently quote that nanotechnology in one of the sciences practiced from antiquity by nature (more than three billion years!). Precise nanomanipulators like ribosomes were revealed by nature which make plethora of proteins by harnessing instructions from the genome. To overcome the obstacles of movement and navigation, elegant molecular motors are found in cells which work at atomic precision to power many biological phenomena, such as DNA replication, Protein trafficking, Neurotransmission, movement of bacteria in aqueous environment and many more to be deciphered. In order to maintain appropriate ionic environment in the cytoplasm, cells employ nanometric tunnels (sometimes Yoctometric also!) for pumping ions in and out with tremendous accuracy (Goodsell, 2004). Departure of waste and entry of raw material, such as glucose, amino acids for carrying out metabolic reactions is also done via specialized transporter proteins present in the cell membranes.

Nanosensors are another evolutionary heritage which optimize internal environment of the cells. Haemoglobin for instance, gets burdened with oxygen in the lungs because of the oxygen high partial pressure and unloads in the regions where oxygen partial pressure is less. During the magnificent evolutionary journey, haemoglobin got modified at molecular level, to work in presence of varying concentrations of oxygen. In foetus, a different isoform of haemoglobin with very high affinity towards oxygen exist in order to clutch oxygen from mother for congregating the essential requirement. This can be very important learning for nanotechnology experts working for devising nanomachines capable of sensing oxgen and delivering it to vital areas of the body.

Nanomolecular logistics is exceptionally sophisticated department perfected by cells over tenure of millions of years. Expert molecular motors march on specialized tracks (microtubules and microfilaments) carrying a cargo containing proteins such as neurotransmitter Acetylcholine, on their heads. Ultra specific binding to a cargo via specific docking proteins and extremely high thermodynamic efficiency, make it the most tactical transportation ever studied. Molecular motors such as myosin are also involved in contraction of muscles via a complex neurochemical signaling.

DNA, the nanoprocessor of cell can store enormous data which can be retrieved via transcription and translation to make efficient proteins. Spatial and temporal expression of information confined within this thread of 2 nm is used to drive the life. Cells can change the shapes, alter metabolisms, and carry out solute transport, replicate into another cell (hallmark of living cells!) and perform almost all the functions by reading these nanoharddiscs. Supercomputers of nanometric dimensions can be made by using DNA as storage device. Enzymes like DNA polymerase exhibits inherent activity of proof reading which prevent addition of faulty base pairs during DNA synthesis. These gifted rectification machineries prevent mutations which is the most hostile phenomenon for a biosystem, if not corrected. Molecular chaperonins are committed to fold a newly born protein in its native conformation by providing hydrophobic environment. These machines can be mimicked for construction of protein rectification devices for curing ailments arising due to misfolding of the proteins. Likewise there are many biomachines toiling for making cell a mightiest empire working with extreme sophistication.

After this brief understanding of functional machineries one may object that:

Why these nanomachines have not been made hitherto? The answer we know is atomic granularity (discrete combinations of atoms that interact through specific atom-atom interactions). Intelligent atomic manipulation perfected during evolution made biological machines to function in highly restricted environment of the cell soup, called cytoplasm, such as balanced pH, temperature, ionic strength and presence of water. Thus, to run a man made device in environment different of cytoplasm, understanding of atomic architecture is an obligation. This demands passionate application of Quantum mechanics for understanding the properties of atoms within biomolecules. *Atoms are glued to each other, in desired level of solidity, by forces called bonds.* These bonds are most of the time covalent bonds, which connect atoms into stable molecules of defined geometry. Apart from covalent bond, stearic repulsion of non-bonded atoms, electrostatic interactions, and hydrogen bonds—allows additional stability of biological molecules. Atomic granularity is not the only hindrance in comprehending nanometric anatomy of biomaterials; instead there are surplus of other complications which make this job exigent.

This chapter is an approach to understand atomic description of life and natural directives to construct molecular machines which function in aqueous environment of biological cells. Methodical understanding of molecular modes of action used by these biological engines within the cells may give us momentum to study the commandments behind construction of functional bionanomachines. Before entering into the intricacies of molecular machines, let's converse about the cardinal parameters which stabilize the functional machines and also the difficulties when we think of shifting to nanometric scale. Later, we will discuss the functioning of life at molecular level and the modifications done by nature during the due course of evolution in order to resolve the difficulties arising at nanoscale dimensions.

2.3 In the Dominion of Biological Machines

Biological machines function in entirely different environment of the cell cytoplasm. At the interface of physics and biology, three major obstacles become severe: *Firstly, the energies and length scales at which biological machines operate are motivating because they are in the regime where the energy and length relations, for a multitude of different phenomenon, congregate* (Phillips and Quake, 2006). *Secondly*, the biological system represents *"many-body"* problems, in particular the orchestrated activity of macromolecular assemblies confined within the cell. Molecular interactions become the major thrust to drive all the mechanochemical reactions in the cell and *Thirdly*, the need for a theory of biological dynamics that represents not only the many-body character of biological systems, but also their *far from equilibrium operation* (Phillips and Quake, 2006). Hence, in order to understand reactions in totality, one has to unlock the basic quantum mechanical and statistical properties of individual atoms and molecules.

The fundamental principles employed in synthesis of biomachines, from proteins, carbohydrates and lipids as building blocks, are completely alien to macromolecular assembly done by contemporary engineering methodologies. For such atomic manipulations, classical physics becomes futile and a giant leap towards Quantum mechanics becomes obligatory. Due to effect of miniaturization (less than 10 nm), movement and navigation becomes a dominant barrier. At nanoscale, collision becomes jitterier due to interaction with the water molecules, thus increasing the Brownian storms (Cavalcanti, *et al.*, 2006 and Curtis, 2005). Water acting as glycerin at macroscale and solid at nanoscale is perfect comparison to comprehend the impact of increased viscosity (Israelachvilli, 1992). This is just like living in an island where a Richter scale 9 earth quake raged continuously. To be more technical, we can say that life in biological system is at very low *Reynolds number* (10^{-4}), as a consequence of high viscosity and low inertial forces. For a man swimming in water, this number is approximately 10^4. A fascinating example of impact of low Reynolds number is movement of flagellated bacteria such as *V.cholerae* or human sperm in liquid medium (Purcell, 1997), given a forward stimulus these creatures will move with the speed of approx. 35 micron/sec. and withdrawing the stimulus will make them to travel 0.1 angstrom before halt within 0.6 microseconds! (Purcell, 1997).

Since biological machineries are made by discrete combination of atoms (atomic granularity), nanomachines of rational size and shape cannot be made (Gauger and Stark, 2006). We cannot imitate modern motors in terms of their smooth design such as rings and shafts, rotating smoothly. Molecular biomotors such as ATPase, kinesins and dyneins perform non-reciprocating movement adopting several discrete rotary states that cycle one after the other unlike macro scale motors. Aquadynamics and frictional properties are not distinct for atomic scale objects (Purcell, 1997; Bhusan, 2002; Bowden and Tabor, 1950). For gluing atoms, as per desired specifications one must understand the nature of chemical interactions such as covalent bond. Directionality of bond plays major roles deciding the strength of bonds. Attractive and repulsive forces should also be understood in order to fabricate atoms mutually.

Inertia plays very central role in dynamics of macroscale object by virtue of its mass. At bulk state, physical properties, such as resistance to walk or slide on a surface, tensile strength, adhesion, and shear strength are equivalent in magnitude to the forces, forced by inertia and gravity. In nanometric objects, centre of mass is not fixed (Roukes, 2001), rather keeps on changing as the particle size decreases. This imparts high friction during movement. This impact of enhanced inertia and viscosity made evolution of cork screw shaped nanopropellers called flagellum used by bacteria to swim in water. Decreased inertia ceases the movement of bacteria is less than an atomic diameter unlike the bulkier object such as submarine in the ocean which traverse little distance before halt.

Another trait at nanoscale is negligible impact of gravity and inertia (Purcell, 1997). Biological systems are totally dependent on melodious communication among macromolecules and respond to external and internal signals. Gravity, due to its less impact on nanometric objects, does not interfere with these fine communications instead makes them robust. The attractive forces between small objects are also stronger than the force of gravity. Exploiting these intense attractive forces flies and lizards can crawl on the wall without getting affected by gravity.

Aqueous environment of the cell also exert great influence on the operation of the nanomachines. Water, being polar in nature forms much favourable interaction, by forming hydrogen bonds. Forming hydration shells around ions and thus regulating their interaction is unique property of water. Macromolecules like proteins and DNA maintain their native conformation in water due to a property called "Hydrophobic effect". Hydrophobic functional groups (carbon rich regions) remain sequestered within the core of the structure and the surface becomes orchestrated with water loving functional groups. This is to maintain the thermodynamic stability of the macromolecules.

Attaining the functional configuration, for protein molecules, is perhaps the most daunting mission happening in biological system. Misfolded protein molecules can create hostile consequences like neurodegenerative disorders caused by prions. The double helical arrangement of DNA is also stabilized in aqueous environment. Nitrogenous bases are sequestered inside the core of DNA exposing negatively charged phosphates. These restrictions, rather principles exploited by biomolecules for their functioning make biological mimicry a difficult job. Non-natural biological machines must be tuned to function in biological systems.

2.4 Biological Engines: Elegant Justifications from Nature

Exploiting the fundamental principles used in modern day engineering, nature has perfected millions of biological machines over tenure of billions of years. These biomachines are completely acclimatized to work inside the biological cell. Tactical movement, at low Reynolds number has unlocked the rudiments for constructing nanorobots capable of using blood vessels as high way, roaming in exploration of the diseases. In the territory of molecular motors, Brownian motion is exceptionally furious. To combat with this jumpy milieu, molecular machines have evolved to take advantage by using the energy of Brownian motion. By converting random *thermal fluctuations into directed forces*, they use chemical energy via intermolecular forces, to attain stable and favorable structure. Molecular motors traverse with extreme thermodynamic efficiency acting as Brownian ratchets (in case of flagella) lowering the effects of inertia and gravity. There is extraordinary discrimination between molecular machines and their man-made macroscopic counter parts (already discussed). The molecular machines fabricated by nature enjoy in the mesmerizing environment dominated by large thermal forces. Consequently, statistical mechanics becomes an essential tool in elucidating the thermal effects on nanomachines*. A molecular machine like ATP synthase operating with 100% efficiency generates about 100 pN-nm of work.

2.5 Nanomotors of Biological Systems: Gluing Pieces Together

Movement is hallmark of biological system. Nature has crafted nanoscale motors such as ATP synthase (also termed as F_0F_1-ATPase), Bacterial Flagellar motors, Myosin, kinesins and proteins like helicases, and the doorway protein in certain viruses that assist in packaging DNA into their capsids. However, evidence that the helicases and viral doorway proteins perform rotatory movement is still a brain child. Motor proteins, for directed navigation, use

---

* The natural units of biological physics is piconewton-nanometrer (pN-nm)

thermal fluctuations and *Brownian movement.* At low Reynolds number (Due to increased viscosity and low inertia), movement becomes non-reciprocating in order to swim in the fluidic environment. Hence, the backward step through the same route turn out to be unlikely, another strategy to work at low inertia. We don't have any general theory for their complex sequestration of atoms in functional conformation *i.e.,* assembly and operation. The fundamental principles of physics are not understood in totality to glue atoms for fabricating the nanomotors. We are lacking the scaffold that combines the elegance of abstraction with the power of prediction. It means we don't have yet, a blue print of making nanomachines with desired structure and function. Molecular motors function at energies and length common to a mass of different processes. This regime adds to the challenges in analyzing cell's molecular machines. *At characteristic size of nanometer, the dissimilarity in energies, such as thermal, chemical, mechanical and electrostatic is lost. In other words, they converge and contribute to movement and function of the nanometric objects.* The richness of molecular motors in short, is due to inexplicable hide and seek between thermal and deterministic forces. For instance, molecular phenomena a such as diffusion, conformational changes, the breaking of hydrogen bonds, and drifting of charges from one molecule to another are permitted by thermal forces. In short, these molecular machines are unbelievably sophisticated and represent the apex of nanotechnology unlike man-made counterparts.

2.5.1 ATP Synthase: A Nanoturbine

MOLECULAR ANATOMY—Molecular understanding of ATP Synthase was instigated in 1960 by Efraim Racker and his contemporaries. They reported the isolation of soluble factor from cow's heart mitochondria. This factor (now known as F_1) had ATPase activity and also could re-establish ATP synthesis in membranes fractions with lost activity for ATP formation. In 1961, Mitchell (Nobel laureate, 1978) theorized celebrated "Chemiosmotic Theory" (1966). This theory clarified the idea of proton gradient and generation of Proton Motive Force (PMF) which drives the motion of most of the molecular motors including bacterial flagella. But the direct link of PMF and ATP synthesis was not proved until A. Jagendorf's "acid base transition" experiment (Jagendorf and Uribe, 1966). The next big heroism was to find out how F_1 exploits the proton flow for ATP hydrolysis and synthesis This mystery was solved by meticulous task of Paul Boyer who illuminated the relation between proton flow at F_0 and ATP synthesis and hydrolysis at F_1. He proposed "binding change mechanism" (Kayalar, *et al.,* 1977; Gresser, *et al.,* 1982) on the basis of studies on O^{18} exchange between H_2O and Pi/ATP (Boyer, 1998). He also proposed that,

ATP, in catalytic site, is in chemical equilibrium with the other two reactants ADP and Pi (Boyer, 1993). This makes ATP synthesis energetically least costly. However, in cellular environment, ATP hydrolysis liberates 12 Kcal/mol. ATP synthase has to pay this energetic price in mechanical work which is a prerequisite for releasing the product ATP. For this unparalleled work Boyer was awarded Nobel Prize in 1997. John Walker (who shared Nobel Prize with Boyer, in 1997) and his group at crystallographic lab, Cambridge University did scrupulous task to decode the fine structure of $\alpha_3\beta_3$ hexamer and most of the γ-shaft (Abrahams, et al., 1994). This was a successful attempt to justify Boyer's "Binding Change Mechanism".

ATP Synthase is perhaps, the most intensively studied molecular motor followed by bacterial flagella. These nanosized (10nm) energy generators are present in the membranes of mitochondria, chloroplast and bacterial membranes for generating biological energy called ATP (adenosine triphosphate)

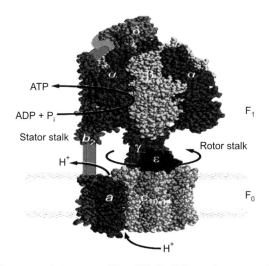

Fig. 2.1. Anatomy of structure of E. coli F_1F_0-ATP synthase; consisting of eight subunits $\alpha_3\beta_3\epsilon\gamma ab_2C_n$. F_1 corresponds to $\alpha_3\beta_3\delta\epsilon\gamma$ and F_0 to ab_2C_n. The rotor consists of $\gamma\epsilon C_n$ and the stator consists of $b_2\delta$. The c-subunits outline a ring, with n=10 preferential in E. coli. a-subunit cooperates with c, and lies outside the c-ring. γ consists of a globular foot, which interacts with c and ϵ, plus a long helical coiled-coil (not shown) which go through almost to the top of the central hollow space within the $\alpha_3\beta_3$ hexagon. The δ-subunit is positioned on top of F_1, most distant from the membrane.

Reprinted from, ATP synthesis driven by proton transport in F_1F_0-ATP synthase, Joachim Weber, Alan E. Senior, FEBSLetters, 545, page no. 62, 2003 © 2003 Federation of European Biochemical Societies, with permission from Elsevier Limited.

using the Proton Motive Force (PMF). In other words, these remarkable nano-energy generators transform electromotive force in to a torque which causes rotation (clockwise rotation) of the central shaft, like a macroscopic electrical motor. In certain circumstances, such as anaerobic conditions (in case of Bacteria), when the proton-motive force is low, the F_1 motor hydrolyzes ATP, driving the F_0 motor in reverse. In this case, the energy is derived from ATP hydrolysis. Cox and his contemporaries proposed a principle for proton transfer in F_0 motor, (Cox, *et al.*, 1986) invoking rotation of subunits.

Intact ATP synthase is composed of eight subunits in a stoichiometry α_3, β_3, δ, ϵ, γ, a_6, b_2 and C_{12} as depicted in Fig. 2.1. These components form two operational domains called F_0, which is membrane spanning complex and F_1, a soluble complex. We can now call this motor protein fabricated as counter rotating rotor and stator assembly as shown in Fig. 2.1. F_0 is called rotor (C_{12} ϵ γ) and F_1 along with its other components ($\alpha_3\beta_3\delta ab_2$) called stator. The F_0 fraction consists of three subunits a_6, b_2 and C_{12}. The 12 C rings are arranged in the membrane circularly into which the central *shaft* γ is inserted. Remaining components of F_0 are a_6 and b_2. There is a contact between b_2 and δ-subunit so that it anchors the subunit to F_1. Another point of contact between F_0 and F_1 is $\gamma\epsilon$. Thus, there are two stalks connecting F_0 to F_1, $b_2\delta$ and $\gamma\epsilon$, (Fig. 2.2).

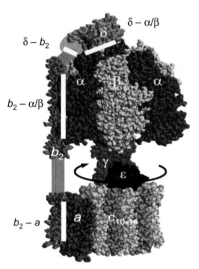

Fig. 2.2. Structure of E. coli ATP synthase viewing stator interactions. White rectangles show stator interactions, established and probable, designed to counteract rotor torque.

Reprinted from, ATP synthesis driven by proton transport in F_1F_0-ATP synthase, Joachim Weber, Alan E. Senior, FEBS Letters, 545, page no. 63, 2003 © 2003 Federation of European Biochemical Societies, with permission from Elsevier Limited.

Head piece of the ATP synthase made up of *soluble heterohexameric complex* α and β with stoichiometry $\alpha_3\beta_3$ arranged alternatively in the head piece in circular manner. It's like cross section of an orange, just to imagine. Each β-subunit harbors catalytic site which show differential affinity to nucleotides. All together there are three catalytic sites present in β-subunit and three non-catalytic sites in α-subunit. Catalytic sites of F_1 motor are best characterized to elucidate the molecular basis of power stroke generation which rotates the central γ-shaft made up of coiled coil structure (Capaldi and Aggeler, 2002; Noji, *et al.*, 1997).

The function of non-catalytic site is still in its infancy, but speculated to be as holder to keep hexameric components intact. *Outstandingly, the reactivity of catalytic sites in F_1 depends on the states of other two sites* (recall that there are three catalytic sites in F_1 subunit) and the position of γ-subunit.

2.5.2 Generation of Power Stroke: Converting Chemical Energy to Mechanical Work

The most noteworthy asset of F_1 ATPase is its thermodynamic efficiency of converting chemical energy to mechanical energy. This remarkable motor protein converts free energy of ATP hydrolysis into elastic strain which is then used to generate rotary torque by string of conformational changes. In this practice, it works with almost 100% efficiency! (Yesuda, *et al.*, 1998). Generation of the torque is initiated in the catalytic sites through sequential increase in the binding strength of ATP via formation of multiple hydrogen bonds between ATP and active site. As the clutch on ATP, of catalytic site, enhances (by formation of 15–20 hydrogen bonds), the power stroke is accomplished by bending of the pivot that rolls the top of the β-subunit down towards the bottom (Oster and Wang, 2003). This bending of β-subunit spins the central γ-shaft (like turning a crankshaft) as represented in Fig. 2.3.

Flow of energy for generating power stroke from ATP hydrolysis consists of following fundamental steps:

1. Attachment of ATP with active site (AS+ATP)
2. Formation of the more hydrogen bonds with ATP (AS·ATP)
3. Hydrolysis of ATP to ADP and Pi (AS·ADP·Pi)
4. Release of ADP and Pi from the active site (AS+ADP+Pi)

Release of Pi takes place before ADP. In step 1, the progression from weak to strong binding takes place through progressive formation of approx. 20 hydrogen bonds. The mechanism of hinge bending movement of β-subunit is exemplified in Fig. 2.3 (G. Oster and H. Wang, 2003). Formation of the more bonds with ATP tightens the active site (or catalytic site) and drags the top portion to bottom as explained earlier. The formation of each bond lessens the free energy that drives the bending of the hinge (Bockman, 2002; Oster and Wang, 2000; Sun Xia. 2002). This progression from weak to tight binding is called "*Binding transition*".

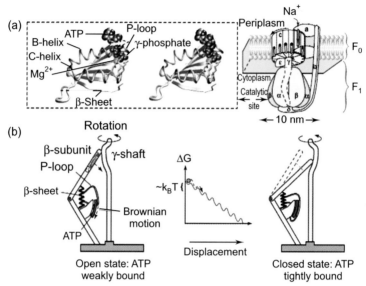

Fig. 2.3 (*a*) ATP synthase is fabricated from two parts: a soluble region (F_1) and a membrane attached region (F_0). As per the function, assembly can be divided into rotor (white) consists of C_{10-14} subunits and stator (shaded) consists of the remainder: the $(\alpha_3\beta_3)$ hexamer with alternating α- and β-subunit is connected via the δ- and β_2- subunits to a subunit. The three catalytic sites lie in the clefts between every other β-α-subunit. Ions in the periplasm enter the inlet channel in a subunit of the stator, where they bind onto the rotor site that is exposed to the inlet channel. Riding on the rotor sites, the bound ions pass out of the hydrophobic rotor-stator interface, where they can dissociate into the cytoplasm through outlet channels (*b*) A cartoon exemplifying how the weak to strong binding changeover of ATP impels the rotary power stroke. The surface of the P-loop slides over the surface of ATP from the proximal (adenine) end towards the distal (γ-phosphate) end. This alters the chemical free energy of the bonds into mechanical force that flows through the P-loop to the β-sheet, and hence outwards, driving the β-subunit to bend. The C-terminal end of the β-subunit bends ~30° with respect to the N-terminal end. The helix-turn-helix motif at the tip pushes on the bulge in the coiled coil β-subunit generating the rotational torque.

Reprinted from, Rotary protein motors, George Oster and Hongyun Wang TRENDS in Cell Biology, Vol. 13, No. 3, March 2003, page no. 116, ©2003 Elsevier Science Ltd, with permission from Elsevier Limited.

As proposed by Oster and Wang, 2000 and Sun Xia. 2002, the characteristic features of binding transition are:

- With *decrease in binding free energy* (hence binding becomes robust), the hinge bending motion of β-subunit takes place and therefore, the γ-shaft rotation takes place. This catalyses *hydrolysis cycle*.

- In reverse direction, the unbending of β-subunit driven by the γ-shaft rotation in reverse direction motorized by F_0 motor.

- When the top of the β-subunit is enforced up and away from the bottom portion, the binding becomes weaker and binding energy increases Fig. 2.3(*b*).

- As a consequence of this, nucleotides get released from catalytic site.
- Around 8–10 $k_B T^*$ of elastic energy gets accumulated in β-sheet (acts like a spring which cause bending during torque generation).

In imprisoning as well as release of the nucleotides, 'Brownian motion plays pivotal role. Generation of the power stroke is by sequential clutching of Brownian motion, which glues the nucleotides in the catalytic domain. Product release is assisted by binding energy of hydrolysis to fade the binding strength so that the thermal fluctuations can hammer them out.' This is the concluding remark of the above dialogues.

We have already honoured F_1 motor for their functional efficiency in terms of thermodynamics and generation of the torque. Let's have a brief account of the fundamental mechanism of functioning of this stunning machine which makes them almost 100 % efficient in coupling energy and torque generation.

Outstandingly high efficiency of F_1 is attributed mainly to its *catalytic cooperativity* and thermodynamic coupling of the events occurring at each site (Fig. 2.4). Hence, the motor proteins must utilize a unique line of attack to convert thermal fluctuations and Brownian movement into navigation and binding forces.

There are three mechanisms of high efficiency in F_1 (Oster and Wang, 2000):

1. Strongly Coupled Mechanical Movement of γ-Shaft: As per the discussion, *the conformational change that drives the central γ-shaft is due to bending of the molecular hinge in each β-subunit (that swivels helixes B and C about 30° with respect to each other).* Thus, power stroke of F_1 motor is bending of each β-subunit. This bending motion is converted into rotary motion of γ-shaft, because shaft is peculiarly bent off-axis. This matchless geometric arrangement makes it an efficient mechanical escapement due to following explanations:

1. The hydrophobic C terminus end of γ-shaft rotates within a hydrophobic covering formed by $\alpha_3\beta_3$ hexamer (Abraham, *et al.,* 1994). This anatomical feature "lubricates" the rotation of the shaft.

2. This formation of "lubricating bearing" that reduces the loss of energy due to friction since no breakage and formation of intermolecular bonds takes place (Tawada and Sekimo, 1991).

3. Each bending of β-subunit is coupled to one-half revolution of γ-shaft. Thus, bending of three can be coordinated to a complete revolution with no hanging up of the motor.

$*k_B T$ measures the thermal energy of Brownian motion and its value is 1 $\kappa_B T \sim$ 4.1 pN nm = 4.1×10^{-21} J at 298 K and 25°C.

4. Bending of β-subunit and rotation of shaft is tightly synchronized which prevent the rattling of the shaft around hydrophobic covering.

The mechanical efficiency of this task was also determined by calculating the average rotational velocity of an actin filament anchored to the shaft (Fig. 2.5) and computing the mechanical work done against viscous drag per rotation (Kinosita, *et al.*, 1998). The ratio of this work and free energy offered from hydrolyzing three ATPs (which it makes in one revolution) was found to be 1. This indicates almost 100% exploitation of free energy to do mechanical work.

2. Multisite Hydrolysis of ATP: Reaction rate increased almost 10^5 times (in comparison with the unisite hydrolysis) when multisite hydrolysis plays the game. Due to infrequent reversal of the steps during multisite hydrolysis, (Noji, *et al.*, 1997; Yasuda, *et al.*, 1998) the hydrolysis cycle at each catalytic site is closely coordinated with rotational position of the shaft. This means that the ATP must be taken to each catalytic site and product formed should be freed in a synchronized manner. *Gating mechanism in catalytic site is controlled by Mechanical signaling* (Al-Shawi, *et al.*, 1997; Ren and Allison, 2000) in following ways:

- Regulating the intake of ATP to catalytic site (ATP gate)
- Regulated release of product (ADP and/or pi gate)
- Reorientation of the catalytic residues

It can be summarized that the mechanical signaling between catalytic sites can be achieved via bending subunit and rotational position of β-subunit.

3. Generation of the Constant Torque by F_0: Surprisingly, the torque generated during *multisite rotational catalysis* and *stokes efficiency* is 40 pN/nm and 100 %, respectively. For this high level of efficiency and accuracy, the points we discussed are not sufficient. The motor must generate a constant torque through out in order to attain 100% stokes effectiveness. We have learned an important lesson from thermodynamics that effectiveness of functioning enhances if the energetic business proceeds in "small steps" (Fig. 2.3). During the rotational catalysis, there are two large energy drops (Senior, 1992; Weber and Senior, 1997):

- $\Delta G = 14$ kT during ATP binding, and
- 10 kT during phosphate release

A characteristic feature of Boyer's binding change mechanism (Boyer, 1998, 2000) has little difference between energy drops during hydrolysis or ADP release. *Power stroke can be efficiently generated by fragmenting large rotational steps into small steps and utilizing free energy change at every step.*

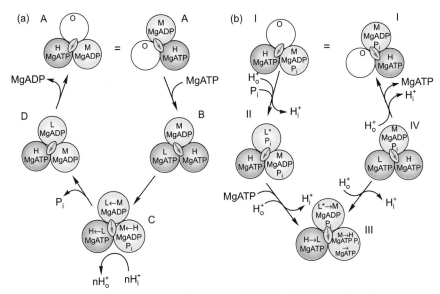

Fig. 2.4 (*a*) Enzymatic mechanisms of ATP hydrolysis and synthesis catalytic site conformations are: O, open (unoccupied); H, highest affinity for nucleotide; M, medium affinity; L, lowest affinity; L*, site with Pi binding pocket present. The central arrow denotes γ-subunit rotation. a: ATP hydrolysis. Binding of ATP to the empty site O (A→B) brings about hydrolysis in the H site by catalytic site cooperativity (B→C). Combined binding and hydrolysis of ATP occur with 90° rotation of γ, which is also associated with switch in conformations of the catalytic sites ('binding change'). Pi is released (C→D) associated with 30° rotation of γ, followed by ADP release (D→A) to regenerate the starting ground state. (*b*) ATP synthesis. Proton-driven γ-rotation generates L* from O (I'!III) so that Pi binds. This allows discrimination so that ADP binds (II→III) despite an unfavorable [ATP]/[ADP] ratio in the cell of >10/1. Next the binding change occurs (II→III→IV) and ADP+Pi condense chemically at the (new) H site. Release of ATP involves transformation of an H site via L to O site (III→IV→I).

Reprinted from, ATP synthesis driven by proton transport in F_1F_0-ATP synthase, Joachim Weber, Alan E. Senior, FEBSLetters, 545, page no. 65, 2003 © 2003 Federation of European Biochemical Societies, with permission from Elsevier Limited.

To understand the jugglery of these energy drops and rotation of the shaft, we must go through the hydrolysis cycle. The summary of overall reaction at catalytic site is as follows:

$$AS+ATP \xleftrightarrow{1} AS\bullet ATP \xleftrightarrow{2} AS\bullet ADP\bullet Pi \xleftrightarrow{3} AS\bullet ADP+Pi \xleftrightarrow{4} F_1+ADP+Pi$$

As we have already discussed, there are two crucial steps of energy drop which make the overall reaction thermodynamically proficient. ATP docks in the active site (AS), forming weak intermolecular interaction (*via* hydrogen bonding) which gradually become strong before the actual process of hydrolysis begin. At steps 1 and 3, free energy drops as mentioned earlier. These free energy drops must be exploited for generating torque in two steps:

1. Fraction of the torque is delivered to γ-shaft (*Primary power stroke*).

2. Remaining is stored as elastic strain energy as the shaft bends away from rest configuration.

This stored elastic energy catalyses the recoil of γ-shaft to its rest state generating a *secondary power stroke*. Each hydrolysis cycle correspond to a free energy change of approx. 24 k_BT. In short, we can conclude that the ATP docking is due to *sequential annealing of hydrogen bonds* as the nucleotide harbors in the docking site (Active/catalytic site).

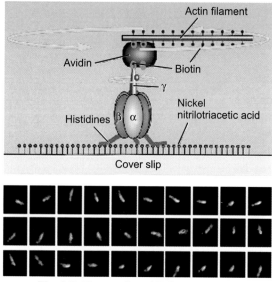

Fig. 2.5 Observation of F_1 Rotation.

Reprinted from F_1-ATPase: A Rotary Motor, Kinosita, Jr., Yasuda, Noji, Ishiwata, and Yoshida, Cell, 93: 21–24, 1998 ©1998 by Cell Press, with permission from Elsevier Limited.

F_0 Motor: the Lords of the Ring: Thanks to Peter Mitchell (Nobel laureate, 1978) for laying the fundamental principles behind functioning of F_0 and many other nanomotors such as bacterial flagella. His most notable *Chemiosmotic theory* explains the formation of proton motive force (Δp) during electron transport chain (ETC). This ionic force accelerates F_0 motors so as to catalyze the synthesis of ATP from ADP and Pi. The two important parameters of proton motive force are (Mitchell, 1979):

- Membrane potential (ΔΨ)
- Transmembrane proton gradient (ΔpH)

In the form of Equation it can be expressed as:

Proton motive force (Δp) = (2.303 RT/F) × ΔpH × ΔΨ   ...(1)

Where, *R* is the gas constant, *T* is absolute temperature and *F* is Faraday's constant.

The involvement of both the parameters varies vastly in different species. In mitochondria, membrane potential (ψ) is the main driving force for ATP formation, while in chloroplasts the same role is played by transmembrane proton gradient (ΔpH) (Graber and Witt, 1976). However, thermodynamically, both parameters exert driving force for ATP formation as per equation 1. Like F_1 motors, molecular world of F_0 is not well explored, mainly because it's a transmembrane protein. However, NMR and molecular modeling studies have exposed high-quality picture of its anatomy in mitochondria as well as *E.coli* (Fillingame, *et al.*, 2002) and sodium driven F_0 motor of anaerobic bacterium *Propionigenium modestum* (Vonck, *et al.*, 2002). In this section, we will understand the importance of Mitchell's hypothesis in generating rotational torque by F_0 motor. This torque is required to release tightly bound ATP form the catalytic site of F_1.

Rotary engines in the car and in the cell

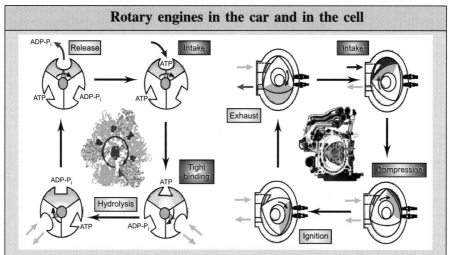

The F_1 motor reminds us of the rotary combustion engine, which was invented by Felix Wankelin in 1957 and was first to be exploited in commercial cars by Mazda in 1967. The rotary engine is small, light, quiet and effortless because the engine can directly translate the fuel energy into rotation of the rotor. It can power the intake of the fuel gas, compression, ignition and exhaust in succession just by a simple rotation of the central rotor, which is quasi-triangular in shape (right panel).The actions occurring on one side of the rotor (green) are explained. The F_1 also carries a central rotor — the γ-subunit — and three reaction chambers (the catalytic β-subunits; left panels).The events occurring in one β-subunit (light red) are annotated according to Boyer's exchange model. The basic principles behind the functioning of these rotors — three reaction sites in turn doing each of three cyclic steps in a 120° phase distinction to cause rotary motion —are outstandingly similar. Reprinted form ATP Synthase — A Marvellous Rotary Engine of The Cell, Masasuke Yoshida, Eiro Muneyuki and Toru Hisabori, Nature Reviews-Molecular Cell Biology, volume 2, September 2001, page no. 673, by permission from nature © 2001 Macmillan Magazines Ltd.

F_0 is transmembrane cylindrical structure fabricated from 10–14 c-subunits, depending upon the species. This cylindrical structure is adjoined by a six helical a (plays important role as stator) subunit. We know that, this subunit forms a connecting link between F_0 and F_1 via b_2 and δ-subunits. Rotary torque is generated by ions flowing through the a/c interface (Fig. 2.6) which is sufficient to spin the γ-shaft and hence ATP release during synthesis cycle. The modus operandi of both the motors (proton and sodium driven) is same with slight disagreement based on certain aspects of geometrical arrangement.

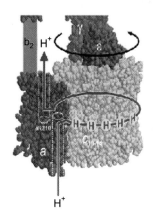

We will, however, concentrate on sodium driven motor owing to its intense molecular studies. Recalling once again molecular structure of motor and stator, the stator consist of subunits *a*, b_2, and δ, and the rotor is made up of 12 c-rings arranged circularly in the membrane. Each c-subunit is a pair of transmembrane α-helices orchestrated with ion binding negatively charged sites in the interior of membrane (Ballmoos, *et al.*, 2002). *In E.coli, the ion binding site is a negatively charged residue (Asp61) and in P. modeustum, it consists of one negatively charged and two polar residues (Gln32 (Q32), Glu65 (S66) and Ser66 (E65) (Fig. 2.5), (Vonck, et al., 2002).* The remarkable feature which makes the discussion more interesting is *existence of positive stator charge Arg210 in E.coli and Arg 227 (R-227)* (Wehrle, *et al.*, 2002).

The rotor site is access to the cytoplasm out of the motor-stator interface (Fig. 2.7). This access is via outlet channels that extend from the rotor site to the cytoplasm. Rotor site is brought into the rotor-stator interface (by diffusion of the rotor) through a polar strip that recompenses for the energy penalty of the rotor site peeling its hydration shell. The polar strip assists in forming *electrostatic interaction* between stator charge (R-227) and rotor site (Fig. 2.8). If the rotor site is loaded with an ion, it gets dislocated into the cytoplasm. Entry of the ion (Na$^+$ or H$^+$) takes place via aqueous inlet channels in a subunit (from periplasm to cytoplasm). A positive ion (Na$^+$ or H$^+$) neutralizes the rotor charge so that it can easily move through hydrophobic rotor stator interface into the membrane

Fig. 2.6 Structure of ATP synthase showing proposed proton transport pathway. Residues cAsp61 and aArg 210 lie in the center of the bilayer, at the a/c interface. Their concerted interaction is required for proton movement. Putative access channels for ingress/egress of protons are shown. The c-ring carries protons around on protonated cAsp61 as it rotates.

Reprinted from, ATP synthesis driven by proton transport in F_1F_0-ATP synthase, Joachim Weber, Alan E. Senior, FEBS Letters, 545, page no.63, 2003 © 2003 Federation of European Biochemical Societies, with permission from Elsevier Limited.

where ions can dissociate through the outlet channels (Fig. 2.7). The outlet channel gets occluded once the rotor site is in the rotor-stator interface. In this state, the ions cannot escape to the cytoplasm. Once the occupied rotor site diffuses though the hydrophobic barrier, the outlet channel reopens facilitating the exit of bound ion in the cytoplasm. Once the site gets free it undergoes hydration.

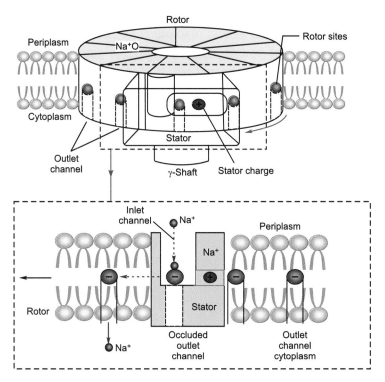

Fig. 2.7 The molecular construction of the sodium-driven F_0 motor of Propionigenium modestum. Ion-binding sites on the rotor have access to the cytoplasm via the outlet channel except in the rotor-stator interface, where the outlet channel is occluded. Ions enter the aqueous inlet channel and bind to a rotor site. The nearly neutralized site loses most of its hydration shell and encounters only a small energy barrier in passing through the hydrophobic rotor-stator interface to the left into the membrane. Once there, the rotor sites regain their access to the cytoplasm via the outlet channels and the bound ions can dissociate, and the rotor site reacquires its hydration shell. The stator charge prevents ions in the periplasm from leaking along the polar strip into the cytoplasm and prevents ions in the cytoplasm from riding on the rotor into the inlet channel.

Reprinted from, Rotary protein motors, George Oster and Hongyun Wang, TRENDS in Cell Biology, Vol. 13, No. 3, March 2003, page no. 117, © 2003 Elsevier Science Ltd, with permission from Elsevier Limited.

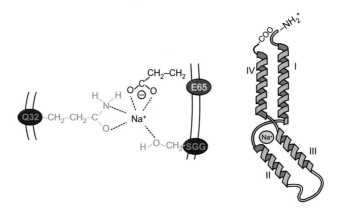

Fig. 2.8 Schematic diagram of Stator-Rotor assembly showing the Stator charge (R227).

This creates large barrier against fluctuating backward (to the right) into the rotor-stator interface (Oster and Wang, 2003). Movement of the motor in forward as well as backward direction, depends upon the binding and dissociation of ions (Fig. 2.7). The stator body is hydrophobic with two valuable exceptions:

1. An aqueous channel (inlet channel) that allows entry of the ions towards the concentration gradient to access the rotor binding site.

2. A polar channel (outlet channel) linking the aqueous channel to the low concentration reservoir.

Presence of stator charge creates an *electrostatic potential well* that attracts the rotor site as we have already discussed. Once captured in the well, the *rotor relies on Brownian storms to getaway*. But the rotor must escape in order to function flawlessly. Membrane potential favours the escape by lowering the left side of the potential well so that the rotor charge is more likely to facilitate movement to left than to right (Oster and Wang, 2000, 2003).

The aqueous channel is at same potential with the high concentration reservoir (periplasm). This facilitates transmembrane potential drop from vertical to horizontal. The combined impact of both the ion concentration difference and membrane potential drives the rotor to left reservoir (Oster and Wang, 2000, 2003).

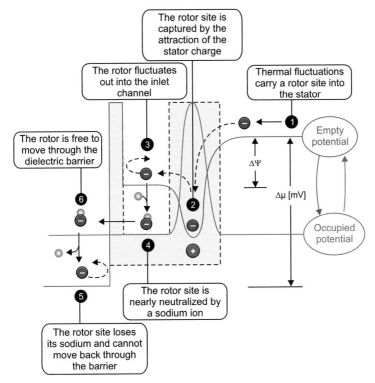

Fig. 2.9 Generation of the rotor torque. The motor is driven by switching between two potentials, one corresponding to empty rotor sites and one to occupied sites. The switch is driven by the binding free energy of ions to the rotor sites (1→2). An empty rotor site fluctuates into the rotor-stator interface, where it is captured by the electrostatic field of the stator charge (2→3). The rotor eventually fluctuates out of the potential well but, because of the biasing influence of the membrane potential, it is much more likely to hop leftwards into the inlet channel. (3→4) Exposed to the periplasmic reservoir, the rotor site quickly picks up an ion. This nearly neutralizes it, where upon it loses its hydration water, switching it to the occupied potential. (4→5) Facing a low energy barrier to leftward movement, the rotor diffuses through the hydrophobic barrier into the membrane domain. If it diffuses to the right, it encounters the stator charge, loses its ion back into the inlet channel and is recaptured by the charge (5→6). Once out of the stator, the rotor site can lose its ion into the cytoplasmic reservoir and rehydrate. This re-erects the barrier to moving backwards, completing the Brownian ratchet step to the left. Computations show that the torque generated by this cycle of events is sufficient to drive the F1 motor in the reverse synthesis cycle.

Reprinted from, Rotary protein motors, TRENDS in Cell Biology, Vol.13 No. 3 March 2003, page no. 117, George Oster and Hongyun Wang, © 2003 Elsevier Science Ltd., with permission from Elsevier Limited.

2.5.3 Flagellar Motors in Bacteria: Turning the Screw

Let's now appreciate the majestic movement of bacteria and many other prokaryotes at low Reynolds number. We have already discussed the obstacles coming across swimming at high viscosity and in absence of the inertia.

To embark upon these two problems most favoured mode of navigation, by biological system, is swimming not crawling. In this way frictional forces are minimized. One of the most sophisticated objects perfected by nature is bacteria equipped with flagellar motor proteins. Like ATP synthase, flagellar motor is fabricated as *motor and stator assembly*. Spinning of motor generates enough force to swim in liquid in desired direction. The bacterial flagellar motor is an exceptional example of extreme bionanotechnological construction. Understanding the construction and modus operandi is one of the first endeavors towards making man-made machines on nanoscale. This motor has the same power-to-weight ratio as an internal combustion engine, spins at up to 100,000 rpm and achieves near-perfect efficiency. Yet at only 50 nanometres across, one hundred million would fit onto a full-stop. The only other natural rotary electric motor is in the enzyme ATP-Synthase.

An *E.coli* cell is driven by tiny helical flagellar filaments (approximately four in number) anchored to the cell membrane. This number may vary in other bacterial species. Each filament is powered, at its base, by a rotary motor fixed in the cell membrane. By changing the bundling properties of helical filament a cell can swim towards a nutrient (counterclockwise) and can tumble away from toxic chemicals or stimulus (clockwise). The stimuli which decide the directions of movement are sugars, amino acids; dipeptides *etc.*, which are sensed by receptor proteins associated with the flagellum. This helps bacterium to find provinces where life is comfortable. *At highest speed, the entire four flagellar filaments rotate counterclockwise forming a bundle that drives the cell in forward direction. During tumbling, one or more filament entwine.* These filaments leave the bundle, and the cell changes direction of movement. Clockwise rotation is optimized by *chemotactic signaling protein, Che Y,* which is regulated by phosphorylation by a chemoreceptor controlled histidine kinase che A. It acts as *"response overseer"*. Phosphorylated *Che Y* binds to the flagellar switch at the base of the flagellar motor (Walch, *et al.*, 1993) and flagellum rotates in clockwise direction (Barak, *et al.*, 1992). Potency of kinase is depressed by addition of chemoattractants which drives the cell in forward direction (Blair, 1995; Falke, *et al.*, 1997; Stock, *et al.*, 1999; Bren and Eisenbach, 2000; Bray, 2002; Bourret and Stock, 2002). Thus, CheY cats as *"molecular switch"* to alter the direction of flagellar rotation. Molecular details, however, which rotate the flagellum, clockwise or anti-clockwise is still not well understood. Apart from swimming, *swarming* (Berg, 2005) is another type of movement performed by bacteria on solid surface. It can be imagined as swimming on a surface! When grown on semisolid medium like agar, *swarming becomes a special type of movement across the surface in a coordinated manner.*

Molecular Structure: Major mass of the flagellum is due to long helical filament formed from thousands of copies of protein called flagellin, arranged circularly to form a hollow tube. Base of the filament is glued to a thick, flexible structure termed as hook. Arrangement of hook is in such a way that it can turn set of rings mounted on a rod (DePamphilis and Adler, 1971). L- and P-rings facilitate the entry of rod through the outer membrane and peptidoglycan layer. This anatomic feature helps the rod to rotate freely. The names of L- and P-rings are according to their locations relative to the cell envelope. *L-ring is present in outer membrane, while P-ring is in peptidoglycan layer* (Fig. 2.10). LP-ring function as bushing for the central rod. MS-ring (M-membrane S-supramembrane) is resident of inner membrane present within and above it. MS-ring is formed from an intact protein FliF (Ueno, *et al.*, 1992). The cytoplasmic domain of MS-ring is glued to C-rings which act as switch regulators (Khan, *et al.*, 1992; Francis, *et al.,* 1994). The C-rings contain three important proteins FliG, FliM and FliN. These proteins play pivotal role in generation of the torque mandatory for spinning the filaments (Khan, *et al.*, 1991; Francis, *et al.*, 1994; Katayama, *et al.*, 1996).

The basal structure forms the rotor and non-rotating part fabricated from circular array protein complexes around the rotor, forms the stator. Basal body which forms the motor consists of a set of rings up to 45 nm in diameter (DePamphilis and Adler, 1971). A unique structural feature of stator (in H^+ driven flagellar motor of *E.coli*) is presence of protein complex MotA *and* MotB. They form a MotA/MotB complex that when oriented properly binds to the peptidoglycan and opens proton channels through which protons can flow (Van way, *et al.*, 2000). Similar types of motors, PomA and PomB are found in Na^+ driven flagellar motor of *Vibrio alginolyticus* and *Vibrio cholera.* MotX and MotY are also present in Na^+-driven motors, but their exact function is not known (McCarter, 1994*a*, *b*).

MotB is postulated to have proton binding site. Asp 33, a highly conserved aspartic residue of *Salmonella enterica* MotB, which corresponds to Asp32 of E.coli invite the proton to form weak interactions (Sharp, *et al.*, 1995; Togashi, *et al.*, 1997). Motor torque is speculated to be a consequence of cyclic interaction of Mot A with FliG coupled with proton pumping (Zhou, *et al.*, 1998; Kojima, *et al.*, 2001). Transport of protons via MotA/MotB complex (stator) drives itself so that the protons bind to the next vacant proton binding site, thereby stretching their connection. The rotor assembly has to rotate by one step, when these connections recoil. Consequently, both MotA and MotB can generate torque after receiving protons from the sink. As per the detailed studies of torque-speed dependence of the motor, the torque range was found to be 2700 to 4600 pN-nm (Berg, *et al.*, 1993; Chen, *et al.*, 2000).

MotA and MotB also function as molecular steering, assisting the cell to change the rotational direction (Yamguchi, *et al.*, 1986). The rotation rate of proton catalyzed flagellar motor is proportional to proton motive force (Gabel, *et al.*, 2003). The C-ring consists of the export apparatus that pump proteins needed to fabricate hook and filament outside the cell (Minamino and Namba, 2004). Monomer amalgamation is strictly regulated by pentameric cap complexes (Yonekura, *et al.*, 2000). Nevertheless, how the proton translocation event is coupled to mechanochemistry of flagellar rotation is still secrecy.

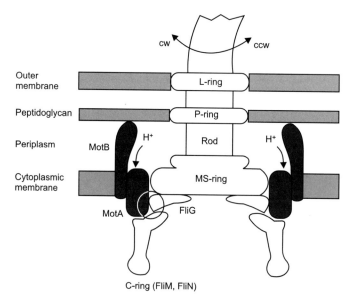

Fig. 2.10 Schematic view of the flagellar basal body imbedded in the cell envelope of a Gram-negative bacterium. The central part (white) rotates, whereas the torque-generating MotA-MotB proton channels (dark shading) are anchored to the peptidoglycan layer of the cell envelope (gray). Torque is generated by the flow of protons (H1) across the cytoplasmic membrane via MotA-MotB and by interactions between MotA (stator) and FliG (rotor) (circled portion). Most flagellar motors reverse the sense of rotation from counterclockwise (ccw) to clockwise (cw), although some rotate unidirectionally cw and modulate rotary speed.

Reprinted from, Helix Rotation Model of the Flagellar Rotary Motor, R. Schmitt, Biophysical Journal Volume 85, August 2003, 843–852, page no. 844, © 2003 by the Biophysical Society, with permission from Elsevier Limited.

The hook and filaments are polymers of single polypeptides, hook protein and flagellin, respectively. The two components are fabricated of 11 parallel rows of slanted protofilaments relative to cylinder axis (Asakura, 1970; Calladine, 1975; Hasegawa, *et al.*, 1998). Under constant and steady rotation of the motor, the filaments behave as rigid propellers. It takes the help of tortionally induced alteration between alternative filament forms with varying numbers of long and short protofilaments. These protofilaments modify the stiffness of filament helix, pushing out the filament from the bundle that drives a swimming cell. Flexibility of the hook is more than the filament and hence, it functions as universal joint to facilitate several filaments from motors all over the cell to rotate the bundle in species with peritrichious flagella. Molecular Dynamics simulation studies based on the filament anatomy deciphered a possible mechanism for switching between long and short form in response to force (Kitao, *et al.*, 2006; Furuta, *et al.*, 2007). The number of revolution by bacterial flagellar motor depends upon number of FliG subunit in C-rings. Due to a recent high resolution study of flagellar motor dynamics it became possible to reveal *number of steps per revolution is 26* (Sowa, *et al.*, 2005).

The rotation rate of flagellar motor protein is radically affected by cytosolic pH. Decrease in cytoplasmic pH bring an *E.coli* to halt or reduces the motility to a great extent. This happens even at constant proton motive force signifying that the absolute concentration of protons in cytoplasm plays pivotal role in determining the rate of flagellar motor rotation (Minamino, *et al.*, 2003). Role of intracellular pH in determining the torque speed of flagellar motor is studied recently by Nakamura, *et al.*, (Nakamura, *et al.*, 2009).

Molecular Physiology of the flagellar motor and stator: As discussed earlier, flagellar motor rotation is catalyzed by electrochemical potential gradient across the membrane. Evidences that flagellar motor is also driven by sodium ions in many species, the sole involvement of protonation in flagellar spinning is ruled out. At lower speeds, the torque appears proportional to the proton motive force and the relation is expressed as:

$$PMF = \Delta\Psi - (2.3RT/F)\Delta pH$$

Where, F is Faraday's constant, T is the absolute temperature, "ψ is transmembrane potential and ΔpH is 'pH difference' across the membrane.

As per the kinetic studies, at normal swimming conditions of ~ 100 Hz, 1200 protons pass through the channel per revolution (Elston and Oster, 1997). The speed of motor is dependent on the number of stators. For 10 stator element, each stator generates rotational torque of 30–60 pN-nm. Thus, motor speed enhances linearly with the number of functional stator. Unlike ATP synthase, in bacterial motors the stator remains fixed (non-rotating). To understand the operational fundamentals of stator, imagine two cylindrical surfaces set in such a way that the inner cylinder is free to revolve with respect to outer (Fig. 2.11). The most fundamental aspect behind the rotation of the motor is hide and seek between equilibrium and non-equilibrium conditions. Consider the cylindrical surfaces be decorated by point charges. The electrostatic potential field defined distribution of charges will have multiple equilibria. In absence of any restriction, inner cylinder will rotate in order to achieve most stable equilibrium condition. Now assume that a positive charge (protons or sodium ions) neutralizes the negative charge present on the surface. In this condition the mechanical equilibrium of the system is no longer maintained. Hence, to achieve a new local equilibrium cylinder will rotate (Elston and Oster, 1997). *Hence, to turn the system into rotary motor, fixed charged must be positioned in such a way that the successive equilibria cause the cylinders to rotate unidirectional.* In flagellar motors in bacteria, the proton trajectory is rather vague. It is generally speculated that the protons interact with α-helices of the MotA/MotB channel complex. But the torque-generating regions appear to be the cytoplasmic domain of MotA which adjoins the FliG component of the rotor. Interaction between acidic and basic residues on rotor (FliG) and stator (MotA and MotB complex) is indispensable for torque generation (Lloyd, *et al.,* 1998; Tang, *et al.,* 1996).

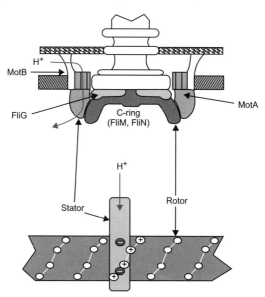

Fig. 2.11 (TOP) Schematic of the rotor and stator configuration. Protons flow downward from the periplasmic space through the MotA/MotB channel complex into the cytoplasmic domain of MotA where they are subject to the composite electrostatic field of the rotor (FliG) and stator (MotA/B) charges. In the model, we assume that protons flow exclusively through the stator. (BOTTOM) The charge distribution rotor stator interface. In this configuration the stator charges are negative and the rotor charges are positive (o). The rotor is divided into $n = 15$ repeating units consisting of 4 charges spaced in a 2 + 2 configuration, giving a total of 60 positive rotor charges.

Reprinted from, Protein turbines I: The bacterial flagellar motor, Timothy C. Elston and George Oster, Biophys. J. page no. 5 73:703-21(1997), with permission from Elsevier Limited.

The actual mechanism of torque generation is brilliantly reviewed by Elston and Oster (1997). Major part of this discussion is adapted from the same review. Let's assume for the time being that acidic residue (negative charge) constitute protonation site in stator and basic residue constitute the gating site in rotor (Fig. 2.11). There is slight slope in the rotor charge with respect to stator charge which is an important prerequisite for torque generation. A positive charge (sodium ion or proton) can enter the stator from extra cytoplasmic fluid to associate with top negative charge of the stator. This happens in absence of blockage due to positive rotor charge during the entrance of positive charge. When the middle rotor charges are not jamming the proton's trajectory, it can push to lower stator charge and when the bottom charge revolve out of the way, the proton can leap out to cytoplasm. The force applied on the rotor by stator is dependent on the habitation of the stator charge sites. We have already discussed the presence of aspartic acid (Asp 33) in proton binding site. Estimating the pKa value of aspartic acid or any other residue present in the binding site, we can represent the stator sites as "Columbic potential wells". The potential sensed by the incoming positive ion is combined effect of field set up by both rotor and stator charges. The transition rates between the stator states can be modeled by Kramer's rate theory, assuming protons in thermal equilibrium in each well (Hanggi, *et al.*, 1990). Important assumption of this study is the time proton spends in transfer between sites is much shorter than the time spent in each potential well. More detailed extension of this concept is beyond the scope of this book. Intersected students should refer review by Elston and Oster (1997).

2.5.4 Linear Molecular Motors: Dancing on the Tracks

Biological cells are equipped with sophisticated logistics, involving intelligent linear and rotary motors engrossed in manufacturing ATP like fuels or generating forces for muscle contraction, marching on the specialized tracks loaded with cargo containing proteins segregating chromosomes during mitosis. About rotary motors we have already devoted several pages of this book and now its linear motor's turn to be highlighted. Motor protein work by converting chemical energy to mechanical force. However, to do this meticulous task they use diverse molecular paths as shown in Fig. 2.12. Although various types of linear motor proteins, such as motors which perform vectorial transport on nanotracks for many ATPases cycles without detaching from the track. Some of these molecular motors are kinesins (Brady, 1985; Vale, *et al.*, 1985; Howard, *et al.,* 1989; Block, *et al.*, 1990; Svoboda, *et al.*, 1993) myosinV (Cheney, *et al.*, 1993; Mehta, *et al.*, 1999; Sakamoto, *et al.*, 2000), dyneins. Others are DNA replicating enzymes, such as DNA and RNA polymerase, helicases, protein folding machineries like proteasomes and many more to be discovered. Understanding these mechanisms is one of the more challenging problems that require concerted efforts by chemists, physicists, and biologists.

Track specific molecular motors walk like us (Kinosita, *et al.*, 1998). The two feet never detach simultaneously from the track (ground in our case). In addition, at least for myosin V and conventional kinesin which known to be *processive*, persuasive proofs exist that these molecular motors toss two feet onward alternately in a hand-over-hand fashion (Yildiz, *et al.*, 2003, 2004; Kaseda, *et al.*, Asbury, *et al.*, 2003; Warshaw, *et al.*, 2005) just like human walks. It is motivating rather to study resemblance of motor movement with our walking mechanism. When a human being walks, he completely relies on mass. Gravity, in this case would help to pull his lifted foot down, instead of pushing it with muscles. Forward movement is assisted by the inertia, without demanding involvement of muscles. If the walking gentleman is attached with a giant helium balloon at his shoulder, and watched from far above, how he goes ahead, the balloon would move forward with little sign of separate stepping, because we exploit inertia of the body to make smoother movement.

Unlike us, for biomolecules influence of gravity and inertia is negligible and mass is also practically nil. A biomolecule, like protein never settles down to bottom of the test tube unless centrifuged at high speed. This is because of negligible mass and the dominant Brownian motion due to water molecules. One of the smartest evolutionary explorations is molecular vehicles walking on specialized tracks. These vehicles must adhere stubbornly to its track; otherwise Brownian motion will badly obstruct their movement. *In absence of any external force, a molecule in watery environment, stops moving vectorially within 0.1 nm and performs completely random Brownian fluctuations. This is due to, as already discussed, swimming at low Reynolds number* (Purcell, 1977). Therefore, a molecular object can not move in a unique direction unless there is a force applied on it.

Another cardinal difference between two-foot motor protein and us is absence of right and left foot in molecular motors. Due to this unique feature, motor would adopt specifically same posture every time it moves not every two steps as human movement. The two legs of molecular motor are glued at the base to fabricate a coiled coil of α-helices (Kozielski, *et al.*, 1997; Li, *et al.*, 2003). When both the feet are bound to the track concurrently, they must be oriented in the same direction dictated by the structure of the binding on the track. Positioning both toes in the same direction is natural mode of walking, but for a molecular motor acquiring this type of posture requires uncomfortable posture of the legs (Kinosita, *et al.*, 2005).

A motor protein like myosins, kinesins and Dyneins interact specifically with certain types of filament along with it moves at the cost of energy derived from hydrolysis of ATP. These nanometric tracks serve as guides or tracks for the movement. The filaments we are talking about are Microtubules and Actin filament. Both the filaments are formed with dimer forming tubulins and monomeric subunits actins, respectively. The most celebrated property of these molecular motors is their vectorial movement along tracks. This is possible because both the ends of microtubule and actin filament are distinct from each other conferring polarity to the filaments. The two ends are called as "plus end" and "minus end" (Fig. 2.12).

Motor proteins are classified into several families: myosin, kinesins and dyneins. Myosins march always along actin filaments and towards the plus end. These are called "plus end directed motors". Kinesins and dyneins move along microtubules. Kinesins move towards the plus end and dyneins towards the minus end. Myosins are dedicated to muscle contraction (Huxley, 1957 and 1969) by generating power stroke at the cost of ATP and regulated by complex neuromuscular signaling. Myosin molecules form linear structure containing many heads interacting with actin filaments arranged in parallel. The stroke of myosin motors then encourages the relative sliding of the two types of filaments. They also play pivotal role in cell movement and organization of actin.

Kinesins form important basis of neurotransmission. In neurons they are found in huge numbers and play important role in transport of vesicles along the axon towards the synaptic cleft. Both myosins and kinesins have two identical heads of a size of about 10–20 nm which are the elementary force-generating elements as well as a tail which is used to attach the motor to another structure. The source of energy for power stroke generation is the hydrolysis of ATP to ADP and Pi. Attachment site (to actin filament or microtubules) in these motors reside in the heads. Heads also have ATP binding site (Let's name it as H). ATP binding site "H" goes through a chemical cycle: it attaches to ATP and hydrolyzes the bound ATP. Subsequently it releases the products ADP and P. Different chemical states during this cycle can be denoted as H, H•ATP, H•ADP•P, and H•ADP, respectively. After completion of the cycle, (H + ATP → H•ATP → H•ADP•P → H ADP + P → H + ADP + P) the motor is unaffected. During this mechanochemical cycle, different conformational changes with typical geometrical features have varying interaction characteristics with respect to filament (Huxley and Simmons 1971). As a consequence of this, motor protein undergoes chemical induced alteration in affinity towards the filament ("attachments" and "detachments"). In other words, motor protein undergoes changes between strong and weak binding states. This coupling between chemistry and binding provide driving force for movement along a polar filament (Hill, 1974; Spudich, 1990; Julicher, *et al.*, 1997). For observing the force generated by an individual motor, the ability of the motor to remain associated with the track or filament is noteworthy. This is called processivity of proteins which are dependent on templates, such as actin filament. Myosin is not processive molecular motor. During the mechanochemical cycle, it detaches from the filament during a considerable period of time. During this instance, it can easily disperse away from the filament if it is not detained in place. *Motor protein induces stochastic displacements of the order of 6–10 nm which lasts for few milliseconds and peak forces of the order of 1 pN* (Ishijima, *et al.*, 1991; Winkelmann, *et al.*, 1995).

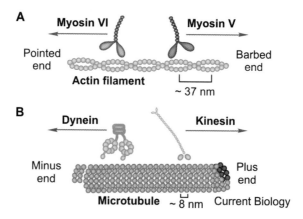

Fig. 2.12 The families of vesicular molecular motors. (*A*) Myosin motors march along actin filaments; the most celebrated myosin that travel towards the barbed end is myosin V, and towards the pointed end is myosin VI. Structurally, the myosin motors share considerable resemblance. (*B*) The two families of microtubule motors, with most of the kinesin-family motors traveling towards the plus end, and cytoplasmic dynein moving towards the minus end. Structurally, kinesin and dynein are very different.

Reprinted from Cargo Transport: Two Motors Review Are Sometimes Better Than One, Steven P. Gross, Michael Vershinin and George T. Shubeita, Current Biology 17, R478–R486, 2007, page no. R479, © 2007 *Elsevier Ltd., with permission from Elsevier Limited.*

Molecular motors like kinesins move directly along stiff nanotracks to transport cargoes as we have already seen. For smaller load forces, each solitary motor step is coupled to hydrolysis of exactly one ATP molecule and dislocates the centre of mass by 8nm towards the microtubular plus end. Alterations in chemical reactions and conformation of a molecular molecule elicit specific chemo-mechanical web. The cycle of this web govern the dynamics of molecular motors. In excess of ATP, kinesins can make about 100 steps in one second with a resultant velocity of about one micrometer. On macroscopic scale, this movement corresponds to a runner who runs 190 meters per second! We have already learned that this movement is in a very viscous and noisy environment, since it steadily undergoes thermally excited collisions with a large number of water molecules.

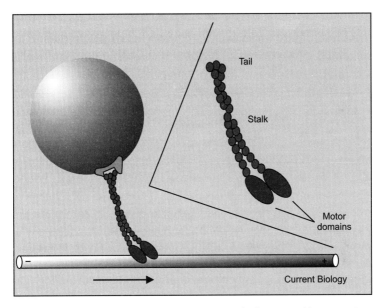

Fig. 2.13 Unidirectional transport. A prototypical motor with two motor domains, a stalk, and a tail domain moves cargo unidirectionally towards the microtubule plus end. The motor domains cooperate with the track, and the tail domain attaches to the cargo via a receptor/adaptor (green).

Reprinted from Bidirectional Transport along Microtubules, Michael A. Welte, Current Biology, Vol. 14, R525–R537, *July* 13, 2004, *page no.* R527, ©2004 *Elsevier Ltd., with permission from Elsevier Limited.*

Let's make this discussion more motivating by discussing more fundaments facets of these magnificent motors. The directed walks of cytoskeletal motors are rather remarkable, since these motors function at nanometric scale. *At this scale, molecular machines become fragile because of low spring constant.* Hence, they can be easily perturbed by random thermal collisions with water molecule which is the major filler of cell. Consequently, the directed movement of these motor proteins directly expose that they are able to escape from the surrounding "molecular chaos". Generally, any nanometric object such as, motors under consideration, in watery environment will undergo unpredictable movement which reflect the underlying thermal collisions and has no preferred spatial direction in absence of the external force. We have already appreciated rotary molecular motors to rectify or bias these molecular fluctuations in order to generate torque. This is talking about Brownian ratchets. A molecular motor can have several more forms such as;

- A single molecular motor.
- A cargo particle which is anchored to a single molecular motor.
- A cargo particle which is anchored to multiple molecular motors.

For a single molecular motor, there are three possible organizations:

- The molecular dynamics of the motor proteins underlying the chemomechanical energy transduction which leads to a single step of the motor particle.
- The directed marching of motor on the filaments.
- The motor walks of the motor particle as it repeatedly unbinds from and rebinds to the filaments.

The molecular organization covers all length scales up to the disarticulation of a single motor protein using ATP as energy source. *Two footed molecules like kinesin walks by distinct steps which results in shifting the centre of mass corresponding to repeat distance of the microtubule* (Svoboda, *et al.*, 1993). Similarly, motors like Myosin V which walks in similar manner with two heads makes steps which lead to a centre of mass displacement of 36 nm close to the helical pitch of the actin filament (Reif, *et al.*, 2000). Stepping time, for kinesin, has been calculated to be faster than 70ms (Cappello, *et al.*, 2003). The time scale for entire motor cycle is considerably longer and depends on the ATP concentration. The cycle becomes essentially independent of this concentration at increased concentration of ATP. The tenure of the cycle is observed to be of the order of 10ms. With decreasing concentration of ATP, the cycle becomes lengthy when it becomes dominated by diffusion-restricted transport of ATP molecules to the ATP adsorption domain of motor heads. This reaction requires assistance of an enzyme in order to proceed with reasonable rates. This enzymatic activity is provided by the catalytic domain of a motor head. There are three cardinal parameters occurring at catalytic domain.

- Adsorption of one ATP molecule at vacant catalytic motor domain.
- The ATP molecule is hydrolyzed corresponding to the forward direction.
- Desorption of ADP and Pi from the catalytic domain.

This was talking about forward reaction, but the backward reaction also has some peculiar characteristics which are as follows:

1. Adsorption of ADP and Pi at the empty catalytic motor domain.
2. Formation of ATP from ADP and Pi.
3. Desorption of ATP from catalytic domain.

How molecular motors move?

As per the hand to hand mode, a processive molecular motor alternates between two-foot and one-foot binding to the tracks (Kawaguchi and Ishiwata, 2001). To move forward, (and to prevent backward movement) a molecular motor, without an intelligent thought and action, must make an acceptable choice mechanically at least two times in a cycle. Thus, when both feet are glued to the track, the motor must use correct foot to bring up. Then, the elevated foot must land on a forward binding site (and not forward site). The existing assumption for this type of navigation is that a strain-dependent mode of action requires the lifting of the trail foot and a lever action in the landed foot biases the motion of the lifted foot forward to assure forward landing.

During a normal walk, both the feet of a molecular motor are landed on the track and their chemical statutes is independent of being associated with ATP, ADP and Pi , ADP, phosphate, or none, may well be different (Fig. 2.14). In other words, chemical states of the two-feet, whether they bind feet in different chemical states should ideally exhibit different chemical affinities for a landing site on track. Thus, in principle, the motor is dependent on the string of hydrolysis reaction to prefer the proper foot to lift, the foot with a weak affinity, because of the differential reaction phase on both trail and lead foot. It means the reaction phase on trail foot is predicted to be ahead of the phase on the lead foot.

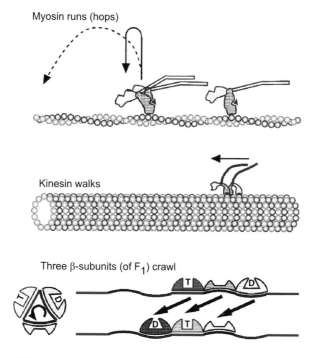

Fig. 2.14 Mode of operation of various Molecular Motors.

Reprinted from F1-ATPase: A Rotary Motor, Kinosita, Jr., Yasuda, Noji, Ishiwata, and Yoshida, Cell, 93: 21–24, 1998, ©1998 by Cell Press, with permission from Elsevier Limited.

To discriminate the phase differences between the chemical reactions, a motor protein exploits the difference in strain in the two feet (Hancock and

Howard, 1999; Mehta, 2001). Since both the feet are articulated to each other through the legs, the trail foot is pulled forward, while the lead foot is pulled backward in the two-foot landing posture.

The difference in the strain could adjust the affinity of the foot for the track, or the preference of the foot for a particular chemical state. Cardinal role of strain in regulated movement of the molecular motor is experimentally verified by Uemura and Ishiwata (2003). In case of kinesin, affinity towards microtubule is known to be feeble when it binds ADP (Hackney, 1994; Uemura, et al., 2002). *According to Uemura, et al., under the physiological concentration of Mg ADP (~10–4 M) strain dependent detachment of kinesin from a microtubule takes place.* Detailed study of this aspect demands writing another book, hence, it is recommended that interested reader should refer advanced review on walking mechanism of molecular motors (Kinosita, *et al., article in press*).

2.6 Summary

Bio-Nanotechnology will boom only after in depth understanding of the molecular machines fabricated by nature over the course of evolution. We must find the tactics to tackle the obstruction coming when an object pierce in the realms of nanotechnology. Upon this magical shift, quantum mechanics become a dominating tool to understand the biophysical characteristics of biological motors. Gravity and inertia almost disappear at nanoscale, giving required competency in biomolecular interactions. Movement is another obstacle need to be solved at nanometric dimensions. Biologicals move in at low Reynolds's number due to increase in viscosity and the movement is purely non-reciprocating. Mimicking the modus operandi of nanomotors, such as ATP synthase, flagella, linear molecular motors like myosin and kinesins, will bring an energetic momentum in nanorobotics and biomimetics. ATP synthase is the smallest known nanomotor functioning with 100% thermodynamic efficiency. Molecular tracks used by myosin can be exploited in fabrication of targeted drug delivery vehicle marching on the tracks. Thermodynamic efficiency of molecular motors, such as ATP synthase can be a cardinal feature embedded in nanorobotics.

□□□

Synthesis of Nanomaterials

> My own judgment is that the nanotechnology revolution has the potential to change America on a scale equal to, if not greater than, the computer revolution.
>
> —U.S. Senator Ron Wyden (D-Ore)

3.1 Introduction

Nanoparticles of less than 100 nm sizes are the bases for constructing nanostructures, nanomaterials, nanomachines and nanodevices. As discussed earlier in Chapter 1, regardless of the composition, every substance when reduced to size of less than 100 nm exhibits new properties. These properties (optical, electrical, mechanical, magnetic and chemical) can be systematically manipulated by adjusting the size, composition and shape of the nanoscale materials. Hence, one of the key aspects of nanotechnology concern is the development of reliable protocols for synthesis of nanomaterial over a range of chemical composition, sizes and high monodispersity.

3.2 Approaches Employed for Synthesis of Nanoparticles

Scientists are fast gaining control over accomplishing the methods for synthesis of desired nanoparticles. Many methods using precursors from all the three phases of matter *i.e.*, solid, liquid or gas phase have been developed. The manufacturing techniques employed may be classified as:

- Bottom-up Approach
- Top-down Approach

Though both approaches play very important roles in fabrication of nanoparticles, they both have advantages and disadvantages. Therefore, one has to choose them as per requirement very carefully.

3.2.1 Bottom-up Approach

As the name suggests it means build up of a nanomaterial from the bottom. Bottom-up approach involves atom-by-atom, molecule-by-molecule or cluster-by-cluster fabrication of desired nanoparticle. Moreover, by arranging molecules to form complex structures, with new and useful properties have also been accomplished by this method.

Advantages of bottom-up approach: This approach creates nanostructures with less defects, more homogenous chemical composition and better short and long range ordering can be obtained. This is because the "bottom-up" approach is driven mainly by the reduction of Gibbs free energy so that Nanostructures and Nanomaterials produced in such way that they are in a state closer to thermodynamic equilibrium state. Whereas, top-down approach is likely to introduce internal stress as well as surface defects.

However, the limitation that has been encountered during synthesis of nano materials using "bottom-up" approach is that it is difficult to make structure large enough and of desired quality.

3.2.2 Top-down Approach

It is just the opposite of bottom-up approach. Here a bulk material is used to generate nanoparticles (top-down) of desired size and shape by employing ultra fine grinders, lasers, and vapourization followed by cooling. The details of techniques used for nano particle fabrication are given later in this chapter.

The main challenge for top-down approach is the creation of increasingly small structure with sufficient accuracy. Moreover, top-down approach usually introduces internal stress and surface defects.

Need for stabilization of Nanoparticles: Nanomaterials, once they are synthesized and especially in the powdered form, they rapidly (sometimes within a few seconds) tend to aggregate through a solid bridging mechanism; and loose their nanoproperties. If the nanoparticles need to be kept without clumping, then they must be prepared and stored in a liquid medium designed to facilitate sufficient inter-particle repulsion forces to prevent aggregation. While using colloidal Nanoparticles keeping them stable in colloidal suspensions a prerequisite, they can be stabilized by various methods. However, encapping nanoparticles have also shown successful avoidance of aggregation. Some of the stabilization mechanisms of nanoparticles are:

(*a*) **Electrostatic Stabilization:** Involving the creation of a double layer of adsorbed ions over the nanoparticles resulting in a Columbic repulsion between approaching nanoparticles; or (*b*) **Steric Hindrance:** Achieved by adsorption of polymer molecules over the nanoparticles.

(*c*) **Encapping of Nanometal:** Khanna, *et al.*, (2005) have stabilized gold nanoparticles with Polyvinyl alcohol (PVA). They prepared stabilized gold nanoparticles in aqueous medium using two different reducing agents, hydrazine hydrate, a stronger reducing agent and sodium formaldehyde sulfoxylate (SFS), a slightly weaker reducing agent. The PVA stabilized gold nanoparticles remained stable over a long period of time with no indication of aggregation.

3.3 Techniques Employed for Synthesis of Nanostructures

Using both, the approaches mentioned above various physical, chemical and biological techniques have been developed for fabricating nanoparticles and nanostructures. Each method has proved their benefit and is being used as per requirement. One of the consideration during nanoparticles synthesis is controlling the chemical composition of surface, because it is surface at which most of the chemistry occurs. The application of nanoparticle is also dependent on the creation of specific surface sites of nanoparticles for selective molecular attachment. It is important for deciding its use in nanofabrication, nano-patterning, self-assembly, nanosensor, bioprobes, drug delivery, pigments, photocatalysis, LED's *etc.*

The synthesis methods for nanoparticles can be done in all the three phases of a matter as discussed here:

3.3.1 Gas-phase Synthesis

Gas-phase synthesis is used for manufacturing inorganic (ceramic and metal) nanoparticles. In gas-phase processes for synthesis of nanoparticles either as gas-to-particles or droplet-to-particle conversion process is followed. Most synthesis methods of nanoparticles in the gas phase are based on homogenous nucleation in the gas phase and subsequent condensation and coagulation. There are few advances in gas-phase synthesis by using solid precursor methods like inert gas condensation, pulsed laser ablation, spark discharge generation and ion sputtering. Chemical vapour deposition (CVD) processes is used to deposit thin solid films on surfaces. Here vapour phase precursors are brought into a hot-wall reactor under conditions that favor nucleation of particles in the vapour phase rather than deposition of a film on the wall. Often Vapour Phase Synthesis word is used for this process. Vapour-phase or gas-phase synthesis is a well-known chemical manufacturing technique for an extensive variety of nanosized particles. For synthesis of nanoparticles, conditions are created where the vapour phase mixture (which is thermodynamically unstable) are converted or condensed to nanosized solid particles. Here several reactor technologies, such as flame, hot-wall, plasma and laser reactors are applied.

Gas-phase synthetic techniques include pyrolysis (where solid bulk material is evaporated and the vapour-phase molecules thus formed is condensed to yield solid nanoparticles. This is done by various methods *e.g.,*

   (*i*)   Combustion

  (*ii*)   Physical vapour deposition with any type of plasma

 (*iii*)   Laser ablation

 (*iv*)   Spray pyrolysis

  (*v*)   Electrospray deposition or

 (*vi*)   Plasma spray

Another synthesis method is **epitaxial growth**. The metal is evaporated onto a textured substrate using molecular beam epitaxy. When the substrate is removed, the particles are freed.

3.3.2 Liquid-phase Synthesis

Liquid-phase synthesis implies chemical reactions in solvents. This leads to colloids, in which the nanoparticles formed can be stabilized against aggregation by surfactants or ligands. Micro-emulsion, reverse micelle process, hot-soap process, spray pyrolysis are the examples of liquid phase nanoparticle synthesis. Liquid-phase synthesis is a better route for making non-aggregating nanoparticles of uniform size and shape, and controlled surface properties. Such desirable properties are difficult to achieve in gas-phase synthesis routes.

3.3.3 Solid-phase Synthesis

Solid-phase synthesis is based on surface growth under vacuum conditions. Diffusion of atoms or small clusters on suitable substrates can lead to island formation, which can be seen as nanoparticles.

3.4 Methods Employed for Synthesis

Both physical and chemical methods can be used for synthesis of nanoparticles using any phase condition.

3.4.1   Physical Methods

3.4.1.1 *Ball Milling*

Ball Milling is an old procedure known since 1870 used for grinding flint for pottery. It has also been used for grinding materials into extremely fine powder for use in paints, pyrotechnics, black powder and ceramics for a long time.

The grinding of solid matters occurs under exposure of mechanical forces that trench the structure by overcoming of the interior bonding forces and creating a changed state of the solid grain size and even the grain shape. A ball mill is a cylindrical device which rotates around a horizontal axis, partially filled with the material to be ground plus the grinding medium consisting of ceramic, stainless steel, high density alumina or even glass balls. High-quality ball mills can grind mixture particles to as small as 0.0001 mm size. The grinding works on principle of critical speed *i.e.,* the speed after which the balls used for grinding of particles, start rotating along the direction of the cylindrical device; thus causing no further grinding.

With the advent of Nanotechnology ball milling is being used for producing Nanosized material. It is an attrition of large particles by simply grounding to nanosized particle. For synthesizing nanoparticles, high energy ball milling with usually tungsten carbide balls and an inert gas atmosphere is used to prevent contamination. This is the only top-down approach for fabricating nanoparticles.

Though the size distribution of the product is high and purity is low, it is preferred for production of large amount of nanoparticles. Ball milling has been successfully used for preparing nanosized particles

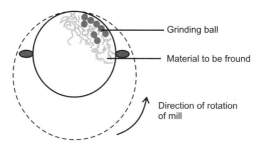

Grinding ball

Material to be fround

Direction of rotation of mill

Fig. 3.1 Schematic representation of working of a Ball-Mill.

However, there is no formula known which connects the technical grinding work with grinding results.

3.4.1.2 *Thermal or Flame Synthesis*

Thermal or Flame synthesis rather than supplying energy externally to induce reaction and particle nucleation, in a flame synthesis heat is produced *in situ* by the combustion reactions. This is used in making carbon black and metal oxides. It is primarily useful for making oxides. In flame spray, pyrolysis process rather than injecting vapour precursors into the flame, one can directly spray liquid precursor into it. Van der Waals and Ticich (2001) have reported Flame synthesis using a pyrolysis flame of CO/H_2 in presence of nanocatalyst an aerosol, which was created by drying a nebulized solution of iron or iron colloid (in the form of ferrofluid). Flame aerosol synthesis is one of the most promising routes to fabricate a range of single and multi-component functional nanoparticles at low cost and high production rate. This is commercially successful approach to nanoparticles synthesis. It is useful in making oxides, since the flame environment is very oxidizing.

3.4.1.3 *Spray Pyrolysis*

In spray pyrolysis, rather than delivering the nanoparticles precursors into a hot reactor as a vapour, a nebulizer is used to directly inject very small droplets of precursor solution. Reaction often takes place in solution in the droplets, followed by solvent evaporation. Hence, it is also known as *Nebulisation method.* For aerosolization of drugs, suspensions of biodegradable and hydrophobic nanoparticles consisting of a polymer and novel comb polymers are nebulized with a jet, ultrasonic and piezoelectric crystal nebulizer.

Recently an ultrasonic spray pyrolysis for nanoparticles synthesis was tried by Tsai, *et al.*, (2004) where they showed the effects of precursor drop size and precursor concentration on produced particle size and morphology.

3.4.1.4 *Laser Ablation*

The word *LASER* is an abbreviation or acronym for *Light Amplification by Stimulated Emission of Radiation.* Lasers are formed by irradiating a material that can generate a large concentration of inversion layer to produce stimulated radiation. This gives a monochromatic radiation. Lasers may be continuous or have pulsing type of frequencies ranging from 5 to 20Hz. The Pulsed Laser Deposition is a versatile technique used to grow or deposit thin-film of CNMs, DLC and amorphous carbon. Various laser parameters that influence the deposition of thin film are the laser's wavelength, the fluence of the laser, the ablation plume and the film quality. A pulsed laser rapidly evaporates a target material forming a thin film that retains target composition. The uniqueness of PLD is that energy source Pulsed laser is outside the chamber. This facilitates a large dynamic range of operating pressure (10^{-10} Torr to 100 Torr) during material synthesis. By controlling the deposition, pressure and temperature a variety of nanostructures and nanoparticles can be synthesized with unique functionalities.

The laser ablation in liquid phase is pioneering method for production for very stable and well-dispersed nanoparticles with very small and narrow size distribution. Pyatenko, *et al.*, (2005) have synthesized silver nanoparticles by laser ablation in liquid phase. Laser pyrolysis photothermal synthesis is an alternate means of heating the precursors to induce reaction and homogenous nucleation in absorption of laser energy. Heating is generally done using an infrared (CO) laser, whose energy is either to absorb by one of the precursors or by an inert photosensitizer, such as sulfur hexafluoride.

3.4.1.5 *Arc Discharge*

The original method of arc plasma heating was reported in 1964 by Holmgren, *et al.*, Later, Kratchmer, *et al.*, (1990) developed mass production method of synthesizing fullerenes by arc discharge. This technique led to the discovery of multi-walled carbon nanotubes (MWCNTs) by Ijima (1991) and Ando and Ijima (1993). In an arc discharge unit, two graphite rods are used as the cathode and anode, between which arcing occurs when DC voltage is supplied. A large quantity of electrons from the arc discharge moves to the anode and collides into the anodic rod. Carbon clusters from the anodic graphite rod caused by the collision are cooled to a low temperature and condensed on the surface of the cathodic graphite rod. The graphite deposits condensed on the cathode contain carbon nanotubes (CNTs) and nanoparticles.

Recently, Kuo-Hsiung Tseng (2009) has developed a method for synthesizing gold nanoparticle in an anhydrous ethanol using the arc discharge method (ADM).

3.4.1.6 *Thermal Plasma*

In thermal plasma, energy required to cause evaporation of small micrometre size particles is given as thermal energy, *i.e.*, thermal plasma temperatures in the order of 10000 K, so that solid powder easily evaporates. Nanoparticles are formed upon cooling while exiting the plasma region. The thermal plasmas are generated by dc plasma jet, dc arc plasma and radio frequency (RF) induction plasmas. The energy is generated by formation of an electric arc between anode and cathode and this energy is used for evaporation of precursor. In induction plasma, energy coupling to the plasma is accomplished through the electromagnetic field generated by the induction coil. The plasma gas does not come in contact with electrodes, which eliminates possible sources of contamination and allowing the operation of such plasma torches with a wide range of gases including inert, reducing, oxidizing and other corrosive atmospheres. The working frequency is usually between 200 kHz and 40 MHz. Since, the residence time of the injected feed droplets in the plasma is very short; the droplet sizes are kept small to get complete evaporation. The RF plasma method is used to synthesize different nanoparticles *e.g.*, ceramic nanoparticles, such as oxides, carbours/carbides and nitrides of Ti and Si.

3.4.2 Chemical Synthesis

Most of the chemical methods use Liquid phase as synthesis medium. Some commonly employed techniques for Nanoparticle manufacturing are discussed as follows:

3.4.2.1 *Sol-gel Methods*

Sol (short for solution) means a solution containing particles in suspension; which is polymerized to form a *Gel*. So it is a simple and inexpensive wet chemical synthesis, based on use of aqueous solution or non-polar organic solvents. In sol gel method, phase transformation of a sol (which acts as a precursor) is obtained from metallic alkoxides or organometallic precursors which is either polymerized, polycondensed or hydrolysed at low temperature to form diphasic gel or colloidal suspension form producing high purity material having controlled composition of chemical. Metal oxide is formed by connecting the metal centres with oxo (M–O–M) or hydroxo (M–OH–M) bridges, thus generating metal-oxo or metal-hydroxo polymers in solution. Often particle density is very low in a colloidal system, hence excess of fluid needs to be removed. This phase separation is accomplished either by centrifugation or by allowing time for sedimentation and then decanting the excess liquid. Further removal of remaining liquid is achieved by a drying process. Nanoparticles synthesized by sol-gel method are finding diverse applications in optics, electronics, space, sensor, for controlled drug release *etc*. It must be mentioned here that sol-gel methods have been traced to be used in as early as 1880 for the formation of SiO_2 fibres and monoliths.

3.4.2.2 *Sonochemistry Methods*

Sonochemistry arises from acoustic cavitations, the formation, growth and implosive collapse of bubbles in a liquid. The advantages of this method are its narrow size distribution and control of particle size. "This methodology has been successfully used for the synthesis of nano size cobalt, iron and many others". They are broadly used in information storage media ferrofluid technology, as a contrasting agent in medical imaging, and for magnetically guided drug delivery.

3.4.2.3 *Photochemical Methods*

Hsan yin Hsu (2004) has reported a photochemical method of synthesis of gold nanoparticle using UV irradiation to form interesting shapes. "Gold salt solution was photo-chemically converted to gold nanoparticles using Xenon lamp as source of light". He could control the production of gold nanoparticles of various sizes and shapes (triangular, pentagonal, and hexagonal). Photothermal size reductions was found very suitable for maximum control of diameter of gold nanoparticles, which existed in the system and photochemical growth controls their size distribution also. They have suggested that during this reaction the gold salt. "Under the influence of UV light the gold salt becomes unstable; this unstable complex further reacts wtih ethylene glycol, which has a dual role to play; both as a solvent and as well as a reducing agent. In the entire process reduction of gold salt takes place. Finally gold atoms are formed due to disproportionation of Au^+. The gold atom thus formed plays a significant role as nucleation site for further growth of gold nanoparticles. The role of PVC in this cardinal reaction is to control the over growth as well as stabilization of nanoparticles formed".

3.4.2.4 *Nanoprecipitation Method*

The Nanoprecipitation technique (or solvent displacement method) for nanoparticle manufacture was first developed and patented by Jessi and co-workers, which involves rapid desolvation of the polymer when it is added to the non-solvent. This is a versatile technique, particularly with respect to the encapsulation of hydrophilic drugs *e.g.,* proteins and modifications of nanoparticles size. The parameters involved in nanoprecipitation method are the solvent and the non-solvent nature, the solvent/non-solvent volume ratio and the polymer concentration. It is extensively used for entrapment of proteins (*e.g.,* tetanus toxoid, lysozyme, and insulin) into poly(D,L-lactic acid) and poly(D,L-lactic-*co*-glycolic acid) nanoparticles.

Due to the plethora of properties exhibited by protein such as its stability, molecular weight, hydrophobicity; the caging of proteins into nanoparticles becomes daunting task. The stability of functional conformation of protein demands a suitable media, pH and a relevant ionic strength of solution to circumvent such problems. So that the proteins can maintain their secondary and tertiary structure (Stevenson, 2000). In a revised methodology by Bilati *et al.,* (2005), the hydrophyllic and hydrophobic compounds were caged within tiny nanoparticles using a relevant solvent and strictly in basence of water, to avoid hydrophobic constraints concerned with the protein stability. In this method the critical issues such as protein stability, high shearing rates, as well as hydrolytic degradation pathways can be reduced.

To overcome such pain staking task for protein solubilization, the solid – in oil – in water method is most widely exploited. This method also surpasses the tedious water organic solvent interface during the emulsification process which is most widely used method.

Stevenson (2000) performed the precipitation of lysozyme and insulin using DMSO as solvent. In this precipitation process, the protein was partially misfolded but its activity was retained once reconstituted in aqueous medium.

In a unique method to encourage the stability of the protein, a protein surfactant complex was synthesized using iron pairing technique. As a surfactant concentration below the critical micelle concentration and at a pH below the isoelectric point of the protein, the confrontation of an anionic amphiphyllic molecule and positively charged protein; the pairing of the ions takes place contributing to the stability of the protein. (Meyer and Manning 1998 and Quintanar – Guerrero et al., 1997).

The ion pairing is a versatile method which makes a protein more hydrophobic and hence increasing its solubility in organic solvents. This improves the loadng of protein onto the nanoparticles.

3.4.2.5 *Hydrothermal Treatment*

Singh, et al., (2006) have demonstrated a water based simple synthesis of re-dispersible silver nanoparticle (of less than 50 nm) using silver nitrate as precursor and tri-sodium citrate as initial surfactant-cum-reducing agent followed by a secondary reducing agent i.e., sodium formaldehyde sulphoxylate (SFS). The citrate ions also created hydrophilic capping to *in situ* generated zero-valent silver, thus leading to surfactant capped particles. Partial re-dispersion of such nanopowder in aqueous medium lead to formation of < 30 nm colloidal silver.

3.4.2.6 *Pyrolysis Including Chemical Vapour Deposition (CVD) Method*

Pyrolysis is an umbrella term wherein; the precursors in vapor form are allowed to pass through a nozzle under very high pressure and temperature. Though this method is widely used for making nanomaterials; it suffers from aggregation and agglomeration rather than monodispersivity of nanomaterials formed. For comprehending the depth of the process, the students are referred to chapter 7.

3.4.2.7 *Colloidal Synthesis*

This method has been known since ancient time. "Chemical synthesis of the magnificent colloidal gold was used for making colored glasses as well as paints from the time unknown. Surprisingly in a process invented by John Herschel, the gold nanoparticles were used to record images on the paper. Thanks to Michael faraday, who examined the physical and chemical properties of gold nanoparticles. This was the incidence of first pure colloidal gold synthesis." It seems Faraday was inspired by work of *Paracelsus* (*Paracelsus* claimed to have created a potion called *Aurum Potabile* (Latin: potable gold). Faraday used *phosphorus* to reduce a solution of *gold chloride. Colloidal gold*, also known as "nanogold", is a suspension or colloid nano-sized particles of gold in a fluid—usually water. The liquid is usually either an intense red color (for particles less than 100 nm), or a dirty yellowish colour (for larger particles). "In this method, the gold salts solution is reduced using a suitable reducing agent. During the reduction process Au^{+3} is converted to gold atoms." Some of the known methods of synthesizing gold nano- particles by colloidal method are Brust and Schiffrin (1990) method and J. Turkevich, *et al.*, method (1951). The CdS nanoparticles have also been synthesized by the colloidal chemical method.

3.5 Structural Uniformity, Monodispersity and Need for Stabilization

During fabrication of nanoparticles, often irregular particle sizes and shapes are formed due to uncontrolled agglomeration caused by Van der Waals forces (Aksay, *et al.*, 1983). Monodispersed powders of colloidal nanoparticles need to be stabilized. Moreover, due to very small size, nanoparticles have a very high surface area/volume ratio resulting in a high reactivity. Therefore, the need for stabilization is all the more. When particles are not stabilized they generally undergo Oswald Ripening; it is a process where the smaller particles have a tendency to merge with one another till a large particle is formed. To avoid this agglomeration, nanoparticles can be stabilized by any of the following two methods:

(*a*) **Sterically** *i.e.*, by coating it with a polymer or protein. Here a ligand is added in the solution to perform capping, or (*b*) **Electrostatically** by attaching ligands with a negatively charged end-group to the particles, hence a repulsive force between the particles is introduced to keep them form coming in contact with one another.

3.6 Self-assembly Technique for Synthesizing Nanoparticles

Along with the physical and chemical methods described earlier; efforts are being tried to synthesize single-nanometer sized nanoparticles in large quantities. Nanotechnologists' interests in mimicking nature to synthesize nanoparticles have given them an impetus to think in the direction of self-assembly techniques. Realization that some molecules can self-assembly (*e.g.,* DNA molecule that has been performing self-assembly since the origin of life on this planet or the tissues involved in healing of wounds in bio-system); have helped the nanotechnologists to develop several methods of self-assembly. Rothmund (2005) has tried to mimic DNA duplication to engineer or create complex components for self-assembly. It is a bottom-up fabrication from atoms to molecules.

Some of the methods of self-assembly of nanoparticles are mentioned below. The mechanism of self-assembly may be by:

3.6.1 Physicochemical Methods

When detergents, fat or lipid molecules are poured on water they self-assemble into a micelle on water surface. Both fatty acids and detergent molecules have a hydrophilic head and hydrophobic tail; the micelle that they form is a spherical structure on water surface (Fig. 3.2). The cytoplasm of a living cell contains a spherical structure known as liposome of about 25 nm to 1 μ which are micelles of triglycerides. Similarly 4 nm wide lipid bilayer a part of cell membrane arranges them.

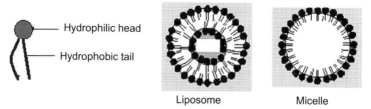

Liposome   Micelle

Fig. 3.2 (*A*) Molecule having hydrophobic head and hydrophilic tail. (*B*) Micelle formation in liposome and (*C*) of detergent on water surface.

3.6.2 Electrodeposition

In microchips, copper atoms are electrodeposited where it self-assemble into an ordered solid material and fill the trench to form nanowires.

3.6.3 By Insulating Holes

IBM uses this technique, where an air gap is formed between copper wires on a computer chip, this air gap allows faster flow of electrical signals. Moreover, it consumes less electricity.

3.6.4 Nanolithographic Self-assembly

Nanolithographic self-assembly is a bottom up approach, in which nanoscale structures having at least one lateral dimension of size of one atom to 100 nm. It is used for fabrication of semiconductor integrated circuits of nanoscale and for NEMS (Nanoelectronic Mechanical Systems).

Various types of nanolithographic assembly are as follows:

3.6.4.1 *Soft Lithography*

It is called 'soft' because it uses elastomeric materials like Poly-di-methyl-siloxane (PDMS).

3.6.4.2 *Photolithography or Optical lithography*

Uses very short wavelengths (usually 103 nm) to produce < 100 nm patterns. It requires use of liquid immersion and a host of resolution enhancement technologies *e.g.,* (*i*) X-ray lithography by using the short wavelengths of 1 nm for the illumination; (*ii*) Extreme Ultraviolet (EUV) lithography (using ultra short wavelengths of 13.5 nm); (*iii*) Electron beam direct write lithography (EBDW)–(that uses a beam of electrons) and (*iv*) Charged particle lithography (*e.g.,* ion- or electron projection lithography, can also form very high-resolution patterns).

3.6.4.3 *Magnetolithography (ML)*

In which magnetic field is applied on the substrate, using paramagnetic metal masks (magnetic masks); to pattern surfaces. Magnetic mask defines the spatial distribution and shape of the applied magnetic field. Then ferromagnetic nanoparticles are assembled onto the substrate according to the field induced by the magnetic mask. These nanoparticles can get assembled by two methods. (*i*) By positive approach where magnetic nanoparticles react chemically or interact *via* chemical recognition with the substrate, causing immobilization of magnetic nanoparticles at selected locations, where the mask induces a magnetic field. It results in formation of patterned substrate and (*ii*) By negative approach, there is no chemical reaction. The magnetic nanoparticles remain inert to the substrate; and after patterning the substrate, they block their binding site on the substrate so that it does not react with another reacting agent, it adsorbs the reacting agent. Then the nanoparticles are removed, resulting in a negatively patterned substrate.

3.6.4.4 *Nanoimprint Lithography (NIL)*

Involves fabrication of nanopatterns by mechanical deformation of imprint resist (imprint resist is a monomer or polymer formulation that is cured by heat or UV light during imprinting). Using nanoimprint lithography, patterns with 3D structures can be fabricated. NIL is being used to fabricate device for Electrical (MOSFET, O-TFT, single electron memory), Optical and Photonic (in fabrication of sub-wavelength resonant grating filter, polarizers, waveplate, anti-reflective structures, integrated *photonics* circuit and plasmontic device) and Biological (in DNA strenching, for shrinking the size of biomolecular sorting device so that they are more efficient) applications.

There are various types of Nanoimprint lithography *e.g.*, Thermoplastic nano imprint lithography, Photo-nanoimprint nanolithography, Electrochemical Nanoimprinting, Full wafer nanoimprint, step and repear nanoimprint *etc.*

3.6.4.5 *Dip Pen Nanolithography or Atomic Force Microscopic Lithography*

In this method, AFM is used for writing nanostructures using molecules or nanoparticles. It is a chemo-mechanical method where tip of AFM forms pattern on a surface (Fig. 3.3). Although, the AFM is an imaging device, it is now being used to either to measure or to apply locally small forces on a sample, thus creating nanopatterns on a desired substrate.

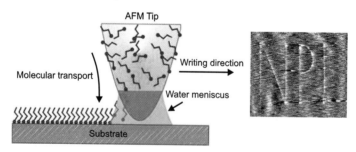

Fig. 3.3 Dip pen nanolithography *e.g.*, structure written with CdSe/ZnS nanoparticles on gold surface (Image size 3.5 micrometer).

3.6.4.6 *Enzymatic Nanolithography*

It is done by using AFM. Here instead of chemicals like *CdSe/ZnS;* immobilizing few molecules of enzyme (*e.g.*, alkaline phosphatase) are used by placing them on the tip of AFM. The substrate is present in the solution and precipitates after enzyme digestion.

Synthesizing nanoparticles using the above mentioned physical and chemical methods has been prevailing for reasonably a long period in most of the industries. But the recent advent of biological synthesis of nanoparticles, is now making every other method to take a back seat. In the next chapter, Biosynthesis of nano scale structures is discussed.

3.7 Summary

In this chapter, Bottom-up and Top-down approach of synthesizing nano-materials is introduced. Moreover, various chemical as well as physical methods of gas-phase, liquid-phase and solid-phase synthesis of nanomaterials is also covered.

4

Biosynthesis of Nanomaterials

4.1 Introduction

A primitive man thought of sky as wonderful, mysterious and awesome, but he could not even dream of what was within the golden disk or silver points of light so far beyond his reach. Man himself, a mysterious and curious object, started exploiting his gray matter to understand the mystery of colloidal solutions when light is impinged on it. The phenomenon of absorption, reflection and scattering of light always fascinated man from ages. No wonder peeping into the world of metal colloids which explicitly deal with this all-pervading phenomena has become a matter of excitement. Material scientists now call metal colloids of size 10–100 nm as nanocrystals formed by simply dicing down the bulk metals.

"The novel properties of the nanoparticles emerge due to their size dependent interaction with light. This property changes the color as well as other spectral properties making it an ideal candidate for optical devices and color coatings. The first evidence of successful synthesis of gold nanoparticles dates back to 1857 by Michael Faraday." His preparation was stable for the next hundred years and was destroyed in the bombardment done by London troops in the World War II. "However, the science of nanoparticles was in practice much before Faraday. For making colorful glases as well as paints Romans used different sized gold nanoparticles as coloring agents."

The absorption and scattering theory of colloidal solutions was propounded by Gustav Mie (1908). In his theory of Colloidal solutions especially of Gold, calculated the allowed states of small spheres of metal and their stability in the presence of a solvent. Hardy (1900) and Schulze (1882) further studied its stability in the presence of ions of different valencies. They studied the coagulation of colloidal solutions in the presence of oppositely charged solvent ions as compared to charge on the metal ions. Robert Brown then observed the phenomenon of Brownian motion in the colloidal solution which was theoretically backed by Einstein (1905). Another milestone in the material world was Smoluchowski's (1916) theory of coagulation process, which explained the diffusion of ions into colloid particle.

4.2 Fundamentals and Applications of Noble Metal Nanoparticles

We are not oblivious of the properties of gold, a precious jewel at bulk level. But if you dice it into smaller and smaller pieces, eventually this crystalline precious metal will enter into the strange realms of the nanoworld—a transition zone or an interface between a bulk metal and a single atom. This nanocrystal has all surfaces and no bulk. This nanocrystal cannot be now called as a jewel anymore because it is halfway between the bulk and the crystal. In such an altered states the nanocrystal show—unusual properties. Gold nanospheres of 5–10 nm show ruby red colour, while nanorods of 15–20 nm exhibit purple colour. Material Scientists are baffled not only about gold nanoparticles, but other metals also exhibit similar transformation in their properties at nanoscale. Copper is harder at nanoscale than at macroscale. PbS acts as an insulator at macrosize, but at nanosize it is a conductor. Iron at macroscale is paramagnetic, but at nanolevel it becomes super paramagnetic. Silicon cermets, calcium carbonate films made of nanometer size are more efficient than the micro-sized films.

Quantum dots are few nanometers in diameter, roughly spherical, fluorescent, crystalline particles of semiconductors like CdS, CdSe and are those excitons (excitons are bound state of an electron with a hole and can be called as that quasi-particle having an overall charge zero) which are confined in all three spatial dimensions. Their potential application in diverse fields can be attributed to the property of quantum confinement. The wavelength of the emitted light can be tuned by using particles of different diameters. The diameter of the semiconductor nanocrystal is smaller than the equilibrium separation distance of the electron-hole pair (exciton) formed in the bulk material upon promotion of an electron from the valence band to the conduction band. The smaller the diameter of the nanocrystal, the more confined the exciton is, resulting in a larger gap between the valence and conduction band. By changing the composition of the nanocrystal and its diameter, the wavelength of emitted light can be varied. Quantum dots can be used as biological tracking agents or fluorescent dyes to monitor the movements of cells and biological molecules.

Fig. 4.1 Gold nanoparticles exhibiting different colours depending on their size (5–50 nm from right to left).

Metallic nanoparticles possess a distinctive optical property—Surface Plasmon Resonance. Surface Plasmon Resonance is due to the collective oscillation of free electrons (Fig. 4.2). Gold and silver nanoparticles are the most extensively studied metal nanocrystals and their surface plasmon resonance is a function of nanocrystal size and composition. The molecules attached to the surface of the metallic nanoparticles and the surrounding medium.

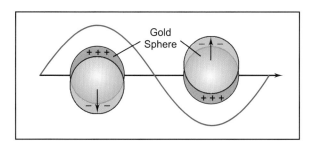

Fig. 4.2 Collective oscillation of free electrons of gold nanosphere.

Nanometals can have various potential applications due to their small size and large surface area; few of the significant ones in the field of biology are: Hyperthermia treatment for malignant cells, Magnetic resonance imaging enhancement, Cell labeling, Cell tracking, *In vivo* imaging, DNA detection, Diagnostics, Gene sequencing, Drug delivery systems, Biomedical sciences *etc.*, (Table 4.1).

The nanometals that have found application and attention include Selenium, Tellurium, Gold, Silver, Cadmium, Lead, Iron, Copper, Arsenic, Chromium and Zinc *etc.* It is interesting to note that most of these metals belong to transition metal group of periodic table. Transition elements are those which possess dual oxidation states because of partially filled or vacant *d*- or *f*-orbital. Though Cd, Fe and Zn do not belong to the transition metal group, but their complexes such as sulfides, amides resemble transition element properties.

Table 4.1. Various applications of Nanometals

Applications	Nanomaterials Used
ABRASIVES The ultra-fine particle size and distribution of properly dispersed products is virtually unmatched by anyother commercially-available abrasives.	• General Abrasives–Aluminum Oxide, Iron Oxide, Cerium Oxide • Rigid Memory Disk Polishing–Aluminum Oxide • Chemical Mechanical Planarization (CMP) of Semiconductors–Aluminum Oxide, Cerium Oxide • Silicon Wafer Polishing–Aluminum Oxide • Optical Polishing–Aluminum Oxide, Cerium Oxide • Fibre Optic Polishing–Cerium Oxide • Jewelry Polishing–Chromium Oxide, Aluminum Oxide, Iron Oxide, Tin Oxide.
CATALYSTS Nanomaterials have enhanced catalytic abilities due to their highly stressed surface atoms which are very reactive.	• General Catalysts–Titanium Dioxide, Zinc Oxide, Palladium • Catalyst Supports–Titanium Dioxide, Zinc Oxide, Palladium • Substrate for Precious Metals–Titanium Dioxide, Zinc Oxide, Palladium • Photocatalytic Oxidation–Titanium Dioxide, Zinc Oxide, Palladium • Oxidation Reduction Catalysts–Palladium • Ammonia Synthesis–Iron Oxide
COSMETICS • Nanoproducts provide high UV attenuation • Transparency to visible light β • Can be evenly dispersed into a wide range of cosmetics to make non-caking products.	• Hydrogen Synthesis–Iron Oxide • Sunscreens fe45b • Colour Foundations with spf fe45b • Lipstick with spf fe45b • Lip balm with spf fe45b–all uses Brown Iron Oxide or T_iO_2 • Foot Care Ointments–Zinc Oxide Powder • otc Topical–Zinc Oxide Powder.
ELECTRONIC DEVICES Nanoproducts can provide unique electrical and conduction properties for use in existing and future technologies.	• High Dielectric Ceramics–Barium Titanate • Conductive Pastes & Inm–Silver, Palladium • Capacitors–Titanium Dioxide, Barium Titanate • Phosphorus Crt Displays–Zinc Oxide • Electroluminescent Panel Displays–Zinc Oxide • Ceramic Substances for Electronic Circuits–Aluminum Oxide • Automobile Air Bag Propellant–Iron Oxide • Phosphorus Inside Fluorescent Tubes–Zinc Oxide

Contd...

Contd...————

	• Reflectors for Incandescent Lamps–Titanium Dioxide.
MAGNETICS Nanoproducts provide unique magnetic properties.	• Ferro-fluids and Magneto-rheological (mr) Fluids–Iron Oxide.
PIGMENTS AND COATINGS Nanoproducts facilitate creation of superior pigments and coatings, High UV attenuation. • Transparency to visible light • More vivid colours that will resist deterioration and fading over time • Can be evenly dispersed.	• General Pigments and Coatings–Iron Oxide, Titanium Dioxide • Microwave Absorbing Coatings–Iron Oxide • Radar Absorbing Coatings–Iron Oxide • UV Protecting Clear Coats–Titanium Dioxide, Zinc Oxide • Fungicide for Paints–Zinc Oxide • Powder Coating–Zinc Oxide • Automotive Pigment–(Demsited on Mica for Metallic Look)–Iron Oxide, Titanium Dioxide.
STRUCTURAL CERAMICS • The nanosize of the particles allow near-net shaping of ceramic parts via super plastic deformation which can reduce production costs by reducing the need for costly post-forming machining.	• Translucent Ceramics for Arc-tube Envelopes–Aluminum Oxide • Reinforcements for Metal-Matrix Composites–Aluminum Oxide, Titanium Dioxide • Porous Membranes for Gas Filtration–Aluminum Oxide, Titanium Dioxide • Net Shaped Wear Resistant Parts–Aluminum Oxide.
BIOSYSTEMS AND HEALTH-CARE	• Hyperthermia treatment for malignant cells – Fe • Magnetic Resonance Imaging Enhancement–Fe • Cell Labeling • Cell Tracking • _In vivo_ imaging–Barium sulfate • DNA Detection • Diagnostics • Gene Sequencing • Drug Delivery Systems–Au, Al • Biomedical Science–Ag • Antimicrobial and for Surgical Devices –Ag, Cu, Cd.
PRODUCTS FOR INDUSTRIAL APPLICATIONS	Titanium Dioxide, Iron Oxide, Zinc Oxide, Cerium Oxide, Yttrium Oxide, Tin Oxide, Copper Oxide, Chromium Oxide, Indium-Tin Oxide, Antimony–Tin Oxide, Molybdenum Oxide, Silver (Metal), Palladium (Metal), Zirconium-Silicate, Barium Titanate.

These transition metals when chisled into nanosize, exhibit many commercially important properties such as magnetic, optoelectronic, semiconductor. These exotic properties of nanometals are explicable due to quantum confinement of electrons at nanoscale dimension.

Production of segregated nanoparticles has unlocked its various applications in optoelectronics, microwave absorption, solar energy *etc*. Amongst plethora of applications; anti-microbial potential of some nanoparticles, such as silver, copper, cadmium *etc*., have drawn attention of material scientists and biologists.

Its applications have also been found in waste water treatment and various other environmental equipments *e.g.,* Fe-nanoparticles can quickly and cost effectively clean up contaminated soil and ground water. Metallic Fe oxidizes organic contaminants; such as tri-chloroethane, carbon tetra chloride, dioxins, DDT; break them down into simple less toxic carbon compounds, while heavy metals (Pb, Ni, Hg, and Ur) are reduced to insoluble forms.

Medical and health care scientists are also not untouched with the applicative potential of nanoparticles. Medical transplants, surgical devices, catheters can be coated with anti-microbial nanoparticles to prevent any cross infection.

The input of nanometals in engineering sector has also been remarkable. It is applied in powder metallurgy, magnetic nanoliquids, coatings and nano-media, for welding joints at 0.4 Tm, in reinforced glues and composites, as additives to engine oil, reusable metal nanofilters and are primary components for porous heat exchangers.

4.3 What is Biosynthesis?

Biosynthesis is a phenomenon wherein chemical compounds are produced from simpler reagents. Biosynthesis unlike chemosynthesis takes place within living organism and is generally catalyzed by enzymes. The process is a vital part of metabolism.

The prerequisite for biosynthesis are:

• Precursor substance
• Energy (usually in the form of ATP)
• Catalyst usually enzymes
• Reduction equivalent (in the form of NADH, NADPH and other).

Important and economically known products of biosynthesis include protein, vitamin and antibiotics. All these components of living beings are a result of this process. But the biosynthetic capacity of a living system does not end at metabolite formation only. For the survival over millions of years, living being have adapted themselves by often biosynthesizing nanoparticles. Adaptability is the most versatile characteristics of living organisms which has helped them to capture the entire earth. Microorganisms, the tiny engineers are the best examples of adaptation in non-conducive environment. One of the most challenging environments for the microbes is the presence of metal ions. Microbes are gifted with many molecular strategies to combat with the metal stress, one of which is decreasing the redox state of metal ion by reducing it with the help of electron shuttlers which may be extra or intra cellular. In due course, the metal ions are converted to nanoparticles of a defined size and shape. This provides a biological means of nanoparticle synthesis by using microbes as nanofactory. It is particularly very important over chemical means of synthesis which generates noxious substances and requires heavy metal expenditure.

4.4 Why Biosynthesis of Nanoparticles?

- Nanomaterials, especially bio-inorganic materials can be very complex and intricate in structure, composition and function. These features of bio-inorganic materials synthesized by living systems are almost impossible to mimic using chemical and physical synthesis techniques in the lab.

- The bio-based protocols for synthesis of nanometals are both environmentally and economically green as they are based on green chemistry principles and are simple and relatively inexpensive.

- The chemical methods available are often expensive, utilize toxic chemicals and are comparatively complicated. Certainly such methods are not eco-friendly and hence cleaner, cheaper, green processes that do not employ toxic chemicals for the synthesis of nanoparticles have to be devised.

- The nanoparticles synthesized by traditional chemical methods are unstable and tend to clump or agglomerate quickly and are rendered useless. The nanoparticles synthesized by various living systems have been shown to be coated with peptides or proteins. This leads to a similar charge distribution all over the surface of nanometal which results in repulsion between them. These inter particle repulsion forces prevent aggregation and so, nanometal solutions synthesized by microbes have been shown to be extremely stable even after a period of six months.

- Functionalization and conjugating the nanoparticles coated with peptides with other therapeutic molecules or other moieties is relatively easy.

These nanoparticles also show increased bioavailability and comparatively low toxicity inside living systems.

- This green chemistry approach for nanoparticle biosynthesis is simple, scaling up is possible, and it is environmentally friendly.

Thus, biological systems fulfill the above objectives and the required conditions as well. Hence, using bio-based protocols for synthesis of nano-materials with desired properties is contemplated.

4.5 Biosystems as Nanofactories

The terrible and demonic clutches of the catastrophic environment led to the survival of only those organisms who have meticulously adopted the survival strategies designed by nature. The efficient natural nanomachineries in the form of enzymes, organic molecules and intra- or extra-cellular shuttlers, have been endowed with special charachteristics for survival. Toxic metals spilled off by human civilization are contaminating the elixir of life *i.e.,* water. Soil also cannot escape this cataclysmic contamination. Eradication of the toxicity of metals is the only other alternative for Biological systems to survive. Hence, living organisms with their proficient nanomachineries transform toxic metal ions into non-toxic ones and moreover, they can even sequester them aside.

Microbiologists from decades were studying the mechanism of tolerance and resistance of metals by microbes. It was an open secret that metal ions were conglutinated together and packed by capping agents present in the microbes. Material scientists happened to get excited by this dexterous candidate which uniquely synthesizes monodisperse, anisotropic nanoparticles. This synthesis is done by using non-toxic chemicals and at enhanced rates. This intrigued them to screen different organisms for nanoparticles production and utilize them for various applications. Bacteria, Fungi, Actinomycetes, Plants and Algae have been extensively studied for their capacities of synthesizing nanometals. They can wither off the deleterious effects of metal ions by changing its redox states via reduction and then glue it up to form a thermodynamically stable nanoparticles. Prokaryotes exhibit tolerance and resistance against metals by four basic mechanisms (Silver, 1992).

1. Modulation of their transport

2. Active efflux

3. Redox changes

4. Sequestration and intracellular compartmentation into detoxified complexes.

Eukaryotes may use all these mechanisms, but the metal resistance is attributable to the intracellular compartmentation of toxic ions in complexes and/or within intracellular organelles. In yeasts and fungi, a major proportion of accumulated Ca^{+2}, Mn^{+2}, Zn^{+2} are located in the vacuoles complexed with polyphosphates. The metal ion sequestration in eukaryotes occurs via three main molecules:

1. Glutathione (GSH) (Coblenz and Wolf, 1994)
2. Phytochelatins (Rauser, 1990, 1995)
3. Cysteine rich metallothioneins (Stillman, *et al.*, 1992).

GSH and metallothioneins occur in animals, several fungi, some prokaryotes and perhaps in plants, which are induced by metals. In this chapter we will be discussing various attempts of biosynthesis of Nanometals and their possible mechanisms.

4.5.1 Bacteria: Awesome Machinery for Synthesis of Nanometals

Bacteria have an utmost capacity to pop in metal ions and conglomerate it within the cell without any harm to its day to day metabolic activities. This phenomenon is known as "tolerance for metal" by microorganisms. Sometimes they can even biomineralize them outside their own system not allowing toxic metal ions to interfere. This survival strategy is known as "resistance for metal" by microbes. Both these phenomena are well-capable of fabricating nano-particles of controlled size and shapes.

Gold: Gold nanoparticles have their applications in almost all the disciplines of science which necessitated scientists to find various means of their production. Beveridge and Murray (1980) for the first time synthesized gold nanoparticles in *Bacillus subtilis*. They noticed that Gold ions accumulated in the bacteria clumping together to form a precipitate. This precipitated complex was then dumped inside the cell-wall. Nair and Pradeep (2002) gave further insights into this work when they themselves inoculated Lactic acid bacteria in gold ion solution. They found that there are two kinds of size ranges in the solution *viz.*, 20–50 nm and above 100 nm which are also called Nanoclusters and Nanocrystals, respectively. Nanoclusters were found to occur within and outside the bacterial contours, while nanocrystals were found outside the contour. Nanoclusters inside the cell-wall actually formed nucleation site for proper crystal growth of Gold. Such gold crystals were then observed to push the bacterial cell wall. Many crystals formed on the surface can then consequently rupture the cell. Coalescence is the phenomenon by which surface area of the crystal is reduced so that it can be effectively protected to avoid any biological damage.

Lengke, *et al.*, (2006) could isolate a sulfate reducing bacteria which can synthesize intracellularly elemental gold using gold thiosulfate complex. The precipitation of gold was due to the formation and release of hydrogen sulfide as an end product of metabolism and occurred via three possible mechanisms. The mechanisms are as follows:

1. Adsorption and reduction of gold on iron sulfide surface.

2. Bacterial electron transport also reduces gold.

3. After the death of the bacteria its metabolic enzymes can also reduce gold ions.

They were successful in the formation of spherical aggregates containing octahedral gold. Konishi, *et al.*, (2006) used mesophilic anaerobic bacterium *Shewanella algae* with H_2 as an electron donor for intracellular precipitation of gold at 25°C and pH 7. The reductive precipitation was a fast process, producing insoluble nanoparticles of 10–20 nm size within 30 min. Extracellular synthesis of Gold nanoparticles is far more beneficial compared to intracellular ones, since there is no requirement of separation of nanoparticles from bacterial cells. When gold ions were added into the supernatant of *Pseudomonas aeruginosa* ATCC 90271, it was found that after some time the solution turns red. This colour is a clear indication of Gold nanoparticle synthesis which also exhibits Surface Plasmon Resonance (SPR) at 540 nm. It was observed that as the nanoparticle size increases the absorption peak position also shifts (Husseiny, *et al.*, 2007). *Rhodopseudomonas capsulata* was also found to be capable of synthesizing gold nanoparticles extracellularly at pH values ranging from 4.0–7.0. Spherical Gold nanoparticles were observed in the range of 10–20 nm at pH value 7 (Fig. 4.3) whereas, a number of triangular nanoplates were observed at pH 4. (He, *et al.*, 2007). The pH plays critical role in the reduction of $AuCl_4^-$ to $AuCl_4^-/Au$. The standard reduction potential of $AuCl_4^-/Au^0$ couple varies with –59 mV/pH unit over the pH range. The variation of pH regulates the proton concentration in the solution consequently leading to controlled size and morphology of gold nanoparticles. The protons affect the functional groups (amino, sulphydryl and carboxylic groups) of enzymes secreted by *Rhodopseudomonas capsulata*. The above functional groups supply electrons to $AuCl_4^-$, thus reducing it to form Au^0. These groups carry more positive charge at low pH values, thus, decreasing reducing power of the bacteria. This deteriorates the gold ion and the bacterial reaction rates, but increases their strength of interaction, thus contributing to nanoplate morphologies. At lower pH values gold nuclei start growing to form big crystals. This is because the surface charge which determines the interaction potential between the nanoparticles is lower. Surface charge varies with the pH of the solution and thus modulating the surface potential. When the pH value increases, correspondingly reaction rate and reducing power also increases, thus contributing to the thermodynamically favoured spherical morphology (Patungwasa and Hodak, 2008).

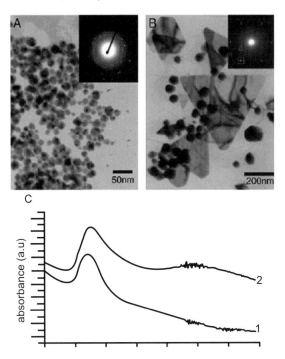

Fig. 4.3 TEM image of the gold nano-particles produced by the reaction of 10–3 M aqueous $HAuCl_4$ solution with bacteria R. capsulata biomass at pH 7. The inset shows their ED pattern. (B) TEM image of the gold nanoparticles produced by the reaction of 10–3 M aqueous $HAuCl_4$ solution with bacteria *R. capsulata* biomass at pH 4. The inset shows a typical SAED pattern of a gold nanoplate. (C) UV–Vis absorption spectra of gold nanoparticles produced at pH 7 (curve 1) and pH 4 (curve 2).

Reproduced with kind permission from Elsevier.

Silver: The antibacterial property of silver has been known for thousands of years with the ancient Greeks who cooked in silver pots. The old adage 'born with a silver spoon in his mouth' referred to more than just wealth. Eating with a silver spoon was known to be more hygienic. Silver cripples the electron transport chain in the bacteria leading to its death. Nonetheless, several bacterial strains are reported as silver resistant and may even accumulate silver at the cell wall to as much as 25% of the dry weight biomass. Thence, Material scientists and Microbiologists started working hand in hand to exploit the bacterial world for silver nanoparticle synthesis. Pooley (1982) noted for the first time biomineralization of silver in *Thiobacillus ferroxidans* and *Thiobacillus thioxidans* obtained from silver mine. Pooley reported deposition of Ag_2S nanoparticles on Thiobacillus when grown in silver containing sulfide leaching system. Later on Klaus, *et al.*, (1999) reported biosynthesis of single crystals of silver on defined compositions in *Pseudomonas stutzeri* AG259 (Fig. 4.4). These biosynthesized silver crystals of 200 nm diameter were often found to be located at the cell poles of the microbes. Nanoparticles of well-defined size, ranging from a few to 200 nm or more and distinct morphology were also deposited within the periplasmic space of the bacteria. The exact reaction mechanisms leading to the formation of silver nanoparticles by this species of silver resistant bacteria is yet to be elucidated. The ability of microorganisms to grow in the presence of high metal concentration might result form specific mechanisms of resistance, such as efflux system, alteration of solubility and toxicity by changes in the redox state of the metal ions, extracellular complexation or precipitation of metals, and the lack of specific metal transport systems (Silver, 1996).

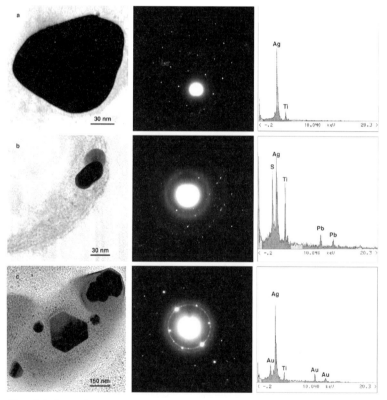

Fig. 4.4 Crystal structure analysis. (a) Regularly shaped nanocrystalline Ag particle taken from a thin, unstained section of a *P. stutzeri AG259 cell,* with a corresponding EDX spectrum (Right) and its electron diffraction pattern (Center) indicating elemental crystalline silver. (b) Second crystal type embedded in the periplasmic space of the cell. EDX spectrum and electron diffraction indicate monoclinic Ag2S. (c) A third type of crystal taken from a whole cell. The crystal structure is not yet clear. The electron diffraction pattern is not consistent with the pattern of elemental silver, whereas the EDX spectrum shows only Ag in considerable amounts.

Courtesy: Proceedings of National Academy of Sciences.

Lengke, *et al.*, (2007) successfully synthesized silver nanoparticles using *Plectonema boryanum* UTEX 455, a filamentous cyanobacterium. They could synthesize spherical nanoparticles and octahedral silver platelets of up to 200 nm in solution. The bioreduction of silver nitrate salt could be associated with metabolic processes that include growth, generation of metabolic energy and redox balancing. The bacterium is supported to reduce Ag+ to Ag0 and nitrate to ammonium which is fixed as glutamine before death. Dead cyanobacteria release certain organic compounds which cause further silver nanoparticle generation. Mokhtari, *et al.*, (2009) have also reported a novel rapid method of silver nanoparticle biosynthesis using culture supernatants of *Klebsiella pneumonia, Escherichia coli,* and *Enterobacter cloacae (Enterobacteriacae).* It was hypothesized that nitrate reductase present in these organisms play an important role in the conversion of Ag+ to Ag0 in the presence of NADH. This was confirmed by using piperitone and menthol which are considered to be inhibitors of nitrate reductase. Once piperitone and menthol were added in the culture media, the organisms lost the potential of synthesizing silver nanoparticles.

Cadmium: Cadmium sulfide and cadmium selenide quantum dots have brought revolution in the facile band-gap engineering of materials. Biological system especially bacteria are trained to transform toxic cadmium ions into insoluble non-toxic cadmium sulfide nanoparticles. Holmes, *et al.*, (1997) gave the first evidence of intracellular formation of cadmium sulfide nanoparticles ranging from 20–200 nm in *Klebsiella aerogenes.* Aiking, *et al.*, (1985) observed that metal-resistant strains of *Klebsiella aerogenes* precipitate lead, mercury and cadmium as insoluble sulfide granules on outer surfaces of the cells. In several bacteria including *Staphylococcus aureus and Bacillus subtilis,* it has been found that active transport of Cd^{2+} depends on the cross membrane electrical potential and this uptake system is highly specific for Cd^{2+} (Belliveau, *et al.*, 1987; Witte, *et al.*, 1986; Trevors, *et al.,* 1986). A common metal induced response in many microorganisms is the synthesis of intra-cellular metal binding proteins which functions in detoxification and also the storage and regulation of intra-cellular metal ion concentration. Metallothioneins, cysteine rich protein can bind metals like Cd, Zn and Cu. Hutchins, *et al.*, (1986) have reported the involvement of Cd binding prokaryotic metallothioneins in *Pseudomonas putida* and even Olafson. *et al.*, (1988) have also found the same in *Cyanobacteria synechococcus.* Inducible Cd binding protein occurs in *E. coli* also, which is larger than metallothioneins responsible for the recovery from the Cd toxicity (Mitra, 1984). Photosynthetic bacteria *Rhodopseudomonas palustris* has also been demonstrated to form CdS nano-particles of an average size of 8.01 ± 0.25 nm. The modus operandi behind CdS nanoparticle synthesis was considered to be the presence of cysteine desulfhydrase in the cytoplasm (Baia, *et al.*, 2009).

Iron: The abundance of iron in the earth's crust and the ability of iron to readily transit between Fe (III) and Fe (II) states has resulted in iron becoming a key metal in environmental microbe metal interaction. (Lovley, *et al.*, 1987) have found, for the first time, magnetite nanoparticles that are naturally synthesized by Magnetotactic bacteria. The dissimilatory ferric reductases (Schroeder, *et al.*, 2003) are essentially an intracellular one. But an extra-cellular one has also been isolated from *Mycobacterium paratuberculosis* (Homuth, *et al.*, 1998). Some bacteria have the capacity to reduce Fe^{+3} oxides by secreting small diffusible redox compounds that can serve as electron shuttle between the microbe and insoluble iron substrates (Newman, *et al.*, 2000). Studies on Fe (III) reduction in Pseudomonas strain 200 (now considered strain of *Shewanella putrefaciens)* suggested that Fe (III) reduction could be linked to electron transport chain involving cytochromes and other electron carriers. Kukkadapu, *et al.*, (2005) have reported for the first time biogenic ferrous hydroxy carbonate when dissimilatory Fe^{+3} reducing bacteria *Shewanella putrefaciens* was incubated with 1:1 mixture of ferrihydrite and nano-crystalline akaganeite under anoxic conditions with lactate as an electron donor and anthraquinone-2, 6-disulphonate as an electron shuttle. The incubation was carried out in a 1, 4-piperazinediethanesulphonic acid buffered medium without phosphate at circum neutral pH. From ancient times, it has been known that metal containing vessels like alloys of copper or iron spoil unpasteurized milk, buttermilk and curd when exposed within the vessel for longer period. Microbiology explains this as the activity of metal ion uptake by microbes, thus leading to toxicity production; but till date we were unaware that this action of the microbes takes place at nanolevel, forming nanometals. These activities (both extra-cellular as well as within the cell) are assisted by the enzymes present in *Lactobacillus*. Here, we are presenting our observations of intracellular as well as extra cellular production of metal (silver, copper and iron) nano-particles by *Lactobacillus sps* (Figs. 4.5, 4.6, 4.7, 4.8). *Lactobacillus* grown in copper, iron or iron alloy vessels could accumulate corresponding nano-particles, whereas those grown in silver vessels did not. However, addition

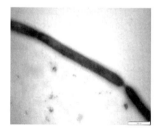 

Fig. 4.5 SEM of *Lactobacillus* showing (*a*) dead cell full of iron nanoparticles and (*b*) bursting of cell wall after over accumulation of iron nanoparticles in it.

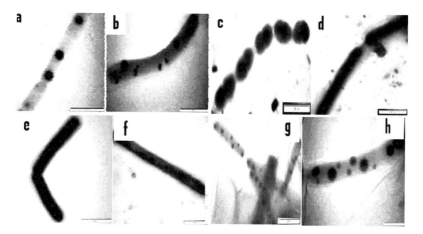

Fig. 4.6 TEM of *Lactobacillus sp.* exposed for 12 hrs. to metal plates of (*a* & *b*) Brass, (*c* & *d*) Copper, (*e* & *f*) Iron, and (*g* & *h*) Stainless steel; showing corresponding nanometals.

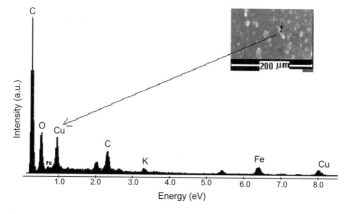

Fig. 4.7 TEM of *Lactobacillus* exposed for 12 hrs. to aqueous metal salt solutions. (*a*) Silver nitrate, (*b*) Copper sulfate, and (*c*) Ferrous nitrate; showing corresponding nanometals.

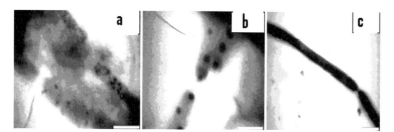

Fig. 4.8 EDAX of *Lactobacillus* grown on Copper sulfate solution. Inserted SEM shows the microbe point where the EDAX was done.

of metal salts of iron, silver and copper all resulted in intra-cellular accumulation of nanoparticles. Hashimoto, *et al.*, successfully studied hollow microtubes/ sheath consisting of amorphous iron oxide nanoparticles produced by iron-oxidizing bacteria *Leptothrix ochracea*. The sheath showed spin glass like magnetic property (Hashimoto, *et al.*, 2006). Watson, *et al.*, synthesized Nanosized strongly magnetic iron sulfide particles using sulfate-reducing bacteria. These nanoparticles were found to be excellent adsorbent for many metal ions. (Watson, *et al.*, 2000).

Palladium: Extensive use of metals in the area of catalysis and electronics has led to the growing interest in their recovery. The routine use of PGM (platinum group metals) is increasing due to their widespread adoption as automotive catalytic converter to reduce gaseous emission in vehicle exhaust for environmental protection (Hoffman, 1998). Platinum and palladium represent 90% of the total demand. To date there is no clean system for PGM processing (Yong, *et al.*, 2002). In 1998 Lloyd, *et al.*, noticed the bio-reductive deposition of palladium (0) onto biomass of *Desulfovibrio desulfuricans*. A preliminary study by Lloyd, *et al.*, has shown that resting cells of sulfate-reducing bacteria *Desulfovibrio desulfuricans* NCIMB 8307 can efficiently reduce Pd (II) to crystalline Pd (0) on the cell-surface at an expense of hydrogen as an electron donor in the presence of oxygen or using formate if oxygen was excluded. The reduction of Pd (II) to Pd (0) was accelerated by using sulfate-reducing bacteria at an expense of formate or hydrogen as electron donor at pH 2–7. Deposited Pd was visible on the cells using TEM and SEM and was confirmed to be 50 nm (Lloyd, *et al.*, 1998).

Platinum: An environmental friendly method using the metal ion reducing bacterium Shewanella algae was proposed by Konishi, *et al.*, (2007) in which resting cells were able to reduce $PtCl_6^{2-}$ into elemental platinum of size 5 nm at room temperature and neutral pH within 60 min. When lactate was provided as electron donor. In another instance, bioreduction of platinum (IV) into platinum (0) nanoparticles was found to be done by two oxygen tolerant/ protected periplasmic hydrogenase enzymes of a mixed consortium of sulphate reducing bacteria (Riddin, *et al.*, 2008).

Zinc and its Complexes: Abundant nanometer scale spherical aggregates of 2–5 nm diameter sphalerite (ZnS) particles formed within natural biofilms dominated by relatively aero-tolerant sulfate reducing bacteria family Desulfobacteriaciae. As discussed earlier, it has been found that *Cyanobacterium sps* and *Pseudomonas putida* have a mechanism of intracellular compartmentation of zinc through metallothioneins, cysteine rich protein (Butt and Ecker,1987; Higham, *et al.,* 1985).

Lead: The most efficient green synthetic process is developed for the synthesis of Lead sulfide nanoparticles. Immobilized Rhodobacter sphaeroides could synthesize PbS nanoparticles of an average particle size of 10.5 ± 0.15 nm (Bai and Zhang, 2009).

4.5.2 Fungi and Actinomycetes as Versatile Fabricators of Nanometals

Fungi and Actinomycetes are the industrial workhorses in terms of productivity and economic viability of desired products compared to bacteria. They can be meticulously exploited for nanoparticle synthesis as they are potent enough to grow wildly covering large surface areas by mycelial growth. Such omnipotent organisms hence are used as Biomimetic agents for the fabrication of metal nanoparticles and their extraction becomes viable due to simple downstream processing and handling of the biomass. Fungi have been found to be efficient secretor of soluble proteins and their mutant strains can secrete up to 30 g/l of extra-cellular protein. It is the character of high level protein secretion that has made fungi as favorite host of heterologous expression of high value mammalian protein for manufacturing by fermentation. Fungi and Actinomycetes stepped in very late in the world of nanoparticle synthesis, but conquered the field due to its versatility in secreting proteins, reducing toxic metal ions and capping them all at enhanced rates.

Gold: Gold nanoparticles exhibit exotic properties and are well-capable to be placed in various applications, such as optoelectronics, catalysis, reprography, single-electron transistors (SETs) and light emitters, non-linear optical devices and photo-electrochemical devices. Sastry's group (Shivshankar, *et al.*, 2004; Ahmad, *et al.*, 2003*a*, *b*, 2005; Mukherjee, *et al.*, 2001, 2002) have extensively worked on gold biosynthesis using different Fungi and Actinomycetes. Their earliest attempt was using an acidophilic fungi *Verticillium sps.* This was isolated from Taxus plant, when challenged with $AuCl_4^-$ lead to their reduction and accumulation as gold nano-particles within the fungal biomass. These gold nanoparticles could be released from the *Verticillium* cells by ultrasound treatment of the *Verticillium* biomass and also by reaction with suitable detergent. The monodispersity of the nano-particles produced intracellularly was not very high and is thus far inferior to that obtained by conventional chemical methods. In their trial for getting mono-dispersed gold nanoparticles, they had success with an extremophilic, *Thermomonospora sps.*

Nanometals have a strong tendency to agglomerate thus loosing the nano-character. To combat this problem nature has devised a simple method of capping the gold nanoparticles with protein right after their biosynthesis. Since, it is well-known that proteins can bind to gold particles through free amine groups or cysteine residue; Ahmad, *et al.*, (2003*a*) have proposed that stabilization of gold nanoparticles may be achieved by surface bound proteins. To confirm their hypothesis they tried to analyze the proteins secreted by *Thermomonospora*. In their effort to standardize the experimental conditions as well as more suitable fungi for biosynthesis they tried an alkali-tolerant fungus *Trichothecium*. They have reported that when biomass of this fungus was cultured on shaker in presence of $AuCl_4^-$; the fungi could assimilate nano-gold intracellularly, whereas cultures grown on static media could produce nano-gold extracellularly. Moreover, when they cultured *Fusarium oxysporum* (Ahmad, *et al.*, 2003*b*) in presence of both $AgNO_3$ and $AuCl_4^-$ they found that nanosized alloy of silver and gold was formed after 96 hrs. of exposure.

Silver: Silver nanoparticles exhibit striking colours from light yellow to brown. The colours are size dependent. Mukherjee, *et al.*, (2001) tried to use *Verticillium* for biosynthesis of silver. They noticed reduction of silver ions; which was lodged onto the cell wall as nanoparticle. Therefore, they suggested that reduction could be either intra-cellular or surface reduction. Recently Duran, *et al.*, (2005) have shown extra-cellular production of silver nano-particles by *Fusarium oxysporum* strain 07 SD. They found that when aqueous silver ions were exposed to *F. oxysporum*, the silver ions got reduced in solution, thereby leading to the formation of silver hydrosol. The silver nano-particles were in the range of 20–50 nm in dimension. They assayed presence of nitrate reductase in the solution and deduced that reduction of the metal ions occurred by nitrate dependent reductases and a shuttle quinone extra-cellular process (Detailed explanation is given later). Vighneshwaran, *et al.*, (2006) used the fungus *Aspergillus flavus* as a nanofactory for synthesis of silver nanoparticles. They also showed an evidence of tryptophan and tyrosine residues as a stabilizing agent of silver nanoparticles. Another attempt to synthesize silver Nanoparticles by *Fusarium semitectitum* was done by Basavraja, *et al.*, (2007).

Kathiresan, *et al.*, (2009) have synthesized silver nanoparticles using marine fungus *Penicillium fellutanum* isolated from coastal mangrove sediment. The biosynthesis of nanoparticles was considered to be the maximum using culture filtrate which was already treated with 1.0 mM $AgNO_3$, maintained at 0.3% NaCl and pH 6.0, incubated at 5 → C for 24 hrs. The culture filtrate, precipitated with ammonium sulphate, was proved to have a single protein band with a molecular weight of 70 kDa using polyacrylamide gel electrophoresis. Similar work was performed with *Penicillium brevicompactum* WA 2315 in which compactin, a compound isolated from the fungus was acting as a nucleating material for silver nanoparticle synthesis. The silver nanoparticles synthesized are of size 23–105 nm (Shaligram, *et al.*, 2009).

Cadmium: Nanometer scale semiconductor quantum crystallites exhibit size dependent and discrete excited electronic states which occur at energies higher than the band gap of the corresponding bulk solid. These crystallites are too small to have continuous energy bands. The onset of such quantum properties sets a fundamental limit to miniaturization in microelectronics. Dameron, *et al.*, (1989) reported the discovery of biosynthesis of the CdS quantum crystallites in yeast *Candida glabrata* and *Schizosaccharomyces pombe* cultured in presence of cadmium salts. Short chelating peptides of glutathione control the nucleation and growth of CdS crystallites to peptides capped intracellular particles of diameter 20 nm. These quantum CdS crystallites are more monodispersed than CdS particles synthesized chemically. According to Ortiz, *et al.*, (1995) formation of CdS nanocrystals in *S. pombe* is a phenomenon consisting of stress protein response. Upon exposure of *S. pombe* to cadmium, series of biochemical reactions are triggered to overcome the toxic effect of the metal. At first, an enzyme Phytochelatin synthase is activated that synthesize Phytochelatins having a basic structure of $(Glu\text{-}Cys)_n\text{-}Gly$. Phytochelatins carry out the chelation of cytoplasmic cadmium to form a low-molecular weight Phytochelatins-Cd complex. Later on, an ATP binding cassette (ABC)-type vacuolar membrane proteins, *viz.* HMT1 transports Phytochelatins-Cd complex across the vacuolar membrane. Within the vacuole, sulfide is added to the complex to form high molecular weight Phytochelatins-CdS nano-crystals (Kowshik, *et al.*, 2001) during their experimental studies noted that formation of CdS is dependent on the time of cadmium addition. During the mid-log phase of growth, maximum production of CdS takes place (Kowshik, *et al.*, 2001 and Williams, *et al.*, 1999). It has been shown by exit gas studies by Bae, *et al.*, (1998) that addition of cadmium in the early exponential growth phase leads to production of CdS nanocrystals. However, the cellular metabolism is affected, which results in efflux of cadmium from the cells. Cadmium addition during the stationary phase does not lead to CdS production. Intensive study of Torres-Martinez, *et al.*, (1999) showed for the first time that the nature of the capping material defines the size and extent of size-distribution of CdS nanocrystals. For example, a GSH-Cap resulted in the formation of CdS nanoparticles that had significantly heterogeneous size and chemical composition. Inducible Cd-binding proteins have been isolated from *Schizosaccharomyces pombe* and were named Phytochelatins. The metal binding peptides are composed of only 3 amino acids L-Cysteine, L-Glutamic acid and L-Glycine (Butt, *et al.*, 1987). Low-molecular weight cadmium binding proteins have been detected in *S. cerevisiae* by Joho, *et al.*, (1986).

Lead, Zinc and Copper Complexes: Kaushik, *et al.*, (2002) have biosynthesized PbS nanocrystallite by *Torulopsis sp.* extracellullarly and have found involvement of the specific enzymes. PbS is considered to be used as semiconductors in various optoelectronics, electronics and catalysis division.

Zinc sulfide nanocrystals synthesized by yeast *Candida glabrata* and *S. pombe* finds its application in the treatment of the specific dermatological conditions. These nanomaterials produce free radicals that can destroy the psoriatic cells upon UV irradiation *in situ*. Further damage to the normal skin could then be prevented by the free radical scavengers, such as glutathione (GSH), which has been proved to be a very important ingredient in the engineering of size controlled production of nanocrystallites, and are shown to be the biomolecules that cap ZnS nanocrystals synthesized by yeast *Candida glabrata* and *S. pombe*. Torres-Martinez, *et al.*, (1999) have recently shown that cysteine and cysteine containing peptides, such as glutathione and phytochelatins can be used *in vitro* to dictate the formation of the discrete sizes of the ZnS nanocrystals. They have proposed and confirmed that there are three steps involved in the biosynthesis of nanosized ZnS. (*i*) Formation of metal-complexes of cysteine or cysteine containing peptides. (*ii*) Introduction of stoichiometric inorganic sulfide into three metallo-complexes to initiate the formation of the nanocrystallites, and (*iii*) Finally size selective precipitation of nanocrystallites with ethanol in the presence of Na^+.

In *S. cerevisiae*, it has been shown that there is a connection between toxicity and energy dependent intracellular uptake of variety of metals including Cu, Co and Zn with decreased influx. Butt and Ecker (1987) have reported that metallothioneins are the intracellular metal binding proteins which function in detoxification of Zn and Cu which have been found in *S. cerevisiae*.

4.5.3 Plants Unveiling the Nature's Nanoengineers

Plants have several cellular structures and physiological processes to combat the toxicity of metals and maintain homeostasis. They also possess dynamic solutions to detoxify metals and hence scientists have now turned into phytoremediation. The modus operandi of detoxification includes immobilization, exclusion, chelation and compartmentalization of the metal ions, and the expression of more general stress response mechanisms, such as ethylene and stress proteins. The ability to tolerate inimical concentrations of toxic metals is found in the plant kingdom from ages. Their ability to accumulate high concentrations of metals was observed for both essential nutrients, such as copper (Cu), iron (Fe), zinc (Zn) and selenium (Se), as well as non-essential metals, such as cadmium (Cd), mercury (Hg), lead (Pb), aluminium (Al) and arsenic (As) (Salt, *et al.*, 1998).

Metallothioneins and Phytochelatins are the cysteine-rich polypeptides present in plants which have a significant role in chelating metals. Exposure to high concentrations of heavy metals leads to induction of these chelating compounds (Rauser, 1999; Cobbett, 2000; Clemens, 2001; Hall, 2002; Cobbett and Goldsbrough, 2002; Rea, *et al.*, 2004). The thiol-group of cysteine-rich polypeptide has very high affinity towards heavy metals. Metallothioneins are sulphur-rich proteins of 60–80 amino acids that contain 9–16 cysteine residues and are found in plants, animals and some prokaryotes (Rauser, 1999; Cobbett, 2000; Cobbett and Goldsbrough, 2002). Phytochelatins (PCs) are a family of γ-glutamylcysteine oligopeptides with glycine or other amino acids at the carboxy-terminal end, in which γ-Glu-Cys units are repeated 2–11 times. They are synthesized from the precursor glutathione (GSH) and its derivates by phytochelatin synthase in the presence of heavy-metal ions (Cobbett, 2000; Rea, *et al.*, 2004). PCs form ligand complexes with heavy metals like Cu and Cd, which are then sequestered into the vacuole. Reduced Glutathione (GSH) also occupies a central role in defense against oxidative stress, heavy metals and xenobiotics. It is synthesized in two ATP-dependent steps that are catalyzed by γ-glutamylcysteine synthetase (γ-ECS) and glutathione synthetase (May, *et al.*, 1998; Noctor, *et al.*, 1998 and Foyer, *et al.*, 2001).

Other low-molecular-weight chelators, including organic acids (malate, citrate), amino acids (*o*-acetylserine, histidine) and nicotinamine, are used in detoxification, sequestration or transport (Cobbett, 2000; Clemens, 2001; Hall, 2002 and Kramer, 2003).

The transformation to less harmful forms is another approach to detoxifying heavy metals, particularly As, Hg, Fe, Se and chromium (Cr), which exist in a variety of cationic and oxyanionic species and thiol- and organometallic forms (Meagher, 2000 and Guerinot and Salt, 2001). The abundance of chelators in plant makes them a suitable candidate for nano-particle synthesis. The above chelators, not only reduce the heavy metals, but also act as capping agent.

Gold and Silver: Gardea-Toresdey, *et al.*, (2002) for the first time exploited Alfa alfa plant as a factory for nanometal synthesis by growing it in aurochlorate rich environment. Nucleation and growth of Au nanoparticles were confirmed by atomic resolution analysis. Neem (*Azadirachta indica*) has been used in extra-cellular synthesis of pure metallic silver and gold nano-particles, and bimetallic Au/Ag nanoparticles (Shivshankar, *et al.*, 2004). On treatment of aqueous solution of silver nitrate and chloroaurate with Neem leaf extract, rapid formation of stable silver and gold nanoparticles at high concentrations was observed. The flavones and terpenoid constituents are the surface-active molecules which play significant role in stabilizing the nano-particles. Synthesis of nanoparticles by reduction of metal ion is possibly facilitated by reducing sugars and /or terpenoids present in leaf broth. Sastry's group has extra-cellullarly synthesized Au and Ag nanoparticles by using root and stem extracts of Geranium. Moreover, they demonstrated that the extracts of Lemongrass plant, when reacted with aqueous chloroaurate ions yield high concentrations of thin flat, single-crystalline gold nanotriangles. The process of nanotriangle synthesis involves rapid reduction of Au^{3+} to Au^0, and sintering at room temperature on liquid-like spherical gold nuclei. The fluidity arises due to nanoparticles surface complexation of aldehydes/ketones present in lemongrass extract (Shiv Shankar, 2004). Ankamwar, *et al.*, (2005) have recently reported extra cellular synthesis of gold and silver nanoparticles using *Emblica officinalis* fruit extract as a reducing agent to synthesize Ag and Au nano-particles. Treatment of aqueous solution of chloroauric acid and silver sulfate with *Emblica officinalis* extract caused rapid reduction of silver and chloroaurate ions, leading to formation of highly stable silver and gold nanoparticles of dimensions 10–20 nm, respectively. Huang and co-workers (2007) exploited sun dried *Cinnamomum camphora* leaf for synthesizing gold and silver nanoparticles ranging from 65–80 nm. They gave an evidence of the involvement of polyol components and water soluble heterocyclic components in the stabilization of nanoparticles. Wang, *et al.*, (2009) exploited the reducing capability of Barbated *Skullcup* herb to carry out green synthesis of gold nanoparticles and understanding its electrochemical significance. The sizes of the nanoparticles were in the range of 5–30 nm. Biosynthesized gold nanoparticles were used to modify glassy carbon electrode to enhance electronic transmission rate between the electrode and *p*-nitrophenol. Extract of *Volvariella volvacea* was chosen as reducing factory by Philip (2009) for making Au, Ag and Au-Ag nanoparticles. Gold nanoparticles of different sizes (20–150 nm) and shapes from triangular nanoprisms to nearly spherical and hexagonal were obtained by controlled reduction under differential parameters. It was found that Au nanoparticles are bound to proteins through free amino groups and silver nanoparticles through the carboxylate group of the amino acid residues. Sathishkumar, *et al.*, (2008) used bark extract and powder of *Cinnamommum zeylanicum to* synthesize silver nanoparticles having bactericidal activity. Zeta potential studies showed that the surface charge of the formed nanoparticles was highly negative.

Ti/Ni Nanoparticles: Small particles of Ti/Ni have been synthesized by a bioreduction method using a suspension of powdered milled *Alfaalfa*. The controlling parameter was the pH of the solution and the smallest size was obtained using pH 4 followed by pH of 7 where over 70% particles have sizes between 2 and 2.5 nm (Schabes-Retchkiman, *et al.*, 2006).

4.5.4 Algae: Dynamic Nanotechnologist

The capacity of Diatoms to take up silica from the surrounding medium and convert into very fascinating intricate patterns and symmetries (Fig. 4.9) has become a field of interest to many Material Scientists and has magnetized Nanotechnologists towards it.

Diatoms are unicellular micro-algae with highly sculpted walls of silica giving fascinating look of a Crystal palaces (Frithjof, *et al.*, 2005).The diatoms have myriad openings (such as pores and slits) through which constant exchange of molecules with the environment takes place. The intricate patterns and symmetries are species-specific and genetically determined Kröger, *et al.*, (2002) shed light on some of the organic molecules that are crucial for the formation of these diatom walls. The high degree of complexity and hierarchical structure displayed by diatom silica walls is achieved under mild physiological conditions. The biological processes that generate patterned biosilica are therefore of interest to the emerging field of nanotechnology. Diatoms successfully process silica from their specific interactions between silaffins and silica. Silaffins are peptides containing lysine residues that are linked to long-chain polyamines. It is due to this modification that Silaffin peptides are able to precipitate silica nanospheres even under slightly acidic pH conditions. Diatom wall formation and silicification occur in the complex environment provided by specialized silica deposition vesicles (SDVs) located in the cell cytoplasm Kröger, *et al.*, first isolated silaffins, a calcium binding protein, from a diatom *Cylindrotheca fusiformis* (Kröger, *et al.*, 1999) and subsequently from a range of diatoms. *In vitro*, silaffins catalyze the polymerization of silica spheres—tiny structures reminiscent of the nanoparticles known to constitute diatom biosilica (Kröger, *et al.*, 2000 and Schmid, *et al.*, 1979). Kröger, *et al.*, have now further defined the structure of the silaffins and discuss their pivotal role in the nanofabrication of diatom biosilica.

The extraction of organic molecules embedded in diatom silica requires harsh conditions that often damage their structure and function. Initial extractions of mature diatom silica by Kröger, *et al.*, (2000) and Van de Poll, *et al.*, (1999) yielded wall-associated proteins that were not localized to the SDV during silicification. However, extraction using anhydrous hydrofluoric acid yielded low molecular weight peptides, allowing the isolation and characterization of silaffins. Silaffins nucleate silica spheres of uniform morphology when added to a solution of silicic acid, although the size, shape, rate of precipitation, and pH of formation differ from those in diatoms. Further refinement of the extraction procedures yielded modified silaffins that could direct silica polymerization via pendant polyamines grafted onto the protein backbone (Kröger, *et al.*, 2000 and, 2001). These modified silaffins dramatically alter the rates of silicate precipitation; the process is accelerated in a mildly acidic environment, a condition thought to characterize developing SDVs. The presence of polyamines on the silaffins not only provides a possible template for nucleation, but might also control the silica colloid size within the SDV.

The globular silica particles observed by electron and atomic force microscopy to constitute diatom silica (Crawford, *et al.*, 2001) may reflect the chain lengths of the polyamines that are used to direct silica deposition.

The discovery of these molecules in a range of diatoms further demonstrates their role in the controlled polymerization of silica. Sumper (2002) has proposed a possible mechanism through which these polymerization determinants could also contribute to the formation of silicified structure. Kröger, *et al.*, further refined the silica extraction procedures to yield silaffins in their native state, in which both the polyamine "tails" and phosphorylation are preserved. The native silaffins are capable of assembling into supramolecular complexes by the intermolecular interactions between the negatively charged phosphate groups and the polyamine moieties. The supramolecular silaffin assemblies therefore nucleate rapid silica formation, and the data suggest that the positioning of silaffin nucleation sites may have a major role in micromorphogenesis (Wetherbee, 2002). A key to the development of nanotechnology will be the ability to make complex nanoscaled three-dimensional structures at low cost and in large numbers. The wide variety of structures in the silicified cell walls of diatoms offer a promising natural source of such materials. Diatom silica can be converted into other materials, with maintenance of detailed morphology. To facilitate the use of diatoms in nanotechnology, specific manipulation of the structure *in vivo* will be desirable (Hildebrand, *et al.*, 2005).

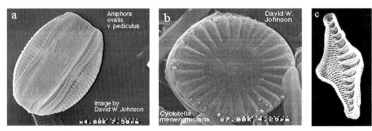

Fig. 4.9 Scanning electron micrographs showing intricate morphology of silica wall of Diatoms (a. Amphora ovalis, b. Cyclotella menenghiniana, c. Nitzchia sps.).

Courtesy Dr Rex Lowe.

Ceramic Nanoparticles: Inspired by the observations of Diatoms, a novel biosynthetic paradigm is introduced for fabricating three-dimensional (3-D) (Dickerson, *et al.*, 2005) ceramic nanoparticles assemblies with tailored shapes and tailored chemistries: biosculpting and shape-preserving inorganic conversion (BaSIC). Biosculpting refers to the use of biomolecules that direct the precipitation of ceramic nanoparticles to form a continuous 3-D structure with a tailored shape. A peptide derived from a diatom (a type of unicellular algae) to biosculpt silica nanoparticle based assemblies that, in turn, were converted into a new (non-silica) composition via a shape-preserving gas/silica displacement reaction. Interwoven, microfilamentary silica structures were prepared by exposing a peptide, derived from the silaffin-1A protein of the diatom *Cylindrotheca fusiformis*, to a tetramethylorthosilicate solution under a linear shear flow condition. Subsequent exposure of the silica microfilaments to magnesium gas at 900°C resulted in conversion into nanocrystalline magnesium oxide microfilaments with a retention of fine (submicrometer) features. Fluid (gas or liquid)/silica displacement reactions leading to a variety of other oxides have also been identified. This hybrid (biogenic/synthetic) approach opens the door to biosculpted ceramic microcomponents with multifarious tailored shapes and compositions for a wide range of environmental, aerospace, biomedical, chemical, telecommunications, automotive, manufacturing, and defense applications.

Silicon-Germanium Oxide: Marine diatom *Nitzschia frustulum* was harnessed to fabricate Si-Ge oxide nanocomposite materials. Germanium was incorporated into the diatom cell by a two-stage cultivation process. The cells assimilated soluble germanium by a surge uptake mechanism. The cell mass was thermally annealed in air at 800°C for 6 hrs. to oxidize carbonaceous materials. The thermally annealed cell biomass was none other than Nanostructured Si-Ge oxides (Rorrer, *et al.*, 2005).

Gold: Singaravelu, *et al.*, (2007) exploited *Sargassum wightii* for extracellular synthesis of gold nanoparticles. They have achieved rapid formation of gold nanoparticles in a short duration. The UV-Vis spectrum showed peak at 527 nm corresponding to the plasmon absorbance of gold nano-particles. Transmission Electron Micrograph showed formation of well dispersed gold nanoparticles in the range of 8–12 nm. An important potential benefit of the described method of synthesis of nanoparticles using marine algae is that they are quite stable in solution which is advantageous over other biological methods. Marine brown algae *Fucus vesiculosus* also has the capacity to reduce Au (III) to Au (0). It was instigated by Mata, *et al.*, (2009) that at pH 7, reduction potential is maximum. Such an environmentally useful process can be used for recovering gold from dilute hydrometallurgical solutions and leachates of electronic scraps.

Table 4.2. Various organisms that have shown capacity to biosynthesize Nanometals

Organism	Biosynthesizing Organisms	Biosynthesized Nanometal	Size(nm)
Bacteria	*Bacillus subtilis*	Gold	50–100
	Thiobacillus ferroxidans		20–50
	Lactobacillus sps.		100–250
	Pseudomonas stutzeri ATCC 90271		10–20
	Rhodopseudomonas capsulate		10–20
	E.coli DH5α		10–20

	T. Terroxidans	Silver sulfide	200
	T. thioxidans		200
	Pseudomonas stutzeri AG259	Silver	200
	Lactobacillus sps.		105–666
	Plectonema boryanum		200
	Klebsiella aerogenes	Cadmium sulfide	20–200
	Mycobacterium paratuberculosis	Iron	
	Shewanella putrefaciens		
	Lactobacillus sps.		76–461
	Leptothrix ochracea		100
	Desulfovibrio desulfuricans	Palladium	50
			50
	Desulfobacteriaceae	Zinc sulfide	2–5
	Cyanobacterium sps. and	Zinc	
	Pseudomonas putida		
	Lactobacillus sps.	Copper	333
	Shewanella algae	Platinum	5
Fungi & Actinomycetes	*Verticillium sps.*	Gold	
	Thermomonospora sps.		
	Trichothecium		
	Fusarium oxysporum		
	Verticillium sps.	Silver	20–50
	Fusarium oxysporum		
	Aspergillus flavus		10
	Fusarium semitectum		10–60
	Candida glabrata	Cadmium sulfide	2–5
	Schizosaccharomyces pombe		
	Torulopsis	Lead sulfide	100–200
Plants	*Alfa alfa*	Gold	
	Azadirachta indica		
	Geranium		
	Lemongrass		
	Emblica officinalis		10–20
	Cinnamomum camphora		65–80
	Emblica officinalis	Silver	10–20
	Azadirachta indica		
	Alfa alfa	Ti/Ni Alloy	2–2.5

4.6 Glutathione, Phytochelatins and Metallothioneins as Nanogetters

The optoelectronic and electrochemical properties of metals are size-dependent. These properties can be fine tuned at nanoscale by efficient regulation of their size. Biosystems are gifted with certain polypeptides like Glutathione (GSH), Phytochelatins (PCs) and Metallothioneins (MTs) which can control size and shape of the nanometals. These cysteine rich polypeptides can bind toxic metal ions (such as cadmium, lead, mercury, copper) and then sequester them in their non-toxic form.

Over-expression of MTs in bacterial cells result in enhanced metal accumulation and thus offers a promising strategy for the development of microbial-based Nanometal synthesis (Kille, *et al.*, 1991; Pazirandeh, *et al.*, 1995; Romeyer, *et al.*, 1990). PCs are short, cysteine-rich peptides with the general structure (γGlu-Cys) (Rauser, *et al.*, 1995, Zenk, 1998). PCs offer many advantages over MTs due to their unique structural characteristics, particularly the continuously repeating γGlu-Cys units. For example, PCs have higher metal-binding capacity than MTs (Mehra, *et al.*, 1995). In addition, PCs can incorporate high levels of inorganic sulfide that results in tremendous increase in the Cd^{2+}-binding capacity of these peptides (Mehra *et.al.*, 1994). However, development of organisms over-expressing PCs requires a thorough knowledge of the mechanisms involved in the synthesis and chain elongation of these peptides. The presence of γ bond between glutamic acid and cysteine in PCs indicates that these peptides must be synthesized enzymatically. PC biosynthesis may proceed by a variety of reactions involving enzymes that transfer γ Glu-Cys from GSH to other PCs (Fordham-Skelton, *et al.*, 1998). Although PC synthase has now been cloned (Clemens, *et al.*, 1999; Ha, *et al.*, 1999; Vatamaniuk, *et al.*, 1999) factors that govern chain elongation of PCs are far from understood.

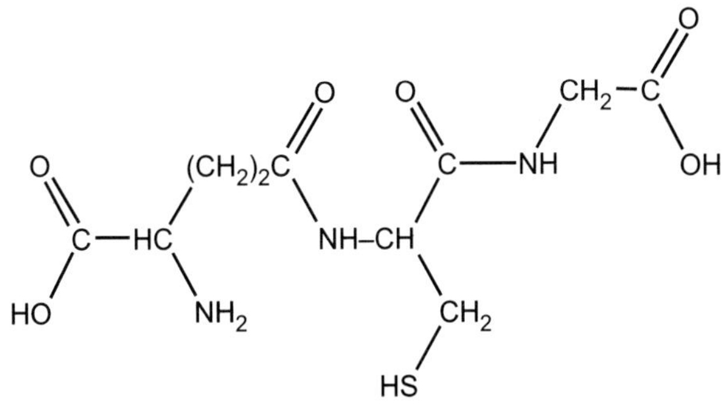

Fig. 4.10 Structure of Phytochelatin.

4.7 Metals as Electron Sink of Excreted Electron Shuttlers

The mechanism of electron transfer by microbes to poorly soluble minerals has been a subject of intense study. (Turick, *et al.*, 2002; and Arnold, *et al.*, 1996), at the cellular level, knowledge of transfer stems from mechanistic study of photosynthesis and respiration both in eukaryotic and prokaryotic system. Today we know in greater detail the structure and function of various membrane bound proteins that are part of electron transport processes. The knowledge of small molecules that participate in extra cellular electron transfer is still in infancy (Stowell, *et al.*, 1995; Gray, *et al.*, 1996). The most efficient route for energy generation involves the proton motive force established across the membrane (Mitchell, 1961). Current work on electron transport pathways that respires by using insoluble minerals as terminal electron acceptor has unlocked the possible importance of small molecules in metabolism that involves extra cellular electron transfer Newman and R. Kolter, 2000; Nevin, *et al.*, 2000. Hernandez and Neumann, 2001, have reviewed the role of excreted compounds in extra-cellular electron transfer. This mode of electron transfer involves small mobile molecules capable of undergoing redox cycling (*i.e.*, an electron shuttle).These molecules serve as terminal electron acceptors and once reduced transfer electrons to metal or metal oxide whereupon it becomes re-oxidized. A single shuttle molecule, in principle could cycle thousands of time and thus have a significant effect on the turnover of the terminal oxidant *e.g.,* iron. Organic molecules which play an important role as electron shuttlers are humic substances, quinones, phenazine and thiol containing molecules like cysteine, but anything that is redox-active and has the right redox potential, could serve the function. Homuth, *et al.*, isolated extracellular dissimilatory ferric reductases from *Mycobacterium paratuberculosis*. Several napthaquinone and anthraquinones with excellent redox properties were reported in *F. oxysporum* that could act as electron shuttle on metal reduction. Duran, *et al.*, found that aqueous silver ions when exposed to several *Fusarium oxysporum* strains are reduced in the solution extracellularly. The silver nano-particles were found in the range of 20–50 nm dimensions. It was a mere speculation earlier that reduction of metal ions is related to a reductases or quinone action. This was corroborated by Duran, *et al.*, in their study on *F. oxysporum* and based on their experiment they have laid down a hypothetical mechanism of silver nano-particles biosynthesis.

4.8 Summary

Microorganisms and plant cells have stupendous mechanisms to act as a nano-factory. Under hostile conditions posed by heavy metal, they have devised both extra-cellular as well as intracellular possibilities to synthesize nanometals. They have superseded the physical and chemical methods for nanometal synthesis. The problem of aggregation has been solved by the intelligent introduction of size regulating, capping agents like Phytochelatins, glutathione, and metallothioneins. There is plethora of physical as well as chemical methods developed for nanoparticles synthesis. But they have many limitations, such as aggregation, polydispersity *etc*. Moreover, as far as the synthesis of nano-particles is concerned, a number of chemical methods exist in the literature that uses toxic chemicals in the synthesis protocol, which raises great concern for environmental reasons. This has led to an exploration of biological means for nanometal synthesis. Moving a step further, biological systems have not only synthesized the nanometal and/or its complexes, but also meticulously defined its dimensions via fetter like proteins, giving it monodispersity. This nanometal synthesized has its applications in various multidisciplinary areas such as optoelectronics, metallurgy, ore factories, Biomedical engineering, textile, photographic imaging, and so on.

5

DNA and Proteins as Templates for Molecular Nanotechnology and Nanoelectronics

> The curiosity remains to grasp more clearly how the same matter, which in physics and in chemistry displays orderly and reproducible and relatively simple properties, arranges itself in the most astounding fashions as soon as it is drawn into the orbit of the living organism.... If it is true that the essence of life is the accumulation of experience through the generations, then one may perhaps suspect that the key problem of biology, from the physicist's point of view, is how living matter manages to record and perpetuate its experiences".
>
> **—Max Delbruck "A Physicist look at Biology"**

5.1 Introduction

A famous cartoonist Walt Disney once said that I try to personify each of our cartoon characters or caricatures—to build up a personality. A good caricature is that drawing or a sculpture or a character which can portray individual characteristics of a person scrupulously. Such strongly individualistic character could be anything from a smiling face or a protruding chin. Its behaviour and characteristics must be beyond reality. We are quite sensitive about the accuracy of the caricature. If an artist or a sculptor is successful in producing specific variation of a common feature (*e.g.*, facial expressions) then we can genuinely say that life has entered into it. All organisms are simply caricatures infused with life. Life remarkably arose from thousands of inanimate lifeless biomolecules interacting in such a way so as to maintain and perpetuate animated life, solely by physical and chemical principles which govern the non-living universe.

A life of scientist with his humble quest to comprehend the physical and chemical characteristics of a living matter has a keen longing for the amalgamation of commonalities and individualities. All living forms have been made from a common well-ordered organization of molecules, extracting nutrients from their environment and above all multiplying in numbers so that an offspring gains all the phenomena akin to that of parents. There is extreme regularity in terms of physical and chemical characteristics of all the organisms or of its cells. There is strict organization and internal order in all organisms. But with all our predilections for commonality we cannot neglect that each individual differs from other individuals of the same species. Thus, within the scaffold of common principles there must be space for individual irregularities.

Deoxyribonucleic acid: DNA is the most celebrated icon of all centuries dictating the common and individual properties of a living matter. DNA is the carrier of genetic information in all organisms excluding few of them. DNA is the connecting link between all individuals, but due to variations within them every individual within same species becomes unique.

The discovery of such a three-dimensional biological riddle was possible due to the unprecedented and ingenious insights presented by Erwin Schrodinger in his most celebrated book "What is life?" Life could be thought of in terms of storing and passing on the biological information which had to be packed in a "hereditary code-script" *i.e.,* DNA studded in the molecular fabric of chromosome. Schrodinger proposed that a gene could be stable for generations if it is an aperiodic crystal—in other words, something with a regular but non-repeating structure. He compared the difference in the structure of a normal and aperiodic crystal giving a beautiful example—ordinary wallpaper having repetitive patterns and periodicity will be a boring view, but a masterpiece of embroidery, such as a Raphael tapestry will be an exciting and lively vista.

Schrodinger's metaphor of a readable book of life, a decipherable genetic code, DNA is a doubly intertwined helical staircase outside of which consists of the phosphates and sugar molecules. Nitrogenous bases are tucked away inside the helix. The nitrogenous bases are adenine (A), guanine (G), cytosine (C) and thymine (T). Schrodinger's aperiodic crystal *i.e.,* DNA is nothing but a regular and non-repetitive structure of the above said nitrogen bases. The different permutation and combinations with which A, T, G and C arrange is known as *aperiodicity* in a DNA crystal. The phosphate groups tend to give negative charge to the double helical molecule which has to be stabilized by an equal amount of positively charged histones. DNA has a diameter of just 2 nm and is wrapped around histones with sufficient compaction and condensation to form a chromosome, a giant fabric. DNA is a spiral staircase which has within its realm a kind of message or a code that needs to be cracked. Molecular nanomachines like RNA polymerases, cracks that information transcribing the code-script to form a messenger RNA. This messenger RNA after joining hands with two counterparts: tRNA and

Ribosomal nanomachineries then translate or decode the information into a protein. RNA is another aperiodic crystal made up of non-repeating nitrogenous bases again (adenine, cytosine, guanine and uracil). Proteins are the nano-machineries of the cell which is capable of performing all the activities like metabolism, reproduction, sensing environmental signals and ultimately embark upon the final destination of a cell *i.e.*, death. Proteins themselves are aperiodic crystals constituted by different arrangements of 20 amino acids.

The assemblage of nucleic acids (DNA and RNA) and proteins represents the pinnacle of *Nanotechnology*. Their assembly is obeying the bottom-up approach and all atoms glue up to form a supramolecular structure in a deterministic fashion. Deterministic feature is the transpiration or emergence of all the atoms determined to perform a given task with the power of prediction. All atoms functionally and individually obey quantum mechanical principles, but as far as Nanomachines like proteins, nucleic acids are concerned, they do not completely follow quantum mechanics. This is because they are Nanopolymers lying in between atomic scale and bulk scale. Such Nanopolymers can often be treated with a blend of continuum mechanics and statistical mechanics. The continuum mechanical behaviour of polymers can be studied only when we consider them to be made from continuous regularly arranged monomers *i.e.*, in continuum and the goal of statistical mechanics or statistical thermodynamics is to interpret the measurable properties of polymers in terms of the properties of their constituent monomers and the interactions between them. This understanding of statistical mechanics has led us to exploit Proteins and DNA as a template for Molecular Nanotechnology and Nanoelectronics. They can be used in microarray technologies, Optical chip designing, Integrated Chip manufacturing, Tweezer designing, Nanodiagnostic tools *etc*…But before dealing with such applications, we need to understand the fundamental properties and thermodynamic stability of such nanomachines at cellular level so that they can be efficiently exploited outside the cell.

5.2 The Renaissance of Forces and Motions at the Nanometer Scale in a Cellular Factory

Venturing into the world of Nanoscale objects in a biological cell needs complete understanding of their behaviour in terms of their stability and dynamism. The cellular machinery is an edifice of Nanometer-scale functional molecules conjoined together to form macromolecules interacting via certain forces and illustrating confluence of thermal, chemical, mechanical and electrostatic energies. We are estranged about the properties of functional molecules at the nanoscale world. The first significant property is that the force of gravity which governs our daily activities is negligible. (This is helpful as most of the time we stand or sit vertically, if cellular machinery would have obeyed gravitational forces, then imagine our condition; all the molecules would have been in the lower part of the body or cell!) The second intriguing property is that the Van der Waals (dispersion) force becomes dominant in the nanoscale. Its attractive force component, which is in the order of piconewtons is about 100μm, but has decaying binding energy potential inversely proportional to the sixth power of separation, while the repulsive component strengthens with the twelfth power of distance. The thermal energy, kT of a molecule moving at a room temperature is something about 4×10^{-21} J (26meV) which is comparable to the Van der Waals binding potential of typical nanometer particles. Consequently, the Van der Waals force binds nanoobjects together. But due to thermal excitation in the cell, these nanoobjects frequently split apart, unless they become entangled by other stronger forces like dipole-dipole interaction, hydrophobic interaction, ionic bond which possess force strength approximately 10 times, 100 times and 1000 times of Van der Waals, respectively. Another imperative force that binds the nano-biomolecules is the residual surface charge generating an electrostatic field that determines their interactions with each other as well as the aqueous environment. This type of force is highly dependent on ionic concentrations and the resultant boundary layer effect. The biggest challenge in nanotechnological applications is *to hold or release nanometer scale objects in a controlled manner.* A biological cell is at par with such challenge.

Isotropy or the measure of asymmetry is a useful property since in the nanoworld the phenomenon of up or down is irrelevant. There is overwhelming difference found near the surface and the bulk volume. The nanoobjects are profoundly and densely embedded in water molecules ($1g/cm^3$ or 33 water molecules in $1nm^3$ of volume). It is quite noteworthy that at room temperature, water molecules continuously form clusters which are responsible for the generation of local density fluctuations. If you peep inside the cell, you will find all the nanoscale objects facing difficulty in their movement since viscous effects become dominant. The reason why viscous forces increase is due to low Reynolds's number (Reynolds's number is the ratio of the inertial force to the viscous force) (Purcell, 1977). Just imagine a condition wherein a fish is swimming in honey instead of water. Since Reynolds's number is proportional to a swimmer's size, its value goes down to lesser than 10^{-4} for all the Nano-objects inside the cell. At such a nanoscale, inertial forces are negligible and a low Reynold's number swimmer has to overcome the viscous forces for its motion by taking the help of thermal fluctuations. This subsequently leads to random bombardment of the molecules, knocking each other at an interaction rate on the order of 10^{12} times per second. Thus, maintaining the direction or trajectory becomes practically impossible without some kind of built-in tethering support mechanisms. The cytoskeletal proteins are very good examples of such support systems. *Controlling the trajectory of nanoscale objects for guided transport is another challenge for nanotechnological applications.*

All objects in the nanoworld behave like a colloidal system and undergo zigzag movement *i.e.*, Brownian motion. This leads to low diffusive transport of molecules. For example, a sucrose molecule (about 1 nm in length) has a diffusion coefficient of 5×10^{-6} cm^2/s in water at room temperature (Chen and Ho, 2006). This indicates that it takes an hour to move 1μm along the chemical gradient direction from its original source. The diffusion coefficients of larger biomolecules are even lesser. *Thus, unraveling an alternative route of transportation of the nanoscale objects is the need of the hour.*

All the above challenges are easily met by a biological cell which is a self-organized system. It is surely one of the greatest triumphs of evolution that a cell can perform all its activities with near fidelity in such a noisy environment. We have a great example of DNA polymerase of just 13nm in size, capable of copying DNA with an error rate of one in one million base pair. Apart from being accurate, molecular machines can also exploit fluctuations as an essential part of their function. For instance, restriction enzymes that recognize and cut specific DNA sequences are extremely efficient at searching through a genome comprising of millions of base pairs. The efficiency is due to the entropic forces at work which is used up by the restriction enzyme leading to the folding of long DNA molecules into a compact coil. The enzymes then easily hop from one strand to another, thus enhancing the search process (Phillips and Quake, 2006).

Interplay between thermal and deterministic force (dipole-dipole interaction, hydrophobic interaction, ionic bond and Van der Waals force) allows processes such as diffusion, conformational changes, dissolution of hydrogen bonds and wandering of charges from their molecular hosts. These processes serve as the basis for the functions of DNA, proteins, Lipids and Carbohydrates. The ability of cell to robustly co-ordinate all nanomachineries within its fluidic capsule, is really awesome. The key aspect which makes a biological cell complete in all senses is its comprehensive capacity to sense and monitor its surrounding, trigger response after processing the incoming signal and finally to actuate its environment towards making it a better place to survive happily in it.

A biological cell can circumvent all the problems posed by Thermal fluctuations, Brownian motion, low Reynolds's number, but exploiting bio-nanomachineries of the cell like DNA and proteins outside it, is an avalanche of challenge in front of a Nanotechnologist. Once these challenges are conquered, then such molecular machines can be readily used for the development of chips, electronic circuits, sensors *etc.*

5.3 The Biology and Physics of Life's Media: Water

Natural bio-nanomachines like DNA and proteins have many limitations for their optimum function outside the cells. One of them is the requirement of the elixir of life *i.e.* water. The nanomachines are quite stable when surrounded by water lest they lose their conformations and become useless to a nanotechnologist. It's just not these nanomachines which exhibit preference, but water is also an unusual molecule with specific predilections for some chemical groups. Water molecules strongly bind each other through hydrogen bonds (Hydrogen bonds are flickering, primarily electrostatic, and weaker than covalent bonds). Water molecules break Hydrogen bonds and form non-covalent interactions with the biomolecules only and only if they offer highly charged regions or regions enriched with Nitrogen or Oxygen atoms. Bio-nanomachines made of simple catenation of carbon would be insoluble in water (hydrophobic), since carbon has very low electro-negativity value compared to oxygen atom. Thus, if a Nanomachine is interested in any conversation with water molecules it has to infuse oxygen, nitrogen, sulfur and phosphorus within its long catenation of carbon thus making it water-soluble (hydrophilic).

Biomolecular structural stability is achieved not only due to the intrinsic covalent and non-covalent bonds within it, but also is a consequence of four types of non-covalent interactions with the universal media of life –Water. They are as follows:

5.3.1 Van der Waals Forces

When two uncharged atoms come very close their surrounding electron clouds have their impact on each other. Variations in the positions of electrons around one nucleus may create a transient electric dipole, inducing a transient opposite electric dipole in the nearby atom. The two dipoles weakly attract each other, bringing the two nuclei closer. These weak forces are called Van der Waals force of attraction. As the two nuclei draw closer to each other, their electrons start repelling each other. The two atoms become stable when the Van der Waals force of attraction cancels the repulsive forces. This is why bio-nanomachines are stable in the vicinity of water molecules.

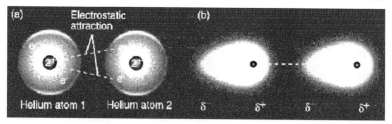

Fig. 5.1 Van der Waals forces or induced dipole-induced dipole forces between two inert Helium atoms.

Exploringberkeley.wordpress.com/tag/berkeley/.

5.3.2 Hydrogen Bonding

Hydrogen bonds behave like a Velcro; they break or make as and when required. They are weak, flickering, and electrostatic in nature. They are formed between hydrogen atoms, already covalent bonded to oxygen, nitrogen and sulfur, of one molecule to oxygen, nitrogen or sulfur atoms of another molecule. The interaction is somewhat directional forming the strongest interaction when the hydrogen atom is aligned directly at the accepting atom. Non-optimal hydrogen-bonding geometries

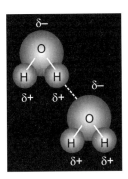

Fig. 5.2 Structure of Water molecules joined by hydrogen bond.
(http://www.ccs.k12.in.us/chsBS/kons/kons/wonderful_world_of.html).

are found in natural bio-nanomachines, since it is fully studded with oxygen, nitrogen and sulfur atoms.

5.3.3 Electrostatic Forces

Electrostatic forces are formed between charged atoms with electronic charges that play an important role in the function and stability of bio-nanomachineries. Opposite charges attract and like charges repel each other. These forces are non-directional acting symmetrically in all directions from each charge centre. The dielectric constant of water is very high which is helpful in the separation of two oppositely charged bio-nanomachines. Water molecules are dipoles: The oxygen atom possesses partial negative charge and the two hydrogen atoms possess partial positive charge. When charged ions are placed in water solution, the surrounding water molecules tend to align their dipole moments to interact with the ion. This reduces the force on other ions, damping electrostatic interactions between the oppositely charged Biomolecular structures by 80-fold. This leads to the stabilization of the bio-nanomachines.

5.3.4 Hydrophobic Interactions

Bio-nanomachines behave in an unprecedented manner when placed in Water. It exhibits the property of *hydrophobic effect*, which is the dominating feature for the stability of the nanomachinery. Water molecules possess flickering hydrogen bonds within it, continuously making and breaking the bonds with their neighbours. The stability of water solution is due to the enthalpic energies, such as forces of attraction like Van der Waals force, hydrogen bonding, electrostatic forces within them and entropic energies leading to more random orientations. Once the hydrogen bonds are broken, water molecules have many other options to bond with, thus favouring entropy. If such options are interrupted, then the entropy of the system decreases, thus making the whole system thermodynamically unfavourable. Hydrocarbon-rich bio-nanomachines interact meagerly with water molecules forming a favourable dispersion interaction, but are incapable of forming any hydrogen bond with water. Thus, these water molecules that are pressed close to the hydrocarbon are unfavourable both in enthalpy and entropy. If in such a system another Hydrocarbon-rich bio-nanomachines are introduced then the situation *analogously turns out to be like that of two unknown persons with a common enemy joining hands and behaving like friends.* Thus, the two hydrophobic bio-nanomachines will come closer to each other and aggregate, thus the total hydrocarbon surface that is exposed to water is reduced when they all associate together. Many water molecules are freed to return to the random, shifting solution. This is known as hydrophobic interaction. Such association of carbon rich molecules without prying in of water molecules is often mentioned to be known as hydrophobic bond. It should also be noted that the bond is not formed due to any intrinsic reaction between the carbon-rich molecules, but is a consequence of removal of water molecules.

Bio-nanomachines like proteins, DNA and lipids are stable and functional because of the hydrophobic effect dominating in the environment. Protein folding is a sort of conformational arrangement of proteins so that they possess least free energy and entropic energies which is possible only due to the sequestration of carbon-rich molecules from water. DNA and RNA strands are stable due to the hidden and sequestered hydrophobic faces of nitrogenous bases. Lipids form stable constituent of cellular and organellar membranes due to hiding of carbon-rich tails. The self-assembly of all the bio-nanomachineries is driven by the Hydrophobicity effect, thus stabilizing the final structure.

Although non-covalent forces are far weaker than the covalent forces, but the cumulative effect of many such weak interactions become dominant for the integrity and stability of the macromolecular structure. The formation of such non-covalent interactions also leads to the decrement of free energy, thus leading to its stability. The pertinence of such non-covalent forces is during dissociation of two biomolecules (enzyme-substrate) when all the multiple interactions have to be disrupted at the same time. As these interactions fluctuate randomly, these simultaneous disruptions are quite unlikely. The macromolecular stability due to 10–20 non-covalent contacts are therefore greater than simple summation of small binding energies. (Van Holde, *et al.*, 1998)

Protein folding occurs due to multiple weak interactions leading to the formation of stable conformations. Enzymes' catalytic power is denoted by the energy released due to the non-covalent gluing of the substrate. Revelation of similar interactions are found when hormones fuse with their specific receptors or antigens merge with their respective antibodies. All such blending are possible due to the exclusion of polar water molecules. At molecular level, complementarity between two nanobiomolecules is significant due to weak interactions between polar, charged and hydrophobic groups. DNA and RNA are also stable due to multiple hydrogen bondings between the nitrogenous bases, hydrophobic interactions and van der Waals forces within them. Carbohydrates also possess weak interactions cumulatively bestowing upon them strength and stability. All the above bio-nanomachines with all weak, but cumulatively strong forces, thus work meticulously in the fluidic soup of the cell.

5.4 Bio-nanomolecular Crowding: Heavy Traffic in the Cell Endorsing Self-assembly

If 25 persons occupy a room having a capacity of 100, then imagine the freedom they will enjoy. Now consider a horrifying situation when a batch of 120 persons are stuffed in and are packed in the same room. People can just stand on their feets, vibrating or rotating within that stipulated space. This overcrowded situation leads to utmost and proficient exploitation of the space provided. Natural bio-nanomachines also experience such an appalling situation and hence function optimally under crowded conditions. Around 20–40% of the cell's volume is occupied by these bio-nanomachines. They perform pivotally under crowded conditions not only inside cells, but also on the surface of the cells and in the extracellular environment. Optimum exploitation of the space provided is the motto of all cellular machineries.

Molecular Crowding has its impact on the assembly and function of the bio-nanomachines. Crowding increases the probability of the molecules to come nearer to each other and like ones will glue each other to form a conformationally functional complex structure on a proper template (for *e.g.*, amino acids from the amino acid pool will be brought down by tRNA on to Ribosomal machinery, aligning them in a proper sequence with the help of mRNA). Mimicking the cellular environment is the most difficult and paramount task for a biochemist. Whenever biochemical studies are done, they are performed under dilute and not under crowded conditions obeying rest all other parameters of the cellular environment like temperature, pH, concentration of ions and redox potential carefully reflecting the physiological state. Hence, optimum activity of a biomolecules can be elucidated only in a crowded milieu.

Eukaryotic cells are complex structures encapsulated by highly folded lipid membranes spanned by integral proteins sensing the extracellular complex signals and passing on the message inside the cells. The gelatinous fluidic cytoplasm is itself overcrowded by cytoskeletal proteins maintaining the integrity of the cells. Such cytoskeletal elements create tracks for the movement of cargos in and out of the cellular milieu. Bionanomachineries like proteasomes and ribosome due to crowding of cytoskeletal proteins will stay in an unperturbed situation performing their jobs with utmost care. Cellular organelles inside cells also stay in an unperturbed state doubly encapsulated via lipoid coat. Of all the other organelles, nuclei are highly heterogeneous and dynamic containing a variety of subnuclear structures, such as the nucleolus, splicing-factor compartments, Cajal bodies, promyelocytic leukemia bodies, replication factors, and transcription factors. It is speculated that the structural stability of the nucleus is entirely due to the stochastic interaction of its components. DNA is wrapped by histones and is compressed to form nucleosomes, is further condensed to form solenoid and later on undergoes compaction to form a giant fabric of chromosome. These enigmatic structures existing in the nucleus also obey the molecular crowding rules laid down by the cell.

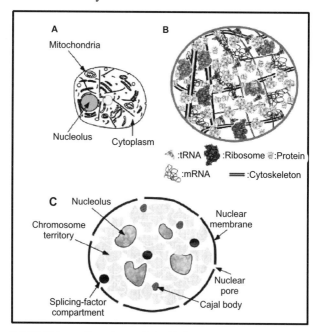

Fig. 5.3 (A) Eukaryotic cell crowded with the organelles and the cytosolic soup (B) Cytosol crowded with Ribosomes, cytoskeletal proteins, tRNA and mRNA (C) Nucleus filled with Nucleolus, Chromosomal elements, Cajal body, Splicing-factor compartment.

Reprinted from Molecular crowding effects on structure and stability of DNA, Miyoshi, Sugimoto, Biochimie 90:1040-1051(2008),by permission from 2008 Elsevier Masson SAS.

It's not only the macromolecules like DNA, RNA, proteins, lipids which stay in crowded environment, but small molecules like urea, salts which correspond to 30% of the volume also create a sort of crowding. They are vital molecules maintaining osmotic balance and homeostasis in all living systems. Hence, aperiodic or heterogeneous crowding both in the cytosol and nucleus greatly affects biochemical equilibrium and kinetics and is responsible for decreasing the entropy of all Biosystems (Miyoshi and Sugimoto, 2008; S.B. Zimmerman, A.P. Minton, 1993; J.R. Ellis, 2001 a & b)

Molecular crowding explains us the consequence of high solute concentrations on chemical reactions. The Biochemical and Biophysical effects of molecular crowding can be explained as follows:

5.4.1 Excluded Volume

The space occupied by any molecule is known as volume. When a single molecule is plunged into a solvent then the volume it occupies is larger. But when many molecules are placed in the same solvent, then the presence of one molecule (or moiety) reduces the volume available for other molecules (or moieties); resulting reductions in their entropy. This reduction in the volume is termed as *excluded volume*. The excluded volume is one of the key aspects of any chemical reaction to happen along with the interactions between biomolecules. Apart from that, a small volume should be excluded to molecules by surfaces of immobile structures, due to which small and large molecules and insoluble components in a small volume suppress molecular movements and dynamics. In fact, each Nanomachine is surrounded by many others and the movement of which is continuously blocked by the neighbouring ones, thus creating a big hurdle for diffusion.

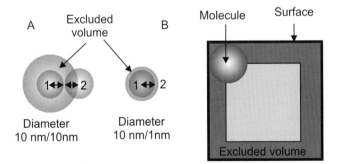

Fig. 5.4 (A) The excluded volume coming from two molecules with 10 nm and 10 nm in diameter (left) or (B) 10 nm and 1 nm in diameter (right) (C) The excluded volume was calculated from the surface of an immobilized and insoluble structure.

Reprinted from Molecular crowding effects on structure and stability of DNA, Miyoshi, Sugimoto, Biochimie 90: 1040-1051(2008), with permission from 2008 Elsevier Masson SAS.

Only those reactions are favoured which lead to the reduction of volume, as free energy of the system increases with restricting conformational entropy of the molecules (S.B. Zimmerman and S.O. Trach, 1991; K.E.S. Tang, V.A. Bloomfield, 2000).

5.4.2 Osmotic Pressure

Water is an integrated part of a living cell occupying 70% of the total weight. Solutes coexisting in such an aqueous environment are deemed to be an osmolyte altering water activity of the system. Water molecules solvate the biomolecules, immobile cellular components leaving very few free water molecules. Small hydrophilic metabolites present abundantly in the cells with meager excluded volumes, perturb the behaviour of water molecules. Osmolyte therefore can have their profound effects on the properties of biomolecules by altering their hydration. This apprehension staunchly demonstrates that osmotic pressure is perturbed by Molecular crowding. Osmotic pressure effects on biomolecular structure, stability and behaviour can be elucidated only in the crowded atmosphere (V.A. Parsegian, *et al.*, 2000).

5.4.3 Viscosity

Molecular crowding augments viscosity of the medium, thus leading to perturbed diffusive rates. This leads to depleted interactions between macromolecules consequently affecting the rates of chemical reactions. Increased Viscous forces also have their effects on folding, recognition and assembly of biomolecules (K. Luby-Phelps, 2000).

5.4.4 Dielectrics

The dielectric constant of pure water is 78.5 and dipole moment is 1.85 Debye. Water being highly polar and possessing high dielectric constant can shield oppositely charged solutes very strongly. Besides its hydrogen bonded structures permit it to form oriented structures that can resist thermal randomization, thus can effectively distribute the ionic charges. Molecular crowding alters the dielectric constant of water gigantically which leads to electrostatic interactions of biomolecules. The strength of such interactions is purely a function of the dielectric constant of the milieu (M. Roca, *et al.*, 2007).

All the above biophysical and biochemical properties perturb diffusion of biomolecules, but once the effective distances decrease then the dispersion forces, electrostatic forces, hydrogen forces and hydrophobic forces marshal two interacting biomolecules in the crowded environment banging them continuously with each other, thus strengthening their bondage (N.A. Chebotareva, *et al.*, 2004).

Such molecular crowded conditions are seen on solid surfaces of an array or an electrochemical sensor developed using DNA and protein templates (Ricci, 2007 and Goodrich, 2004). DNA and proteins are immobilized on a solid support of metal, glass or synthetic polymer and are used for microarrays, biosensors and electronics (Mirkin, *et al.*, 1996; Schena, *et al.*, 1995; Park *et al.*, 2002 and Akamatsu, 2006). The arrays and sensors with DNA are based on the sequence-specific hybridization of DNA or the specific recognition between DNA and target molecules. It has been found that there is a vast difference between the binding behavior on solid support and in solution. The reason for this behaviour has been speculated that molecular crowded conditions affect the functioning of DNA which is immobilized on the support (Goodrich, 2004). Consequently, immobilized DNA density on the support also leads to enhanced efficiency of hybridization and rate of charge transfer (Ricci, 2007).

5.5 DNA and Proteins—Dynamic Bionanomachines

In this chapter, we will now explore the two most dynamic Bionanomachines of Living systems *i.e.,* DNA and Proteins. Towards this endeavour, we will have an apprehension of their three-dimensional structural stability, functionality and dynamism in the cellular context. Cellular Intelligence cannot be mimicked with the same efficiency outside the cells, but this should not prevent us from exploiting such efficient engines. Information stored in DNA can act similar to the circuitry of the Integrated Chips. There are plethora of canonical as well as non-canonical (explained later) DNA nanostructures that can be integrated together to design a conducting chip. Similarly, Proteins which are efficient charge transporters can also be exploited for designing a Protein Nanochip. After giving a brief account of the mechanism of charge transfer and transport in DNA and Proteins, we will venture into their applications in diverse fields of Bionanotechnology like molecular electronics, sensing and microarray technology.

5.6 Unveiling the Electronic Properties of the Readable Code of Life-Deoxyribonucleic Acid

DNA is a blueprint or architecture's plan to design the whole organism. All the information is stored in that piece of coded-script and just sitting inside the control-room named nucleus, master-minds the whole organization of the body via RNA and proteins which are akin to supervisors, transforming that information into definite function. DNA runs very definite program for the cell via its workhorses named proteins, thus navigating the functions of each cell towards successful development of the whole organism.

Recently, DNA has been heralded as the cornerstone of a new generation of electronic devices and computers, since it possesses electron transfer properties within it. Just imagine a condition when we are running fastest computers, mobiles, communicators all made of DNA chips, we are owning a car having circuitry systems constructed out of DNA, all the motorized machines controlled by DNA molecular devices, all the Diagnostic sensors and navigating controllers for surgeries made out of DNA devices. This remarkable vision can be transformed into reality in the near future, since different DNA nanostructures possess specific properties of their own. But before getting into the deeper intricacies of the electronic properties of DNA, it is necessary to understand the basic constituents of DNA and their versatile nanostructures which play pivotal role in charge transport.

5.6.1 The Double Helical Structure of Deoxyribonucleic Acid

DNA is simply a linear polymer of Deoxyribonucleotides which possess phosphate groups connecting 3′ and 5′ positions of successive 2′-Deoxy-D-ribose sugar residues and nitrogenous bases. At physiological pH, phosphate groups of phosphodiester bonds are acidic and hence, they are polyanionic in nature. DNA possesses four types of Nitrogenous bases: Adenine (A), Guanine (G), Cytosine (C), and Thymine (T). According to Watson and Crick base pairing pattern, A always pairs with T and G always pairs with C via Hydrogen-bonding.

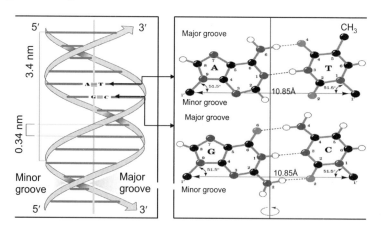

Fig. 5.5 Schematic representation of Watson-Crick model showing double helical structure of DNA, having 20 Å diameter, 10 base pairs per helical turn and a pitch of 34 Å. Watson and Crick's Hydrogen bonding pattern between Adenine and Thymine as well as Guanine and Cytosine is also shown.

However, there are other Hydrogen bonding patterns that have been observed, including Hoogsteen (Fig. 5.6), mismatched base pairs, base triplets, base quartets and reverse base pairs (show bonding patterns in the structure). The nitrogenous bases are planar, aromatic, heterocyclic molecules. Resonance among atoms in the ring gives most of the bonds in the bases, partial double-bond character. This leads to *delocalization of π-electrons consequently conferring electronic properties to DNA*. Nitrogenous bases are tucked inside the DNA polymer being insoluble in nature at physiological pH. Sugars and phosphates are highly soluble in water, but nitrogenous bases are insoluble structures which pose conformational constraints when in water. Hence to make the overall DNA structure thermodynamically stable, there is a need to have double helical structures which keeps phosphate and sugar residues exposed to the aqueous environment and bases hidden inside the spiral or helix.

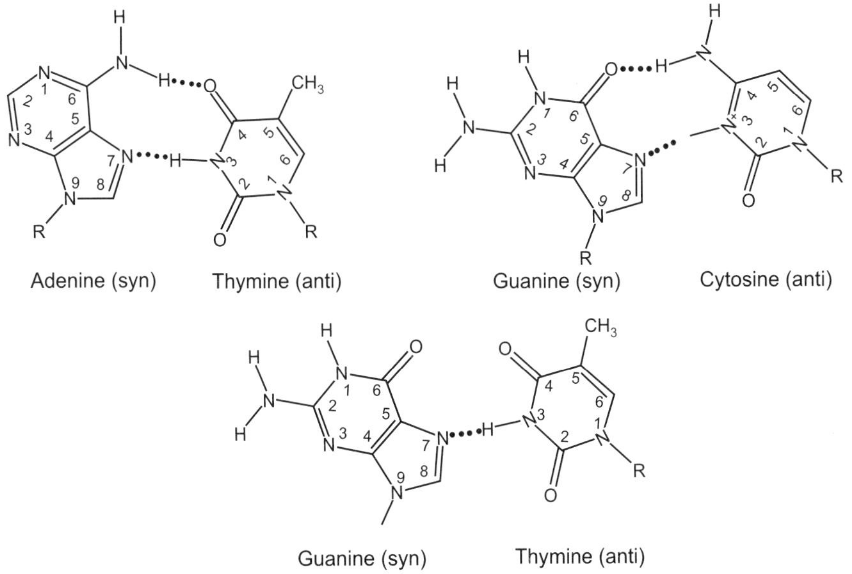

Adenine (syn)   Thymine (anti)   Guanine (syn)   Cytosine (anti)

Guanine (syn)   Thymine (anti)

Fig. 5.6 Hoogsteen base pair formation between adenine and thymine, guanine and cytosine, and guanine and thymine. Reprint from Biochemical evidence for the requirement of Hoogsteen base pairing for replication by human DNA polymerase.

Courtsey: Johnson, Louise and Satya Prakash, Proceedings of National Academy of Sciences 102, 10466–10471 (2005) permission © 2005.

Watson and Crick model also denoted as B-DNA consists of a right handed double helix whose two anti-parallel sugar-phosphate chains wrap around the periphery of the helix. The planes of the aromatic bases are perpendicular to the axis of the helix. Neighbouring base pairs, whose nitrogenous bases are 3.4 Å thick, are stacked in Van der Waals contact, with the helix axis passing through the middle of the base pairs. B-DNA is 20 Å in diameters and has two deep grooves, the narrow minor groove and the relatively major grooves. Canonical or ideal DNA has a helical twist of 10 base pairs per turn and hence a pitch of 34 Å, (Lehninger, 2005 and Van Holde, 1998).

**Table 5.1 Tabular representation of complete details of
Canonical forms of DNA**

	A-form	B-form	Z-form
Helical sense	Right handed	Right handed	Left handed
Diameter	26 A	20 A	18 A
Base pairs per helical turn	11	10.5	12
Helical rise per base pair	2.6 A	3.4 A	3.7 A
Base tilt normal to the helix axis	20	6	7
Side view			
End view			

When the relative humidity is reduced to 75%, then B-DNA undergoes a reversible conformational change to the so called A-DNA. A-DNA forms wider and flatter right handed helix than does B-DNA. Similarly, at high salt concentrations B-DNA which is a right handed helix flips to form a Z-DNA, a left handed helix. This Helical transition from B-to A or from B-to Z-DNA can be used to generate a nanomechanical device. The table gives complete details of all the canonical forms of DNA.

5.6.2 Mastering the Complex DNA Nanostructures

The DNA-based nanostructures are fabricated from both canonical (regular) and non-canonical (irregular) forms of DNA. Watson and Crick B-model (canonical structures), along with Triple helix and quadruple structures (non-canonical DNA structures) have been found to possess charge transport properties. The branched intermediates like Holliday Junction and double crossovers formed during DNA recombination can also be engineered to synthesize topologically and geometrically defined DNA branched nanostructures which possess plethora of Nanobiotechnological applications.

Elley and Spivey (1962) were the first visionaries to give the idea that regularly stacked B-form DNA might serve as a pathway for charge transfer processes. But as DNA was difficult to be obtained from natural sources in large quantities, hence charge transport studies using DNA had to be halted. This hypothesis was then later experimentally proved by Barton and colleagues in 1990s. Since then, the subject has grown to an enormous research field and whether B-form DNA is conducting or non-conducting has become a matter of scientific controversial dispute. This has led researchers to study other non-canonical structures like G-Quadruplexes also. G-Quadruplexes are formed by the intermolecular or intramolecular association of guanine-rich oligonucleotides with four Hoogsteen-paired coplanar guanines, whose structure is called a G-quartet (Alberti and Mergny, 2003). G-wire structure, a nanostructure based on G-Quadruplexes, is promising for nanoscale bimolecular electronics (Porath, *et al.*, 2000). Such structures can be exploited in designing DNA chips for diagnostic purpose. They can also act like a piggyback to load on them various macromolecular agents.

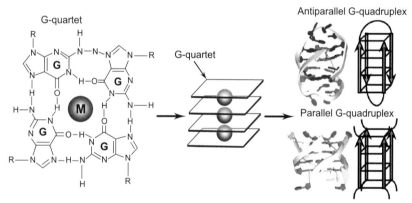

Fig. 5.7 Schematic representation of mechanism of G-quadruplex formation (Left) Chemical structure of G-quartet with a cation shown by the M. (Centre) Scheme for cation coordinations and stacking interactions among G-quartets. (Right) Three-dimensional structures of antiparallel and parallel G-quadruplexes.

Reprint from Molecular crowding effects on structure and stability of DNA (Courtsey : Miyoshi, Sugimoto, Biochimie, 90: 1040-1051 (2008) © 2003 Elsevier Science Ltd.).

Apart from canonical and non-canonical structures, most of the researchers are well inspired by the branched intermediates of Nucleic acids formed during recombination. Particularly, immobile Holliday junctions and other junctions with 3–6 double helical arms can be exploited to design networked DNA nanostructures in which several DNA chains are mechanically coupled through multiple junctions. Such junctions have become a paradigm for Structural DNA nanotechnology, which are capable enough to assemble in such a way that it is possible to obtain arbitrarily complex nanostructures of virtually any shape. These complex DNA nanostructures have been predicted as templates for immobilization on solid support. Such supramolecular nanopatterns have their role in nanoelectronics, nanorobotics, DNA microarray chips *etc.*

Structural DNA nanotechnology requires that unusual DNA motifs can be glued together by specific structurally well-defined cohesive interactions (primarily sticky ends) to produce highly predictable 3D geometries. Sticky ends are typically four to eight bases long, cohering with utmost fidelity via complimentarity methods. The second most important point is that the sticky ends form B-DNA only, so any DNA branched structure with its sticky end will have the same local geometry as is predicted. (Qiu, *et al.*,1997; Mirkin, *et al.*, 1996; Alivisatos, *et al.*, 1996). This relieves us from the headache of crystal structure determination. There are different ways to develop a 4 arm branched junctions. Such junctions should be flexible enough to develop into a periodic and aperiodic array (Qiu, *et al.*, 1997). One of the mechanisms that biological systems use to develop a Holliday junction is reciprocal exchange or double crossovers (Seeman, 2001).

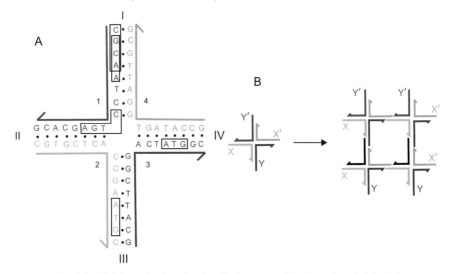

Fig 5.8 (A) A branched molecule with four arms. The four strands labeled with Arabic numerals combine to produce four arms, labeled with Roman numerals. Arrowheads indicate strand polarity. (B) Formation of a two-dimensional lattice from a four-arm junction with sticky ends. X is a sticky end and X′ is its complement. The same relationship exists between Y and Y′. Four of the monomeric junctions on the left are complexed in parallel orientation to yield the structure on the right. Note that the complex has maintained open valences, so it could be extended by the addition of more monomers.

Reprint from At the Crossroads of Chemistry, Biology, and Materials: Structural DNA Nanotechnology (Courtsey: Seeman, Chemistry & Biology, 10: 1151–1159, 2003, © 2003 Elsevier Science Ltd.).

Such double crossover strands can cross link with the other strands via same polarity or opposite polarity. (Lilley and Clegg, 1993; Churchill, *et al.*, 1988). The process of reciprocal exchange leads to the generation of DNA Motifs which is exploited in structural DNA nanotechnology. Two strands of opposite polarity undergo reciprocal exchanges (Fu and Seeman, 1993). This DX molecule can be combined with a DNA hairpin, an extra helix roughly perpendicular to the plane of the other two, to synthesize DX+J motif (Li, *et al.*, 1996). DX when combines with another double helix consequently leads to TX motif (LaBean, 2000). The PX motif is derived by performing reciprocal exchange between two helices at all possible positions where strands of the same polarity come together. The JX_2 motif is similar to the PX motif except that reciprocal exchange is omitted at two adjacent juxtapositions. (Seeman, 2001).

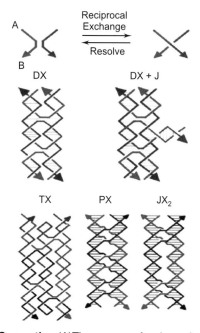

Fig. 5.9 Motif Generation (A)The process of reciprocal exchange. A red strand and a blue strand exchange to form a red-blue strand and a blue-red strand. (B) Motifs used in structural DNA nanotechnology. Two reciprocal exchanges between strands of opposite polarity yield the DX molecule shown. The DX+J motif, usually made with the extra helix roughly perpendicular to the plane of the other two, is made by combining a DNA hairpin and a DX molecule. The TX motif results from combining the DX molecule with another double helix. The PX motif is derived by performing reciprocal exchange between two helices at all possible positions where strands of the same polarity come together. The JX_2 motif is similar to the PX motif except that reciprocal exchange is omitted at two adjacent juxtapositions.

Patterned 2D array can be synthesized using DX, DX+J, and TX motifs (LaBean, *et al.*, 2000; Winfree, *et al.*, 1998) and DX, PX, and JX$_2$ motifs have been used as components of nanomechanical devices (Mao, *et al.*, 1999; Yan, *et al.*, 2002). These motifs are rigid structures which make them suitable for different applications. As we have already seen in the Fig. 5.9, two reciprocal exchanges between two strands of opposite polarity yield DX motif. Now, this DX motif can give rise to two component and four component arrays which are then examined by atomic force microscopy (Winfree, *et al.*, 1998). The combination of these constructions with other chemical components is expected to contribute to the development of nanoelectronics, nanorobotics, and smart materials. The capacities of structural DNA nanotechnology are just beginning to be explored, and revolutionary three-dimensional geometries are synthesized from variety of different motifs.

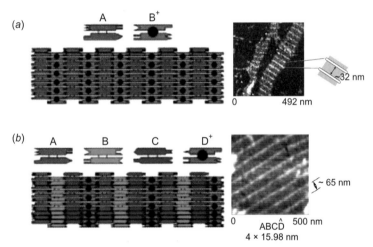

Fig. 5.10 DNA double crossover arrays. (*a*) A two-component array. The two component tiles, A and B*, are shown. The extra DNA domain in B* is shown as a filled black circle. The sticky ended complementarity is represented geometrically. The two-dimensional array that they form is shown below the molecular schematics. With tile dimensions of 4 x 16 nm, stripes should arise every ~32 nm, as shown in an AFM image on the right. (*b*) A four-component array. There are four tiles shown here, A, B, C, and D*. The same conventions apply as in (*a*). The design should lead to ~64 nm stripes, as shown in the AFM image on the right.

Courtesy: Reprint DNA Nanotechnology, Seeman, Materials Today, 24–29, 2003, © 2003 Elsevier Science Ltd.

DNA can be designed synthetically and different kinds of 4 arm junctions can be created with the help of various restriction enzymes like DNA ligases, endonucleases and topoisomerases. This effort has generated various ligated products from flexible DNA components, such as DNA sticky cubes (Chen and Seeman, 1991), sticky truncated octahedrons (Zhang and Seeman,

1992, Zhang and Seeman, 1994), Borromean rings and polyhedral catenanes (Mao, *et al.*, 1997).

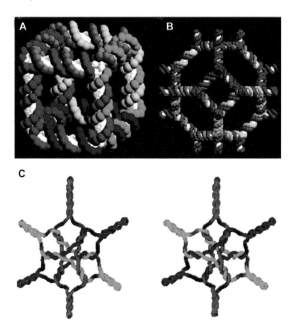

Fig. 5.11 (A) a stick cube and (B) a stick truncated octahedron. The drawings show that each edge of the two figures contains two turns of double helical DNA. There are two turns of DNA between the vertices of each polyhedron, making them, respectively, a hexacatenane and 14 catenane. (C) is a stereo view of Borromean rings. A right-handed 3 arm junction is in front, and a left-handed 3 arm junction is at the rear; if any of the circles is cleaved at one of its equatorial hairpins, the other two circles dissociate.

Reprint from At the Crossroads of Chemistry, Biology, and Materials: Structural DNA Nanotechnology (Courtsey: Seeman, Chemistry & Biology, 10, 1151–1159, 2003, © 2003 Elsevier Science Ltd.).

5.7 Charge Transport in DNA Nanostructures

Material scientists, physicists and chemists are becoming increasingly curious to comprehend the electronic properties of Schrodinger's metaphor of decipherable code of life, DNA. The variety of DNA nanoarchitectures described above can be navigated towards different purposes. At times they can act as conductors, semiconductors and above all can even behave like a superconductor. DNA has been represented as a special medium in terms of energy and charge transfer processes. Barton and Kelly (1999) have reported long range electron tunneling in B form of DNA to occur between intercalated reagents separated by 40 Å. This shows that electron tunneling processes in DNA are rather independent of number of interspersed bases. But there are other research groups which have reported that DNA is an appropriate medium for fast electron tunneling limited to a few base pairs. Electron Tunneling in biological systems is well explained by Marcus theory. This section will purely deal with the basics of Marcus theory and then we will deal with the underlying mechanisms behind charge transport in DNA so that it can be potentially exploited in the field of Nanoelectronics.

5.7.1 Basics of Marcus Theory

The process of electron transfer was the most fundamental quest for an electrochemist during 1950's. When two molecules in a solution exchange an electron, then the one which loses an electron is oxidized and the other which accepts is reduced. There are many such reaction systems where single electron transfer takes place. But it was really a matter of great curiosity that the rates of all such single electron transfer reactions were different. For example, single electron transfer rate of ferrous to ferric was very less as compared to single electron transfer rate of Fe $(CN)_6{}^{3-}$ to Fe $(CN)_6{}^{4-}$. This

Rudolph Marcus

difference in rate of reactions was as great as that of between a tortoise and a rabbit.

Why variation in reaction rates exist, was an unsolved puzzle. Rudolph Marcus gave a successful attempt to solve this mystery for which he received a Nobel Prize in the field of Chemistry in the year 1992. He developed a theory which says that *in simple chemical reactions during single electron transfer between two molecules, no chemical bonds are broken, but there are changes that take place in the molecular structure of the reactants and the nearest neighbouring solvent molecules.* This change leads to temporary rise in the energy of the molecular system, thus enabling the electron to jump from one molecule to another. Therefore, there is a need of activation energy for the electrons to traverse the energy barrier. The size of the energy barrier thus determines the speed of the reaction. Marcus assumed that the reacting molecules must first be loosely bound to each other for quantum mechanical purpose and secondly the solvent molecules in the immediate vicinity must change their position and orientation, thus increasing the energy of the molecular system. Only then electron can jump between two states that have same energy. The practical consequences of his theory extend over all areas of chemistry and biology, and makes predictions of widely differing phenomena as the fixation of light energy by green plants, photochemical production of fuel, the conductivity of electrically conducting polymers, corrosion, *electron transfer in DNA and proteins* and many more.

According to the Marcus theory, the potential energy of the entire system, reactants plus solvent, is a function of the many positions and orientations of the solvent molecules (and hence of their dipole moments) which constitute *the outer sphere electron transfer* reactions, and the vibrational co-ordinates of the reactants, particularly those in any *inner coordination shell or sphere* of reactants. The outer spheres of solvent molecules which are in the immediate vicinity of the reactants, during electron transfer, change their configurations due to thermal fluctuations thus energizing the whole system. The vibrational motion in any inner coordination shell that takes place in the molecular structure of any reactant can be explained by considering a diatomic molecule in which the bond joining the two atoms vibrate so that the energy of the bond varies as the length of the bond varies. This vibrational motion or harmonic oscillation can well be explained by Hooke's law in which a bond behaving like a restrained spring undergoes three cycles, negative compressed state, zero relaxed state and positive stretched state. Hooke's law states that the restoring force of a spring (bond) is proportional to the displacement x.

$$\text{Restoring force} = -kx$$

Where k is the force constant. The potential energy V of a bond subjected to this force increases as the square of the displacement.

$$V = 1/2\ kx^2$$

The variation of V with x exhibits the shape of the parabolic curve and the bond undergoing harmonic motion has a "parabolic potential energy". Hooke's law description is useful in discussion of the energy levels in more complicated molecules. The displacement x is replaced by a nuclear co-ordinate, which lumps together all the distances in all the bonds, and a single representative parabola is used to represent the parabolas of all the bonds which can be viewed in the Fig. 5.12. (Atkins, 2001).

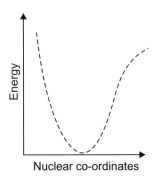

Fig. 5.12 Potential energy curve showing nuclear co-ordinates with change in energy.

Marcus described the rates of electron transfer between weakly coupled donor and acceptor states when the potential energy depends on a nuclear co-ordinate, *i.e.,* non-adiabatic electron transfer (explained later). Let us understand the theory with a simple illustration of electron transfer taking place between donor and acceptor.

$$D + A \rightarrow D^+A^-$$

For electron transfer in solution, we most commonly consider electron transfer to progress along a solvent rearrangement co-ordinate in which solvent reorganizes its configuration so that dipoles or charges help to stabilize the extra negative charge at the acceptor site. This type of *collective* co-ordinate is illustrated in the Fig. 5.13. The external response of the medium along the electron transfer co-ordinate is referred to as "outer shell" electron transfer, whereas the influence of internal vibrational modes that promote electron transfer is called "inner shell". (Marcus, Nobel Lecture, 1992).

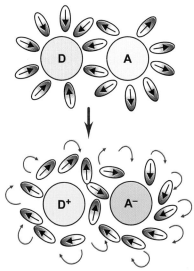

Fig. 5.13 Reorganization of the configuration of solvent surrounding the charged solute.

The Arrhenius expression for the rate constant k for an electron transfer is given as,

$$k_{et} = A \exp(-\Delta G^*/k_B T) \qquad ...(1)$$

Marcus theory builds on the above equation giving a correction to Ä G*

$$\Delta G^* = \lambda/4(1 + \Delta G^0/\lambda)^2 \qquad ...(2)$$

The pre-exponential factor, A, is often expressed as an exponential term itself, dependent on the distance, r, between the molecule or group that donates the electron and the one that accepts it

ΔG^0 is the standard Gibb's free energy of reaction, k_B is the Boltzmann constant, T is the temperature of the system, and λ is a "reorganization term" composed of solvational (λ_0) and vibrational (λ_i) components

$$\lambda = \lambda_0 + \lambda_i \qquad \qquad ...(3)$$

The pre-exponential factor, A can further be calculated as follows

$$A = \exp(-\beta\Delta r) \qquad \qquad ...(4)$$

1. *Where, β is also known as damping factor (attenuation factor) which depends on what is between the donor and acceptor of the electron. Basically, electrons go through some materials like Guanine in DNA much more easily than others and β just reflects this. (Marcus and Sutin, 1985; Renger and Marcus, 2003).*

The value of ΔG^* was derived by considering the thermodynamic surfaces of weakly coupled reactants solving it quadratically. Marcus theory can be represented graphically for non-adiabatic electron transfers between weakly coupled reactants. (Marcus, 1956)

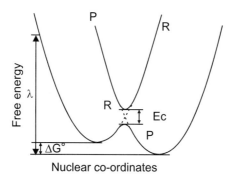

Nuclear co-ordinates

Fig. 5.14 Graphical representation of non-adiabatic electron transfers between weakly coupled reactants (R) to form Products (P). Ec is the electronic coupling site or cross-over point at which electron transfer occurs.

The system is represented in two states, before electron transfer (R, the reactant state or D + A), and that after electron transfer (P, the product state or $D^+ + A^-$).

The above graph of nuclear co-ordinate vs. free energy can be explained as follows:

1. The above graph is parabolic because nuclear vibrations are harmonic in nature and obey Hooke's Law.

2. Marcus theory discusses about non-adiabatic electron transfer between weakly coupled reactants. So why Non-adiabatic only? Non-adiabatic or Diabatic processes are the ones in which electron transfer is a quantum jump from one curve to the other as is observed in the graph. Electron jump at Ec which is also known as electronic coupling points or crossover points can occur only in a non-adiabatic reaction. Now,

let's discuss about adiabatic (impassable to heat) processes in little detail. In thermodynamics, an adiabatic process is one in which no exchange of heat with the environment occurs. In the Carnot cycle of an ideal gas engine, the steps in which expansion or contraction occur without exchange of heat are adiabatic. In the electron transfer context, an adiabatic process is one in which no quantum jump occurs, the electron lingers at the barrier, and the curves representing the two states smooth to form a continuum, with a quasi-state at the top of the activation barrier.

3. All Marcus did was realize, that he could solve for the point of crossing between the parabolas and derive the pre-exponential factor A as shown in Eqn.4 which is based on the electronic coupling Ec between the initial and final states of the electron transfer reaction (*i.e.,* the overlap of the electronic wave functions of the two states). He also gave a correction factor to ΔG^0 (Standard Gibb's free energy change) because according to him, activation energy of any reaction system does not just depend on Gibb's free energy but needs an additional parameter called the reorganization energy, λ. *The reorganization energy is that energy required to distort the nuclear configurations of reactants into the nuclear configurations of the products without considering any electronic transfer taking place.* The most pertinent prediction is that the rate of electron transfer will increase as the electron transfer reactions becomes more exergonic *i.e.*, ΔG becomes negative, but only to a certain point known as crossover point where

$$\lambda = -\Delta G$$

it is at this point where two opposite signed components meet. Now, just go back to Eqn.4. After expansion of the equation we get a term $(\lambda + \Delta G)$ which becomes zero when reorganization energy becomes equal to Gibbs free energy change, but with opposite sign. *The maximum rate constant for electron transfer occurs when the free energy is equal and opposite of the reorganization energy.* The activation energy required for the reactants becomes zero at this point and electron transfer can occur very easily. But after this value of Gibbs free energy change, the electron transfer rate decreases as the driving force $(-\Delta G)$ increases (In spontaneous reactions, when free energy of the system is negative or is exergonic, then that acts as a driving force for a reaction). The conditions under which this dropping-off of rate with increased driving force occurs is known as the Marcus inverted region (Marcus, Nobel Lecture, 1992 and Marcus, 1956).

4. The electron exhibits *Quantum Mechanical Tunneling Effect* at the cross-over point. *Quantum mechanical tunneling is the phenomenon in which electrons move over a barrier through which it is impossible to move classically.* In the sense that when electrons behave like a particle (classical theory) then it is impossible for it to pass through a wall or a barrier, but when they behave like a wave (modern theory) then it can easily tunnel through the barrier (Analogously speaking, if you see those ghost movies, you will find ghosts can easily cross the walls of a house which is rather difficult for a living human body!). In the quantum mechanical world, electrons have both particle as well as wave-like properties. If electrons are moving with enough momentum and with particular amplitude, and if they come across a thin barrier of few nanometres then they easily travel through it on the other side, but with decreased amplitude. When an electron moves though the barrier in this fashion, it is called tunneling. (Griffith, 2004).

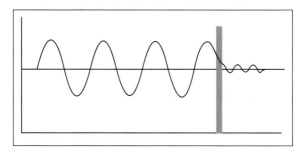

Fig. 5.15 Quantum mechanical tunneling effect: Electrons tunnel through the barrier.

Here, in this context, when we talk about the barrier in terms of electron transfer reactions, then the solvent systems as well as intramolecular or intermolecular vibrational energies pose a barrier to an electron.

5.7.2 Mechanisms Involved in Charge Transport in DNA

The structure of DNA possesses within itself π-bonded ring structures in the form of nitrogenous base stacked in the core. The DNA structure is ideal for electron transfer because some of the orbitals belonging to the bases overlap quite well with each other along the long axis of the DNA. Thus, the possibility of long distance electron transfer had intrigued many researchers to perform some experiments to prove DNA is a conducting material. It has now become a well-known fact that DNA-mediated charge transport processes can occur on fast and ultra fast time scales. They follow different mechanisms depending on the DNA structures and can yield chemical reactions over distances in the nanometer range. The three most successful and dominant mechanisms for photo induced electron transfer in DNA are as follows:

(*i*) The molecular wire model

(*ii*) The super exchange or single step electron tunneling mechanism/model

(*iii*) Thermal hopping model.

In all the above models, there is a requirement of a charge donor that can be photoexcited in order to initiate the charge transfer process. A very simplified picture on the relative energies in any synthetic donor-bridge-acceptor system shows that the level of the bridge medium in relation to the energetic levels of donor and acceptor dictates which model is to be selected. In the case of a *molecular wire*, the bridge states are thermodynamically comparable to the level of donor. When electron donor injects its electron, it gets localized within the bridge and moves incoherently to the acceptor.

In the case of *super exchange mechanism*, the bridge states lie above the level of the donor. Thus, electron tunnels into an acceptor within one coherent jump and is never localized within the bridge. Hence, rate of such reactions decrease exponentially with the distance between donor and the acceptor.

In another instance, if the bridge state of the electron transfer medium is energetically comparable to the photo excited donor, an electron transport via *thermal hopping* could occur as an alternative mechanism. In this incoherent process, the electron is localized on the molecule and exchange energy with it. Electron transfer occurs in a multistep fashion from donor to acceptor. Such hopping process can transfer charge over far longer distances than coherent tunneling process, and the motion can be thought of as diffusive.

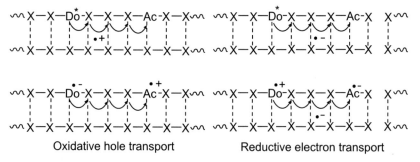

Oxidative hole transport   Reductive electron transport

Fig. 5.16 Charge transport in DNA can be either Oxidative or Reductive.

In principle, DNA-mediated charge transfer processes can be divided into oxidative (electron) hole transfer/transport or reductive electron transfer/transport processes. During hole transfer, the electron is transferred from the DNA to the photoexcited charge donor, thus generating free radical cationic nitrogenous base and the hole is transferred, while in the case of an electron transfer, the photoexcited electron of the donor is injected into the DNA generating free radical anionic nitrogenous base and the electron is transferred. Hence, in both these cases charge transport occurs.

DNA-mediated charge transfer occurs due to nitrogenous bases which are important constituents of DNA. From a great body of experimental data, it has been found that amongst all the four bases present in DNA, only Guanine can effectively transfer electrons through the molecular machine. This is because Guanine has lowest ionization potential (0) as compared to Adenine (0.45), Cytosine (0.66) and Thymine (0.66), thus Guanine possesses lowest oxidation potential also. Therefore, only Guanine is supposed to transfer electrons through DNA (Lakhno and Sultanov, 2003). This proves that DNA can act like a molecular wire only when it has a sequence having only Gs in it which is possible synthetically and not naturally. Thus, the most successful mechanisms to date are Super exchange and Thermal hopping models. In both these cases, electron transfer occurs from donor to acceptor via intermittent bridge.

In super exchange model, charge transfer occurs from donor to acceptor in a single coherent jump (coherent tunneling), since the intermittent bridge energies are sufficiently high and does not allow the charge to get localized. The coherent transfer processes cannot obviously get very far at room temperature, because the orbitals in which the electron density is found do not extend effectively over long distances. So, if the bridging states (*i.e.*, the intervening base pairs in DNA) are high in energy compared to the initial and final states, one should anticipate a super exchange pathway, characterized by a rapid exponential decay of the electron transfer rate or yield as a function of distance. (Fink and Schönenberger, 1999; Bixon, *et al.*, 1999 and Jortner, *et al.*, 1998).

In thermal hopping mechanism, the bridge states that lie in between the donor and acceptor, exhibit sufficiently low energies. The Guanines are responsible to transfer the charge, hence G to G hopping occurs in either of the strands and if the distance between two Gs exceed four AT base pairs then the charge transfer can occur via Adenine also, since it possesses little higher oxidation potential than Guanine. (Berlin, *et al.*, 2000–2001 and Ratner, 1999).

(*An analogous example for such a motion can be, leaping across a stream. It is very simple to cross a narrow stream in a single jump, but it is impossible to leap across a wide river unless there are stepping stones along the way. A person can simply hop from one stepping stone to another to cross the river!*) (Dekker and Ratner, 2001).

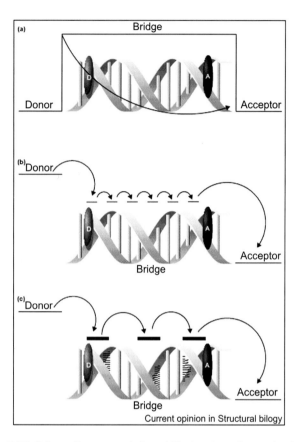

Fig. 5.17 Schematic representation of Electron transfer mechanisms (*a*) The super exchange or single step electron tunneling mechanism/ model, (*b*) The molecular wire model and (*c*) Thermal hopping model.

According to *Marcus theory* (already discussed exhaustively), the rate k_{et} of an electron transfer from a donor D via a Bridge B to an acceptor A in DNA depends upon the driving force ΔG between the donor and acceptor, the electronic coupling (Ec) mediated by the bridge, and the reorganization energy λ of the system. The driving force for the injection of a positive charge was increased by a single Guanine G by a GGG sequence. In all cases, the electron transfer rates increased demonstrating their abidance to Marcus theory. The electronic coupling and as a consequence electron transfer rate decreases with increase of the bridge length (the distance between electron donor and acceptor). The effect of the distance Δr on the electron coupling is already given by the equation 4, in which β is an attenuation factor for the electron transfer rate k_{et} (Marcus and Sutin, 1985; Renger and Marcus, 2003).

$$A = c \exp (-\beta \Delta r) \qquad \qquad ...(4)$$

Large values of β indicate higher resistance exerted by the medium for the electron/hole transfer to occur from donor to acceptor. The medium here refers to solvent, surface *etc.*, in case of an intermolecular charge transfer and to the intervening bonds (covalent/non-covalent) between the donor and the acceptor in an intramolecular situation. Thus, three important observations were done after various DNA-mediated charge transfer studies:

(*a*) The hole transfer via the super-exchange mechanism is limited to short distances (<10Å).

(*b*) Short-range hole transfer reactions occur on a very fast timescale $(k_{HTsss} = 10^9 – 10^{12} \, s^{-1})$.

(*c*) The typical β value of DNA-mediated hole transfer is 0.6–0.8 Å^{-1}.

However, in experiments using charged injection systems, such as acridinium ions, enol ether radical cations, electron transfer rates at short distances are much faster than predicted by Eqn. 4. This effect was explained by the solvent reorganization energy λ_s. If a charge is shifted from a highly solvated ion to a neutral base, the solvent reorganization energy should be a function of the distance. The solvent reorganization is small for a very small shift in the charge; it increases with distance and reaches a plateau for long charge migrations.

5.7.3 Transforming Theories into Reality

In 1960s, Daniel Eley and D I Spivey were the first to give a remarkable vision that DNA could serve as an electronic conductor. This long-awaited vision of Eley and Spivey was then turned into reality by Barton's group in 1980s. The Barton's group studied the problem of 'long-range electron transfer mediated through DNA base stack' by adopting two approaches. In the first approach, the donor (D) and acceptor (A) molecules were fixed at known distances on the duplex via intercalation or covalent attachment, and photo-induced electron transfer between them was investigated. In the second approach, measurements were made of either the oxidative damage at the G (guanine) sites or repair at the dimerized TT (T = thymine) sites, both initiated by photoexcitation of a metal complex bound to DNA at a site far away from the damaged G or TT site. (Murphy, *et al.*, 1993; Arkin, *et al.*, 1996 and 1997; Hall, *et al.*, 1996; Dandliker, *et al.*, 1997).

Two relevant results emerged out of these experiments:

(*i*) DNA can transport electrons over large distances (> 40 Å) using the core π-stack of its base pairs, called π-ways, and

(*ii*) Value of β (Eqn.4) was as low as 0.1 Å^{-1}.

But this result was not consistent as β values many a times approached even more than 1.5 Å^{-1}. (Brun and Harriman, 1992; Atherton and Beaumont, 1995; Fukui and Tanaka, 1998; Meggers, *et al.*, 1998). This has given rise to many schools of thought saying DNA is a semiconductor, superconductor or just a conductor. Many experiments also prove that the value of β is dependent on the structure of DNA and also depend on the position of Gs.

To support this idea, a simple experiment was performed by Meggers, *et al.*, (1998) and Harriman (1999). A positive charge was created on Guanine base (designated as G+ in Fig. 5.18) by a photochemical event in a DNA strand and was allowed to reach a target site (GGG site) separated by varying sequences. The energy of the hole was lower on Guanine as compared to other because Guanine has lowest oxidation potential. Thus, electron localizes itself only on Guanine. Further measurements were done to find out the relative rates of hole transfer between G$^+$ and GGG, when the intervening AT base pairs increased from two to four. It was found that there was 100-fold reduction in the rate of hole transfer. Further keeping the distance between G$^+$ and GGG same as that in c an intervening GC base pair enhanced the rate constants as is exhibited in d.

Table 5.2 Summary of few results of experiments performed on long range charge transfer mediated through DNA

Results of Experiments	Research Groups
Efficient fluorescence quenching over 40 Å proving coherent tunneling and thermal hopping mechanisms.	Wan, *et al.*, 1999 and Fiebig, *et al.*, 1999.
Oxidative damage in DNA results in charge transfer, DNA binding enzymes influence this charge transport.	Giese, *et al.*, 1999; Voityuk, *et al.*, 2001; Hess, *et al.*, 2002; Hess, *et al.*, 2001; Davis, *et al.*, 2002.
Measurement of hopping mobility in DNA was done proving hopping mechanism.	Ratner, 1999
Ultra-fast electron over 17–36 Å with $\beta < 0.1$ Å–1.	Kelly, *et al.*, 1997
Guanine doublets oxidized upon triggering from distances > 30 Å.	Schiemann, *et al.*, 2000 and Stemp, *et al.*, 1997.
Long-range oxidation of Guanine doublets across 17–34 Å.	Hall, *et al.*, 1996 and 1997.

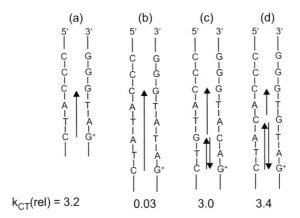

Fig. 5.18 Relative rates (kCT(rel)) of hole transfer between G+ (donor) and GGG (acceptor) along the strand, consistent with the random walk model (Giese, 1999).

5.8 Applications of DNA Nanostructures in Molecular Nanotechnology and Nanoelectronics

DNA nanostructures which have been derived from canonical/non-canonical structures and from vast array of cross-over junctions like DX, PX, JX motifs, can be exploited for different purposes. Base mismatches and DNA lesions lead to oxidation of nitrogenous bases in DNA and are responsible for charge transport. In the last 10 years, genomic research has demanded highly meticulous approach towards diagnostic analysis, mechanical devices and integrated circuits (electronics). The various applications of DNA nanostructures in the area of Molecular Nanotechnology and Nanoelectronics are discussed as follows:

5.8.1 Electron Transfer in DNA Microarray Technology and for Base Mismatch Detection

DNA chip designing for different purposes has become a prime motto of structural and material scientist. In principle, DNA chips are segmented, planar array of immobilized DNA fragments and are used in a wide range of applications from expression analysis to diagnostic microarrays. (Niemeyer and Blohm, 1999; Lovrinovic and Niemeyer, 2005).

For this purpose, the microarray has accommodated subsets of probes containing the fully complementary oligonucleotides for the detection of cancer, AIDS and other viral diseases. The idea is very simple; the mismatch detection is based on the observation that the binding energy of a mutated test sequence to the single strand probe on the microarray chip is lower than the binding energy of a completely complementary strand yielding. Hence, mutations are detected due to diminished emission by the chip. The principle is the same as studied in the previous sections of charge transfer in DNA. It is clear that these processes show an extreme sensitivity towards perturbations and interruptions of the base-stacking which are caused by base mismatches or DNA lesions. Thus, charge transfer in DNA should be suitable in order to obtain a highly sensitive electrochemical read out on DNA microarray chips. The critical gene is immobilized as single strand oligonucleotide on an electrode or chip, and contains a redox active probe that is intercalated and /or covalently attached. DNA which is added to the chip forms intact DNA duplexes leading to an efficient electron transfer between the electrode surface and the distant redox active probe. Base mismatches and DNA lesions significantly interrupt charge transfer in DNA, and as a result, the electrochemical response is lacking. (Drummond, *et al.*, 2003).

The most efficient technique for the deposition of DNA on solid phase surfaces are the self-assembled monolayers (SAM) (Kumar, *et al.*, 1995 and Porier, 1997). Based on this technique, Barton and coworkers attached the alkyl thiolate linker to the 5'-terminal hydroxy group of the oligonucleotide which then interacts with the gold electrode to form DNA films. (Kelley, *et al.*, 1998 and 1999). The DNA was labeled with daunomycin, a redox active intercalator which forms covalent crosslinks with the N2-unit of Guanine. The cyclic voltagramms of the prepared DNA films showed a significant electrochemical signal for the reduction of daunomycin. Variation of the distance between daunomycin and the electrode showed that the rate of electron transfer was essentially unchanged. Using this assay, single base mismatches and biologically important DNA lesions could be detected, even small and local distortions turn off the electron transfer process completely (Drummond, *et al.*, 2004). *Heller* and *coworkers* have also used DNA-modified gold surfaces to investigate electron transfer processes. They attached the oligonucleotides to the gold electrode via a short two-carbon linker which has been covalently attached to the 3'-terminal phosphate group (Hartwich, *et al.*, 1999). Additionally, pyrrolo-quinolinequinone (PQQ) has been covalently tethered to the 5'-terminal T of the oligonucleotide and serves as a redox active cap of the DNA duplex. Electron transfer from the electrode to the PQQ probe was investigated by cyclic voltammetry. The insertion of two adjacent mismatches (GA and GT) resulted in a threefold decrease of the rate of electron transfer. (Hartwich, *et al.*, 1999). Houlton and coworkers used a different surface than gold for the immobilization of DNA (Pike, *et al.*, 2002). The alkylated surface of silicon devices was used as the solid-phase for the automated oligonucleotides synthesis. After hybridization with the complementary DNA strand, the binding of methylene blue to the Si–DNA device yielded an electron transfer which could be detected by cyclic voltammetry. The ability to combine automated solid-phase synthesis with semiconductor electronic materials, such as silicon, provides further scope for the development of advanced DNA chips.

It can be concluded based on the work summarized here, that the efficiency of electron transfer through DNA films offers a new and suitable approach for the development of sensitive DNA sensor, arrays or chips. Using this method, a broad range of single point mutations and DNA lesions can be detected without the context of certain base sequences.

5.8.2 DNA-based Nanodevices for Molecular Electronics

With the introduction of π-conjugated systems in electronic devices, construction of supramolecular architecture, out of such individual π-conjugated molecules have become the subject of very intense research. The central issue in nanoelectronics is the miniaturization of integrated circuits which drives the search for efficient molecular devices at the nanometer scale. The basic idea of molecular electronics is to use individual molecules as wires, switches, rectifiers and memories. As DNA can attain different structures with efficient electronic properties, it acts as a scaffolding structure for molecular electronics. DNA represents a superior material for research as a supramolecular architecture for building up of various electronic devices.

In principle, both types of charge transport (oxidative and reductive) could be applied to molecular electronics, since they can occur over long distances. We are all aware of n-type and p-type materials and their joining forms the basis of all semiconductors, such as diodes, triodes and transistors. (G+C)-rich DNA shows p-type properties and (A+T) sequences show *n*-type ones (Lee, *et al.*, 2002). If we combine such DNA molecules, they can create logic elements that will be more powerful than any silicon-based device because, in theory, just a short sequence of DNA base pairs may be enough to create all the combined *n*-and *p*-type properties. In fact, DNA based single-electron transistors and quantum-bit elements have already been proposed (Ben-Jacob, *et al.*, 1999).

Gordon Moore from Intel Corporation had formulated Moore's law in the year 1965 stating that the number of transistors on a chip would double every 18 months, but this doubling rate would drop to every 4–5 years between 2010 and 2020. DNA-based nanoelectronics now proclaim to end the conventional microelectronics. DNA based chips labeled as "Genes Inside" is the futuristic goal which a material scientist has kept. But to design DNA based electronic components, the electronic properties of DNA and its modus operandi further needs complete assurance from researchers. A DNA-based computer chip, even if it becomes reality, may not be totally based on DNA. We are actually heading towards a hybrid technology, in which the transistor could be made out of DNA molecules that are connected by carbon nanotubes, and other parts would be made out of silicon. (Bhalla, *et al.*, 2003).

Further electron capabilities of DNA can also be enhanced using metals. Metals like copper can be incorporated in the interior of DNA duplex. Copper ions interact with the nucleoside electronically, thus this type of interaction is known as *Metal-mediated base-pairing*. Electron transport efficiency of DNA can also be enhanced by its co-ordination to divalent ions like Zn^{2+}, Co^{2+}, Ni^{2+} along the DNA helix. Further efficiency can also be enhanced by metals coating the duplex completely.

5.8.3 DNA-based Nanodevices based on the B-Z Transition

Seeman and co-workers reported for the first time a nanomechanical device based on DNA which utilized the conformational change of dsDNA from its right-handed B-form to the left-handed Z-form. This process known as B-Z transition is known to synthesize molecular machine based on DNA (Mao, *et al.*, 1999). B-DNA is a right handed canonical structure at neutral pH and moderate salt concentration, but at high salt concentration in particular, in the presence of multivalent cations, such as cobalt hexamine ($[Co(NH_3)_6]^{3+}$), B-DNA undergoes transition to form Z-DNA which is a left handed helix. This transition from B-DNA to Z-DNA can be utilized to induce motion on the nanometer scale. To this end, Mao, *et al.*, combined double-crossover (DX) structures serving as stiff lever arms with an 'active' structure undergoing a B-Z transition upon a change in the buffer conditions. In the B-Z device, two DX structures (red and blue strands) are connected by a DNA strand containing a sequence d(m5CG)10 (with cytosine methylated at the C-5 position) which is particularly prone to a B-Z transition. A change of the concentration of $[Co(NH_3)_6]^{3+}$ from 0 mM to 0.25 mM (accompanied by a change in $[MgCl_2]$ from 10 mM to 100 mM) promotes this transition in the middle section of the device.

For each d(CG) dinucleotide, this corresponds to a change in twist of approximately 128. = .10%. .50. 2 × 34., or –3.5 helix turns. From this, an overall change in the conformation of the device as depicted in Fig. 5.19 can be expected. Most importantly, the two lever sections are rotated with respect to each other by one half turn. The inner edges of the lever arms labelled pink and green in Fig. 5.19 have a distance of 7.0 nm in the B-form and 8.9 nm in the Z-form, as derived from fluorescence resonance energy transfer (FRET) experiments.

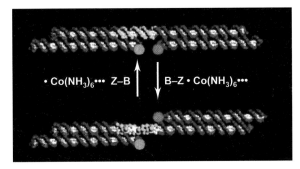

Fig. 5.19 The B-Z device consists of two levers constructed from double-crossover junctions (blue and red strands) and an active region (yellow strand) which performs a B-Z transition when $[Co(NH_3)_6]^{3+}$ is added to or removed from the buffer. The device is depicted in the B-form in the upper part of the figure, and in the Z-form in the lower part. The green and pink dots symbolize two fluorescent dyes forming a FRET pair. The B-Z transition effectively leads to a half-turn of one handle with respect to the other. The larger distance between the dyes in the Z-form makes energy transfer between them less efficient.

Reprinted from Mao, et al., Nature 39:, 144 (1999) *by permission from Nature,* ©1999 *Macmillan Publishers Ltd.*

5.8.4 DNA Tweezers

DNA tweezers are little different from B-Z conformational transition which involved change in the buffer condition to drive the molecular machine. Instead, DNA tweezers are run by changes in their mechanical states by hybridization of their fuel strands. DNA tweezers are DNA based nanoscale devices which involve both branch migration and Watson-Crick base pair complementarity. (Yurke, *et al.*, 2000). The tweezers could be used as a component in nanomachines or to build molecular scale electronic circuits. They are formed from three DNA strands A, B and C. They can be closed by adding another strand of DNA as fuel and opened again by adding still more DNA as anti-fuel strand. The first strand A contained 40 bases, with the middle four bases acting as a hinge. Strand B is bonded with 18 bases on one side of the hinge, and strand C with the bases on the other side. Since two strands bonded together are much stiffer than a single strand, the tweezers, which are open to begin with, consist of two rigid arms with a flexible hinge in between, and the loose ends of strands B and C dangling freely. The tweezers can be closed by adding a 'fuel' strand which bonds with the loose ends of both B and C, and so pulls the two rigid arms together. The tweezers can be opened again by adding an 'anti-fuel' strand, to which the fuel strand will bond in preference to B and C. Both the fuel and anti-fuel strand bond together forming waste product that floats away from the tweezers making it functional in an open state. The tweezers could be used to investigate chemical interactions, by attaching chemical components to its arms, or to control nanoscale machines. Electrically conducting molecules can also be attached to such tweezers so that molecular scale electronic circuits can be built up.

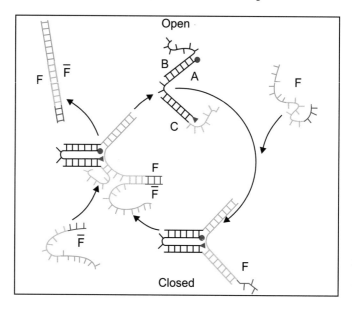

Fig. 5.20 Design and operation cycle of the tweezers: The open tweezers are assembled from three strands A, B, and C (top structure). As indicated by the colours, a fuel strand F can hybridize to the arms of the tweezers, thereby closing the device (bottom). A short overhang section (red) serves as a toehold for the removal strand F⁻ which can wrest the fuel strand F from the device via three stranded branch migration. This restores the open state of the tweezers. In each operation cycle one waste duplex FF⁻ is produced. The red circles and triangles symbolize the two dyes of a FRET pair. In the closed state, the two dyes are very close to each other, leading to maximum energy transfer. In the open state, energy transfer is less efficient .

Reprinted from Yurke, et al., Nature 40:, 605 (2000) by permission from Nature, © 2000 Macmillan Publishers Ltd.

5.8.5 DNA Actuators

The DNA actuator is similar to the DNA tweezers, but with the two arms connected by a single-stranded loop to form a closed motor section allowing the device to perform both pulling and stretching movements. Hybridization of the loop with fuel strand F stretches the device. The relaxed state can be restored by branch migration using the removal strand R. A different kind of fuel strand can be used to close the structure analogously to the DNA tweezers (Simmel and Yurke, 2001).

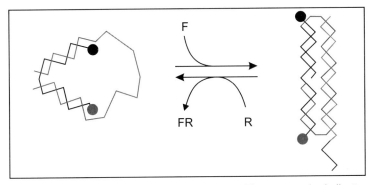

Fig. 5.21 Stretching motion of the DNA actuator: The actuator is similar to the DNA tweezers, but with the two arms connected by a single-stranded loop. Hybridization of the loop with fuel strand F stretches the device. The relaxed state can be restored by branch migration using the removal strand R. A different kind of fuel strand can be used to close the structure analogously to the DNA tweezers .

Adapted from Simmel and Yurke, Phys. Rev. E 6:, 041913 (2001).

5.8.6 DNA Scissors

Another variation of DNA tweezers are 'DNA scissors'. DNA scissors are capable of transducing the hybridization-driven closing motion in one section of the device (the 'handles') to another section (the 'jaws'), 13 nm away. The resulting motion resembles that of a pair of scissors. This particular motion is achieved by introducing short C3 carbon linker sections (*i.e.*, CH_2) into the device. The C3-spacers provide full rotational freedom, while still acting as stiff hinges rather than flexible connectors.

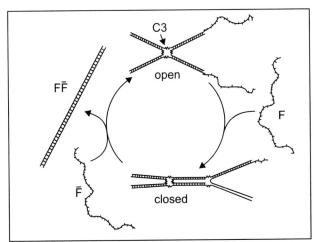

Fig. 5.22 Design and operation of DNA scissors: The double helical sections of the device are connected by C3 spacers (indicated by red zigzag lines). Closing the right pair of arms of the device (the'handles) with the fuel strand F also closes the left pair of arms (the'jaws). Addition of F⁻ restores the open state.

5.9 Protein Nanomachines: The Thinking Robots of Nature

Proteins are the three-dimensional nanostructures derived from an aperiodic array of 20 different types of amino acids. These entire amino acids assemble to form a thermodynamically stable conformation capable enough to perform plethora of functions for the cell. Such Nanorobots are the centre of action for all the biological processes. They play multitude of roles for the survival and maintenance of the cell. As DNA is the fundamental genetic information storage of the cell, similarly proteins are the gluttonous structural and functional Bionanomachines of cellular nanofactory. Protein acts as a scaffold of cell (cytoskeleton) giving structural integrity and strength to it. They function as sensors (receptors), sensing the extracellular environmental signals, transducing them inside the cell. They function as enzymes which are simply nanomachines of the metabolic pathways enhancing the rates of chemical reactions with an unpregnable precision. They also modulate other cells of the organism by sending chemical messengers in the form of hormones (endocrine system) or in the form of cytokines (immune system). They piggyback O_2, glucose, lipids and many other molecules and are responsible for their transportation and storage. Supramolecular architectures of proteins developed with the help of various nanoassemblers perform various co-ordinated mechanical motions of numerous biological processes, such as separation of chromosomes (microtubules), movement of the whole organism, movement of the eyes. They also have important roles in bones, tendons, ligaments (collagen) which provide characteristic tensile strength. They also perform the function of energy generation, hence are also called as Nanomotors.

The structural and functional descriptions of these nanorobots can be classified into four levels of organization:

(a) Primary Structure: Proteins are conglutination of amino acid sequences with their dangling side chains stable in an aqueous environment of the cell. The gluing of amino acids occurs on a Nanoassembler called Ribosome and then move on for such nanostructures which are thermodynamically stable. All proteins, whether from the most traditional lines of bacteria or from the most complex forms of life, are constructed from the same ubiquitous set of 20 amino acids, covalently linked in characteristic linear sequences via peptide bond. Analogously speaking, amino acids are like alphabets of a language named protein. It is really an astonishing quest for a Nanotechnologist, how aperiodic arrangement of 20 amino acids can give rise to myriad of different nanostructures and functions. Thus, the sequence of amino acids in a protein is characteristic of that protein and is called its primary structure.

(b) Secondary Structure: Secondary structure is the regular spatial arrangement of amino acid residues in a segment of a polypeptide chain. All amino acids are sealed by peptide bond which has many geometries of its own to give a functional conformation to the protein. They have different folding patterns such as helices, pleated sheets and turns. The comprehension of peptide group is necessary before getting into the intricacies of different covalent structures of proteins. The peptide group has a rigid, planar resonating structure giving double bond character to the peptide bond. The peptide bond's resonance energy has its maximum value (85 kJ/mol) when the peptide group is planar since its π-bonding overlap is maximized in this conformation. The resonance energy becomes zero as the peptide bond is twisted to zero. The peptide group will possess allowed torsion angles or rotational angles or dihedral angles about the C_α—N bond (Φ) and the C–C bond (ψ) of each of its amino acid residues.

(c) Tertiary Structure: Tertiary structure is the complete three-dimensional structure of a polypeptide chain. There are two general classes of proteins based on tertiary structure: fibrous and globular. Fibrous proteins, which serve mainly structural roles, have simple repeating elements of secondary structure.

Globular proteins have more complicated tertiary structures, often containing several types of secondary structure in the same polypeptide chain. The first globular protein structure to be determined, using x-ray diffraction methods, was that of myoglobin. The complex structures of globular proteins can be analyzed by examining stable substructures called supersecondary structures, motifs, or folds. The thousands of known protein structures are generally assembled from a repertoire of only a few hundred motifs. Regions of a polypeptide chain that can fold stably and independently are called domains.

(d) Quaternary Structure: Quaternary structure results from interactions between the subunits of multisubunit (multimeric) proteins or large protein assemblies. Some multimeric proteins have a repeated unit consisting of a single subunit or a group of subunits referred to as a protomer. Protomers are usually related by rotational or helical symmetry (Lehninger, 2003; Van holde 1998; Voet & Voet, 1995).

5.10 Self-assembly of Supramolecular Protein Nanoarchitectures

Proteins have the capacity to self-assemble forming regularly sized (monodisperse) nanostructures that can act as templates for the build-up of required architectures, so that they can be used in molecular electronic devices. Proteins are capable enough to produce complex nanoscale devices that consist of interlocking modules which is rather impossible in a lab, but Nature does this with ease. They can play an important role in coating medical implants, or acting as detectors in a biological-silicon hybrid devices. They can be produced in large amounts via bacterial expression systems, hence relatively cheaper compared to inorganic materials. Complex Protein nanomachines are considered to be the best candidates for producing artificial nanoscale devices. But the biggest hurdle is its complexity itself and manipulation at such dimensions is rather tricky job. This problem can be solved by using a protein building block approach. In this approach, protein components are used that naturally or artificially can have affinities for other building blocks such that, nanodevices can spontaneously self-assemble upon mixing of the individual components. Protein cages, rings and tubes, all offer useful starting points for further modification and have been engineered to produce supramolecular nanoarchitectures which may be used as basic components for electronic devices, Nanoarray chips *etc*.

5.10.1 Protein Cages

Protein cages are three-dimensional spherical protein nanostructures made up two surfaces: inner and outer one. The cage possesses a central hollow space in which all types of reactions can occur. Such cages self-assemble from monomers in solution and the biggest advantage is that the central cavity is completely enclosed sequestering all the reactions happening inside from the bulk solution. There are different types of Protein cages:

(*a*) Dps (DNA-binding protein from starved cells, from *Listeria innocua*,) with an external diameter of 9 nm, is possibly the smallest. Dps is a bacterial protein produced in response to starvation and oxidative stress which has numerous reported functions including a role in protection of DNA from oxidative damage (Mann, 1993). It can also biomineralize iron , cobalt oxide (Allen and Willits, 2002) in its central cavity. Dps is an extremely small cage protein, with a central core approximately 4.5 nm in diameter (Stiltman, *et al.*, 2005). It has also been used as part of a "ball-and-spike" supramolecule in which the C-terminal of gp5 protein (gp5c), part of the cell puncturing apparatus from bacteriophage T4 (Kanamaru, *et al.*, 2002) was fused to the N terminus of Dps via a short (22 residues) linker peptide. Gp5c assembles as a trimer into a triangular prism. Since Dps is a dodecamer, it has four 3-fold symmetry axes meaning that with a linker of the appropriate length (22 residues), four full gp5c trimers can be assembled with equal spacing around the surface of Dps, at positions corresponding to the vertices of a tetrahedron. Dps can be filled with metal or semiconductor, which can function as a quantum dot. If gp5c can also be modified to act as a template for biomineralization, then it is hoped that biomineralized gp5c "spikes" could act as electrodes around a central Dps quantum dot, forming the basis of extremely small electronic components.

(*b*) Ferritin can also be utilized as cage proteins since it is found in all prokaryotes and eukaryotes and is highly conserved in nature consisting of 24 identical protein subunits. It is a store-house of ferric-oxide naturally. Ferrous ions enter the cavity from the solution via acidic amino acid residues. Once inside, biomineralization begins at specific nucleation sites on the interior surface to form Ferric oxide. Mann and colleagues could biomineralize manganese oxides in the central cavity of purified ferritins. Ferritins have this fantastic property of biomineralization only inside the cavity and not on its outside surface (the so-called Janus effect) (Mann, 1993). Further Yamashita and colleagues have biomineralized nickel hydroxide, cadmium selenide, zinc selenide and gold sulfide which are potent materials to be used as semiconductors. They have used Ferritin based quantum dots as the basis of prototype electronic devices and thus anticipations are made that they can be an important component of future electronic devices. (Yamashita, *et al.*, 2001, 2004 and 2006).

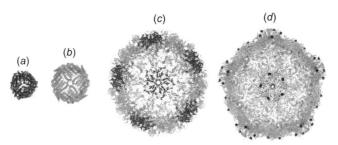

Fig. 5.23 Crystal structures of a range of cage proteins. (*a*) Dps protein (pdb 2bjy (Ilari *et al.*, 2005), diameter 9 nm; (*b*) Ferritin (pdb 2za6 (Yoshizawa, *et al.*, 2007)) diameter 12 nm; (*c*) Cowpea chlorotic mottle virus (CCMV, pdb 1cwp (Speir, *et al.*, 1995)), diameter 26 nm; and (*d*) Cowpea mosaic virus (1NY7 (Lin, *et al.*, 1999), diameter 28 nm. In (d), red spheres show the N-terminal glycine that points into the central cavity. Dark blue spheres show the C-terminal lysine on the external surface.

(*c*) There are other major class of protein cages used in bio-nanotechnology, such as Virus capsids of size 18 to 500 nm in diameter. Cowpea chlorotic mottle virus (CCMV), cowpea mosaic virus (CPMV) and Simian virus 40 are some of the viruses whose capsids are best candidates to act as Protein cage and can be utilized in electronic devices.

5.10.2 Protein Rings

Protein rings are squat, three-dimensional tubes whose length is less than their diameter. Ring structures have a central hole and generally, four distinguishable surfaces; the surface lining the central hole, the corresponding outer surface and two "end" surface orthogonal to the hole axis. There are many Protein rings existing in nature. One well-known example is Rad52, which is involved in homologous recombination in DNA. The tube is 12 nm across with a central hole ranging from 2.5 to 5 nm in diameter. (Stasiak, *et al.,* 2000; Singleton, *et al.,* 2002; Kagawa, *et al.,* 2002).

Another ring protein is the phage recombinase known as β which is a 30 kDa protein that achieves single stranded DNA-annealing, working in tandem with a partner protein Exo. B forms 12-membered ring in, nsolution but forms 12 to 18 membered ring if ssDNA is present. (Court, 2002, Passy, *et al.,* 1999).

The holes within ring proteins can be used to capture inorganic materials and array them on surfaces. The beta subunit of the HSP 60 chaperonin protein from *Sulfolobus shibatae,* forms a barrel-like ring structure that has been successfully used to capture gold nanodots over the central core and arrange them into a well-ordered 2D array (McMillan, *et al.,* 2002).

TRAP (trp RNA-binding attenuation protein) found in Bacillus, which is involved in regulation of Trytophan synthesis, is a ring structure consisting of 11 monomers, with a central hole of 2.5 nm in diameter. (Babitzke, *et al.,* 1994, 1995; Babitzke, 1997, 2004; Gollnick, *et al.,* 2005).

TRAP being a thermostable element (Baumann, *et al.,* 1997; Heddle, *et al.,* 2006), is tolerant to mutation and can be used as an engineerable building block to construct complex nanostructures. The central hole can be lined with cysteine residues in order to capture gold nanodots. (Heddle *et.al.,* 2007). Gold nanodot arrays can be constructed when gold bound rings were placed on titanium or silicon oxide surface, a prototype MOS capacitor. (Sano and Shiba, 2003).

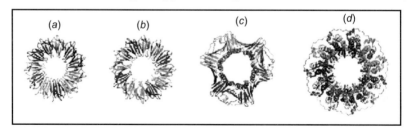

Fig. 5.24 Crystal structures of various ring-shaped proteins. (*a*) wild-type TRAP protein (pdb 1qaw (Chen, *et al.,* 1999), diameter approximately 8 nm); (*b*) mutant 12-membered TRAP protein (pdb 2zd0 (Watanabe, *et al.,* 2008); (*c*) PCNA (pdb 1axc (Gulbis *et.al.,* 1996); (*d*) RAD52 (pdb 1kno (Kagawa, *et al.,* 2002), (e) GROEL (pdb 1grl (Braig, *et al.,* 1994). All proteins shown approximately to scale.

5.10.3 Protein Tubes

Protein tubes such as microtubules, flagella and pili are elongated rings having number of potential uses in controlled release drug formulation, materials and electronics (Bong, *et al.*, 2001; Son, *et al.*, 2006). They have potent role in nanotechnology in such a way that individual subunits of the protein tube can undergo programmed self-assembly and further can be modified as per the requirement. Ribonucleoprotein is also such a tubular shaped protein made from 2130 copies of the coat protein of Tobacco Mosaic Virus (TMV) which assemble around a single-stranded RNA core. The inner cavity has been used as a template for mineralization with nickel, Cobalt, Cobalt-platinum and Iron-platinum nanowires (Knez, *et al.*, 2003; Tsukamoto, *et al.*, 2007b). The external surface has also been used as a template for the biomineralization of metals (Dujardin, *et al.*, 2003; Górzny, *et al.*, 2008) and the insertion of cysteine residues on the surface via genetic engineering allowed the chemical attachment of fluorescent chromophores to form a self-assembling light harvesting system which was able to collect light over a wide spectrum with high efficiency (Miller, *et al.*, 2007) and may be the basis for future components of optical devices. This exciting new area of research may result in new ways of producing nanowires, biosensors, drug delivery systems and structural components of future complex nanodevices.

5.11 Electron Transfer Reactions in Protein Nanostructures

Electron transfer reaction in proteins depends on the structure of the polypeptides. Depending on the structures of intervening polypeptides between a donor and acceptor, the electron transfer mechanisms can be divided into two types:

5.11.1 Super Exchange Mechanism

In case of intramolecular electron transfer, the protein matrix provides the solvent whose dielectric constant is less inside as compared to outside. Electron transfer is possible due to overlapping of the π-orbitals of the conjugated structures present in proteins. In an intermolecular situation, electron transfer occurs due to tunneling or super exchange mechanism as deeply explained in section 5.7.2 with respect to DNA. Such a tunneling process can be well explained by Marcus theory where electron transfer rate between the donor and acceptor held at fixed distance depends on the orientation of the solvent as a function of temperature, reaction driving force($-\Delta G°$), nuclear reorganization energy (λ), and electronic coupling matrix element. The reorganization parameter reflects the changes in the structure and solvation associated with electron transfer from donor to acceptor. A balance between nuclear organization and reaction driving force determines the transition state configuration and hence, the height of the energy barrier for the electron tunneling process. At Ec, the energies of both the reactants are equal and are in similar vibrational states and hence tunneling becomes favourable (Marcus, 1999).

In 1961, McConnell described the super exchange coupling across a bridge made from identical subunits. The tunneling pathway was well explained by Beratan, Onuchic, and Hopfield (1987). This pathway reduces the complex array of interactions in a folded polypeptide to just three types: covalent bonds, hydrogen bonds, and through-space contacts. The tunneling pathway is extremely sensitive for the structure of the intervening polypeptide between the donor and acceptor. Dutton and co-workers (Moser, *et al.*, 1992 and 1997; Farid, *et al.*, 1993; Page, 1999) suggest that the manifold contacts in a folded polypeptide, like those in a rigid solvent matrix, would create a uniform barrier (UB) to electron tunneling (Moser, *et al.*, 1992). Analysis of a variety of ET rates, especially those from the photosynthetic reaction center, produced a universal distance-decay constant of 1.4 Å–1 for protein ET (Moser, *et al.*, 1992). The tunneling pathways can mediate long range coupling reactions to occur. Tunneling pathways have been proposed for several natural protein ET reactions (Mei, *et al.*, 1999; Cheung, *et al.*, 1999; Roitberg, *et al.*, 1998; Sevrioukova, *et al.*, 1999); ET from the binuclear CuA site in subunit II of cytochrome-c oxidase, to cytochrome-a in subunit I have electron transfer rate constant of with a rate constant of >104 s–1, at a driving force of just 90 meV (Regan, *et al.*, 1999). CuA to cytochrome a3 ET (R = 21.1 Å) does not compete with the CuA–cytochrome-a reaction. Rapid CuA–cytochrome-a ET has been attributed to low reorganization energy (< 0.3 eV) and strong electronic coupling.

5.11.2 Multistep Electron-tunneling

The structures of enzymes and enzyme complexes reveal that in some instances electrons must be transferred over very great distances (>25 Å). It is now clear from our previous discussion, that very-long-range tunneling, even when driving-force optimized, will be too slow to sustain biological redox activity. In these instances, electron transport can be effected by multistep tunneling reactions, in which real rather than virtual electron or hole states of the bridge are involved. Evidence for multistep electron-tunneling in biological ET reactions is beginning to emerge. Ulstrup and co-workers (Iversen, *et al.*, 1998) have analyzed the possible role of real intermediates in long-range reactions, and Jortner, *et al.*, (1998) have suggested that some of the controversy associated with ET in DNA might be attributable to multistep tunneling reactions. Dutton and co-workers (Page, *et al.*, 1999) have employed a steady-state kinetics model to argue that multistep tunneling among redox cofactors can be effective even when formation of intermediate species involves highly endergonic steps ($\Delta G° \rightarrow 0.5$ eV). If bridge electron or hole states can be real intermediates in multistep tunneling processes, then they should also be more effective at mediating superexchange couplings (Iversen, *et al.*, 1998). Theoretical analyses illustrate that superexchange tunneling and multistep tunneling can operate together (Kimura and Kakitani, 1998; Zusman and Beratan, 1999). It can be quite difficult to determine whether an ET process proceeds via superexchange or multistep tunneling; the controversy surrounding the primary charge-separation events in the photosynthetic reaction centre is a classic example (Volk, *et al.*, 1998). Multistep tunneling in proteins with high-potential redox sites can involve amino acid sidechains; the reduction potentials of tyrosine, tryptophan, and histidine radicals are near 1 V compared with the normal hydrogen electrode (NHE) at pH 7 (Navaratnam and Parsons, 1998).

5.12 Applications of Protein Nanostructures in Nanotechnology and Nanoelectronics

Protein nanostructures have emerged as high-throughput screening tool for a variety of diagnostic assays, such as tissue engineering, pharmacology, and proteomics. (Mazzola and Fodor, 1995; Ekins, *et al.*, 1997; Bieri, *et al.*, 1999; Emili and Cagney, 2000; MacBeath and Schreiber, 2000; Zhu and Snyder, 2001; Jenison, *et al.*, 2001; Knezevic, *et al.*, 2001; MacBeath, 2002; Schweitzer, *et al.*, 2002).

Miniaturized protein biochips/protein nanoarrays have been fabricated using Nanoimprint lithographies for various diagnostic purposes. Semiconductor nanowires can also be developed using proteins as natural structural motif and their exploitation can be done in various fields ranging from electronics to photonics. Nanowire electronic devices enable a detection and sensing modality—direct and label-free electrical readout (*i.e.,* without the use of bound dyes or fluorescent probes)—that is exceptionally attractive for many applications in medicine and the life sciences also. Proteins can also act as molecular motor, mechanical joint, transmission element or sensor. If all these different components were assembled together, they could potentially form nanodevices with multiple degrees of freedom, able to apply forces and manipulate objects in the nanoscale world, transfer information from the nano to the macroscale world and even travel in a nanoscale environment.

The advantages in developing Protein Nanobiomachines include:

(*a*) Energy efficiency due to their intermolecular and interatomic interactions.

(b) Low maintenance needs and high reliability due to the lack of wear and tear and also due to nature's homeostatic mechanisms (self-optimization and self-adaptation).

(c) Low cost of production due to their small size and natural existence (Dubey, *et al.*, 2004).

Various applications of Protein Nanobiomachines are as follows :

5.12.1 Protein Nanostructures as Data Storage Devices

Biological macromolecules like Proteins perform all the different functions with dexterity. One of the most important properties of proteins is their ability to store and transport charges both within a protein and between proteins. This forms the basis of several elementary and essential processes, such as vision, respiration, general oxidation-reduction, all of which are indispensable for life. All these processes are finely tuned and are coupled with conformational change of a protein. There are two very important properties of proteins which make them best candidates in the design of computer devices, such as memories-ability to store/transport charge and accompanying conformational change of protein. This has led to the development of prototype computational devices including processing units and three dimensional memories. The most dynamic protein which can handle such a challenge is bacteriorhodopsin (bR) found in bacteria that grows in salty marshes with temperatures exceeding 65° C. The protein is heat-stable and photo-stable; instead it uses light energy to transport charges. In addition it can self-assemble to form thin films. Moreover they can be easily mass produced. All these properties make it an ideal device to be used in data storage. The protein has a chromophore which is the main driver of charge transport. It absorbs a particular wavelength of light and concomitantly trigger a photocycle which consists of a series of structural and chemical changes. These structural or conformational changes can be stabilized, thereby leading to the assignment of the binary code to any two long-lasting states or indeed to the assignment of some new ternary code to three such conformational states. The transition between stable states can be controlled by lasers of appropriate wavelengths; in particular the two-photon method of reading/writing data is found be the most efficient. The next stage saw the exploitation of the three-dimensional plasticity of proteins and their coupling into cubic matrices in polymer gels for building of high-density, high-speed, truly three-dimensional memories (Birge, 1992). This increased the storage limit from ~108 bits/cm^2 in two dimensions to 1011–1013 bits/cm^2. While the working prototypes only store information of ~18 Gbytes, the theoretical limit of storage is as high as 512 Gbytes and all of it in a volume of just 5 cm^3. The properties of bR-like molecules that enable them to be used for computational devices can be summarized as: long-term thermal and chemical stability; high quantum yields; accurate and reproducible states; ability to generate thin films; ability to construct efficient three-dimensional assemblies (Ortiz-Lombardía and Verma, 2006).

5.12.2 Nanowire Field-effect Sensor

Semiconductor Nanowires can be configured as field-effect transistors (FETs) for detection and sensing modality. The mechanism involved in this sensing technology is that it exhibits a conductivity change in response to variations in the electric field or potential at the surface of the device. In such a field-effect transistor, a third gate electrode capacitively coupled via a thin dielectric layer to the semiconductor Nanowire modulates the conductance of the semiconductor between on and off states. In the case of a p-type semiconductor, applying a negative gate voltage leads to an accumulation of majority charge carriers (positive holes) and a corresponding increase in conductance. This relationship between conductance and corresponding charge at the gate electrode/dielectric interface makes FETs good candidates for electrical based sensors. Binding of a charged or polar biological or chemical species to the gate dielectric is analogous to applying a voltage using a gate electrode which can be exemplified by binding a protein of net negative charge to the surface of a p-type FET which can lead to an accumulation of positive hole carriers, thus leading to increase in device conductance.

Nanowire based sensing devices can be configured from high performance FETs by linking specific receptor groups to the surface of the nanowire which is either silicon or some other semiconductor. (Patolsky and Lieber, 2005; Patolsky, *et al.*, 2006; Cui, *et al.*, 2001; Hahm and Lieber, 2004; Wang, *et al.*, 2005). When these surface-modified devices are exposed to a solution containing a macromolecule species, such as a protein, specific binding to the receptor will lead to an increase or decrease in the device conductance depending on the net charge of the biomolecule and the semiconductor type (p or n). Fig. 5.26 illustrates an integrated microfluidic/nanowire sensing platform that incorporates nanowires with well-defined p- and/or n-type dopants, specifically tailored surface functionality, source and drain electrodes that are insulated from the fluid environment by a dielectric passivation layer. Also, and a polymer-embedded microfluidic channel that enables rapid solution delivery to arrays of nanowire devices. In addition, the small size of the nanowire devices not only enables very high-sensitivity detection, but also allows for tens to hundreds of individually addressable nanowire devices to be defined within a single microfluidic delivery channel. Significantly, defining distinct surface receptors on different nanowire elements open up the potential for multiplexed, real-time assays of multicomponent solutions, such as the simultaneous detection of proteins, DNA, viruses and small molecules. (Patolsky and Lieber, 2005; Patolsky, 2006, Patolsky *et al.,* 2004, 2006; Zheng, *et al.,* 2005).

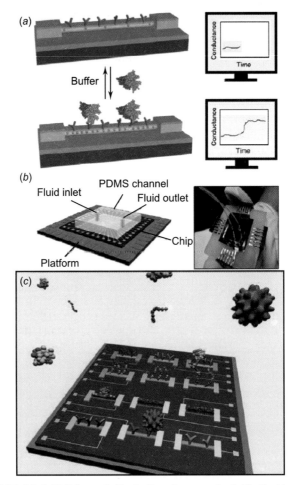

Fig. 5.25 (*a*), (left) Schematic illustration of a nanowire field-effect transistor configured as a sensor with antibody receptors (blue). (right) Binding of a protein with a net negative charge to a p-type nanowire yields an increase in conductance. (*b*), (left) Schematic and (right) photograph of a prototype nanowire sensor biochip with integrated microfluidic sample delivery. (c) Schematic of a nanowire device array for multiplexed, real-time sensing of multiple biological species. The size of the chip is about 15 μm * 15 μm.

5.12.3 The VPL Motors and its Types

Viral Protein Linear motors are basically actuators which change their 3D conformation depending on the pH of the environment. We are all aware of the role of envelope glycoproteins of different viruses which fuse with the membrane of the host cells. During membrane fusion, the glycoprotein undergoes conformational change so that it can readily attack and infect the cell. This conformational change is due to the change of pH in the vicinity of the cell. This mechanism of viral envelope glycoproteins can be exploited to produce actuators. The different viral peptides which can be used as motors are as follows:

1. The Influenza virus protein Hemagglutinin (HA) peptide HA2
2. The Human Immunodeficiency Virus type 1 (HIV 1) peptide gp41
3. The Human Respiratory Syncytial Virus (HRSV) protein subunit F1
4. The Simian Immunodeficiency Virus (SIV) protein gp41
5. The Human T cell Leukemia virus type 1 protein gp21
6. The Simian Parainfluenza Virus peptide unit SV5
7. The Ebola virus protein gp2.

All the above viral peptides can lead to the formation of a different VPL motor and can possess vast range of properties based on weight, volume, motion, force and speed capabilities. Common characteristics in all the above viruses are that they all change the structure of a portion of the surface protein (envelope glycoprotein) and they also possess similar mode of infection (Dubey, *et al.*, 2004).

Stepping towards synthesizing an able VPL motor for nanoactuation, Hemagglutinin (HA) from influenza virus was studied in details and then exploited for various purposes. The main advantage of HA is that it performs repeatable motion which can be controlled by variation of the pH. HA exhibits pH-dependent conformational change that leads to membrane fusion. HA comprises of two polypeptide chain subunits (HA1 and HA2) linked by a disulfide bond. HA1 contains sialic acid binding sites, binding sialic acid of the target cells, thus helping the virus to recognize a cell (Sauter, *et al.*, 1989). The spring loaded conformational theory is the most widely accepted one to explain the process of membrane fusion (Carr and Kim, 1993). As per the theory, there is a specific region (sequence) in HA2 that has a propensity to form a coiled coil. In the original X-ray structure of native HA, it has been found that this region is simply a random loop. Upon activation, a 36 amino acid residue region makes a dramatic conformational change from a loop to a triple-stranded extended coiled coil along with some residues of a short α-helix that heralds it, thus locating the hydrophobic fusion peptide (and the N-terminal of the peptide) by about 10 nm. Biomimicry of 36 amino acid residue, long peptide was a challenge for material scientist for designing a VPL motor. This was successfully achieved by cutting out the loop 36 from the

VPL motor that led to the formation of a peptide having closed length of about 4 nm and an extended length of about 6 nm, thus giving it an extension by two-thirds of its length. After the characterization of the peptide, it will be subjected to a reduced pH thus resulting in a conformational change which can be monitored by fluorescence tagging techniques and the forces can be measured by using Atomic Force Microscope. (Dubey, *et al.*, 2004).

Fig 5.26 a and b shows a schematic view of the VPL motor supporting a moving platform. The motor is in its initial contracted phase similar to the virus native state (Fig. 5.26a) and at its extended, fusogenic state (Fig. 5.26b). The total outward protrusion or extension of the motor is measured to be 10 nm. Just to enhance the force capabilities of the VPL motors, several VPL actuating elements could be attached in parallel as its exhibited in Fig. 5.26. This consequently results in the generation of extremely powerful micro-, meso-, or even macro-actuators that will be able to apply ultra large forces, while their dimensions are extremely small. To augment the displacement capability of the VPL motors, several VPL elements could be connected in a series as seen in Fig. 5.27. (Dubey, *et al.*, 2004).

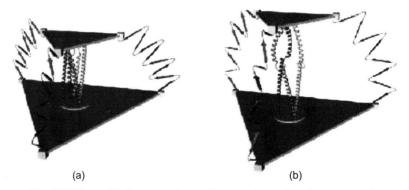

(a)   (b)

Fig. 5.26 Three titin fibres can be used as passive spring elements to join two platforms and form a single degree of freedom parallel platform that is actuated by a Viral Protein Linear (VPL) actuator (center). (*b*) The VPL actuator has stretched out, resulting in the upward linear motion of the platform. The three titin fibers are also stretched out.

The next step is to develop nanomechanical assemblies that will be powered by the proposed VPL motors that can be interfaced with other machine components so that they form multi-degree-of-freedom bionanodevices. Towards such an initiative, a futuristic model of a bionanomachine is designed which has a three-degree of freedom, with three-legged manipulator. The top and bottom platform are made from carbon nanotubes which are connected to each other using three legs made from the VPL motors, while the joints between the VPL motors and carbon nanotubes are formed by DNA-based universal joints. (Dubey, *et al.*, 2004).

Fig. 5.27 Several VPL motors placed in parallel (left) and series (right) to multiply force and displacement, respectively.

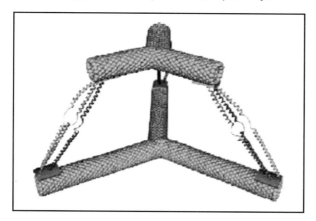

Fig. 5.28 Three-degree-of-freedom parallel platforms can be formed using controllable VPL actuators attached at the three legs of the platform.

5.13 Summary

Beside a variety of more detailed questions, a pretty clear picture about the phenomenon "Charge transfer processes in DNA and proteins" has emerged now. The extreme controversy has been solved by a very differential interpretation of the applied DNA and protein systems and the description of several mechanistic pictures. In conclusion, it turned out that excess electron transport occurs via a hopping mechanism over long distances. It looks also very probable to assume that the DNA dynamics, such as base fluctuations, influence or even gate both types of charge transport processes. Similarly, hopping mechanism also leads to charge transport in Proteins. Such macromolecules can be constructed for their exploitations in Nanotechnology and Nanoelectronics.

Introduction to Carbon Nanomaterials

"Just as silicon transistors replaced old vacuum tube technology and enabled the electronic age, carbon nanotube devices could open a new era of electronics."

—**Margaret Blohm**
GE's Advanced Technology Leader for Nanotechnology

6.1 Introduction

In the periodic table, carbon is a singular element which stands at the sixth place and is the first element of column IV of periodic table. Out of six electrons of carbon, four electrons of valence shell play a significant role of forming three hybridizations *viz.*, sp, sp^2 and sp^3, which has led to the formation of many stable compounds of carbon. Carbon is one of the most abundant elements on this earth. It is fundamental for the living world giving rise to the organic chemistry, the biochemistry and the miracle of life on earth.

Identification of the different forms of solid carbon progressively came to light with the diamond and graphite as natural form of carbon. In 1985, the discovery of basic molecule C_{60} by Kroto, *et al.*, has added new excitements which led to the discovery of carbon nanotubes and many more interesting forms of carbon. Though carbon nanotubes were first reported by Iijima, but this material was prepared by Prof Y Ando of Meijo University, Japan who handed over to Prof Iijima for a deeper analysis as his TEM was not good enough to observe the fascinating structure of nanotubes. Some of the important, but basic characteristic of carbon is shown in Table 6.1.

Table 6.1. Classification of the different forms of carbon

Types	Diamonds	Graphites	Carbynes	Nanotubes
Hybridization	sp^3	sp^2	sp^1	mixture of sp^3 & sp^2
Coordination number	4	3	2	mixture of 3 & 4
Bond length (nm)	0.154	0.140	0.121	~0.133

6.2 Why Pure Diamonds Show p-type Character?

Detailed properties and preparation of diamond and graphite has been known to us very well since our school days and hence, there is hardly any need to devote any time on the various aspects of diamond and graphite. However, it is necessary to understand why pure diamond shows p-type character, while pure graphite shows n-type character. This is interesting because the conventional solid state physics suggests that pure materials should not show any specific characters like n-type or p-type unless they are doped with some suitable dopants.

Crystal of diamond is formed by coordination of sp^3 carbon leading to four σ-bonds. In such type of bonding, each electron of each carbon atom constituting diamond structure, is shared by covalent type of bonding. On the other hand, graphite structure is formed by sharing of sp^2 type carbon. This type formation means, a formation of a mixture of σ-bonding and π-bonding. We know from our school days that π-bonding is formed due to overlap of p-orbital electrons which did not participate in the formation of σ-bonding. If we examine these two types of structural situation, soon it will appear that in three dimension in space there are two places: one where there is no free electrons (*i.e.* the place where σ-bondings are present) and others where non-covalent electrons are present (*i.e.*, the place where π-bonding are present). In diamond type structure, the space which is to be occupied by π-bonding is empty where as in graphite type structure space occupied by π-bonding would be filled with electrons (Fig. 6.1). The conventional theory of physics suggests that if an orbital is empty of electron would mean it contains holes and such materials should be classified as p-type materials. On the other hand, if orbital is filled with electrons and making it available for conduction then the material should be classified as n-type materials. If this concept is considered to be correct, then naturally diamond whose π-bonding is empty should show p-type character. Diamond thus is found to show p-type character. Whereas, graphite which has large concentration of free electron in it's π-bond should behave like n-type, which is the case with graphite. As a matter of fact, theoretical calculation suggests that if material contains less than 33% of sp^3 type carbon, the material will show n-type character and if it is more than 33% sp^3, the material will show p-type character. Most of the carbon materials like carbon nanotubes, nanobeads *etc.*, show intrinsically p-type character. It is interesting to conclude that if carbon material contains exactly 33% of sp^3 carbon, it will show pure intrinsic behaviour.

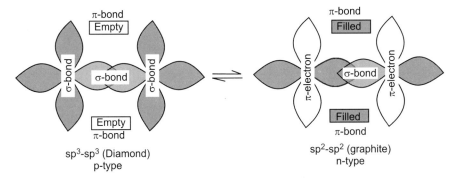

Fig. 6.1 Schematic diagram showing formation of σ-bonding and π-bonding and its effect on showing diamond as p-type and graphite as n-type material.

6.3 Allotropes of Carbon

Before the discovery of fullerenes family, we knew that there are only two types of allotropes of carbon: diamond and graphite. After C$_{60}$ was discovered, third allotrope of carbon was added "fullerene". Does that mean that as we keep on discovering new forms of carbon, number of allotropes of carbon will also keep on increasing? Or should we not develop a classification of carbon such that no matter how many forms of carbon are developed, the number of allotrope does not change? If structure of most of the stable carbon materials are examined, they contain either 100% sp^2 carbon (graphite type structure), or 100% sp^3 carbon (diamond type structure) or carbon containing both sp^2 and sp^3 carbon (carbon nanotubes, fullerenes *etc.*). Hence, if this logic is accepted the classification of carbon should be done as: (*i*) diamond type (containing 100% sp^3 carbon), (*ii*) graphite type (containing 100% sp^2 carbon) and (*iii*) intermediate type (which contains both sp^3 and sp^2 carbon). This classification suggests that almost all type of carbons which have been discovered so far or will be discovered in future can be from one of these three types of allotropes only.

6.4 Generalized Shapes of Carbon Nanomaterials

Before we discuss the various forms of carbon nanomaterials, it may be useful to analyze the possible basic shapes of such materials, so that one could decides the type of carbon material to be used for the specific application in biological systems.

Since discovery of newer forms of carbon keeps on emerging now and then, it would be extremely difficult to give all possible shapes of carbon nanomaterials which have been synthesized and reported. However, it may be possible to classify them in such a fashion that almost all forms fall within some limited types of morphology. These are shown in Fig. 6.2.

From this generalized morphology of various shapes of carbon, it appears that we may encounter either spherical shapes carbon (either hollow or solid beads), or cylindrical shape (either hollow, branched type). Each of these shapes will be useful for a specific purpose. Therefore, before launching on any selection of carbon nanomaterials, it is desirable to have some preconceived idea about the shape which should be most appropriate for the application in question.

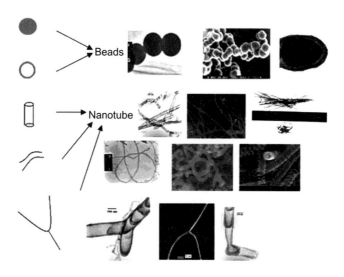

Fig. 6.2 Various basic shapes of carbon nanomaterials.

6.5 Various forms of Carbon Nanomaterials

It will be futile efforts to discuss all types of carbon nanomaterials which have been discovered so far, for the reasons discussed in the previous section. Hence, we shall discuss here only those forms of carbons which have found some specific applications. Carbon nanomaterials that has been discovered so far includes fullerenes, carbon nanotubes, carbon nanofibres, carbon nanobeads, carbon materials with porous structure having pores of nanoscale and channel type carbon nanomaterials. Some of the popular forms of carbon which are expected to show engineering applications are discussed in this section.

6.5.1 Diamond

In *Diamond,* all the four valence electrons orbital (one s and three p) of carbon intermixes with each other and give four sp^3 hybrid orbitals of equal energy, which upon overlapping with neighboring carbon hybrid orbital, create isotropically strong diamond structure. In other words, diamond is found in a cubic form in which each carbon atom is linked to four other carbon atoms by sp^3 bonds in a strain-free tetrahedral array. Diamond also exists in a hexagonal form (Lonsdaleite) with a Wurtzite crystal structure with a C–C bond length of 152 pm. Natural and synthetic diamonds contain various impurities. Nitrogen and boron are found as substitutional impurity atom in the crystal lattice. Significant quantities of hydrogen and oxygen are found in diamonds, especially at surfaces where they stabilize dangling bonds. Diamond like Carbon (DLC) though has great similarity to diamond are, amorphous or partly crystalline phase in the lattice of carbon structures.

6.5.2 Graphite

By sharing three sp^2 electrons with three neighbouring carbon atoms, carbon forms a layer of honeycomb network of planar structure called *Graphite* and the fourth π-electron of carbon is delocalized over the whole plane of graphite. As π-electrons are mobile, graphite conducts electricity in the plane. This conduction of electricity can occur only in plane, but not from one plane to another. This sp^2 carbon build layered structure. Graphite has been used as lubricant and is considered a soft material.

Crystal structure of graphite is the graphene plane or carbon layer plane, *i.e.,* an extended hexagonal array of carbon atom with sp^2 bonding and delocalized bonding. The commonest crystal form of graphite is hexagonal and consists of a stack of layer planes in the stacking sequence ABABAB. In-plane C–C distance in graphite is 142 pm, *i.e.,* intermediate between C_{sp3}–C_{sp3} and $C_{sp2} = C_{sp2}$ bond lengths, 153 and 132 pm, respectively. Principal uses of natural graphite are in the foundry and steel industries as well as in the refractory and electrical industries. Most synthetic graphite used for engineering applications are granular composites consisting of a filler (usually a coke) and a binder carbon formed from pitch. Well-graphitized synthetic graphites are produced by hot-pressing pyrolytic graphite (HOPG grade).

6.5.3 Carbynes

Carbynes are a form of carbon with chains of carbon atoms formed from conjugated C(sp')=C(sp') bonds (polynes) (Fig. 6.3). In 1968, El Goresy and Donnay discovered a new form of carbon which they called *white carbon.* Despite many publications on carbynes, their existence has not been universally accepted and the literature has been characterized by conflicting claims and counter claims. Large number of chemical and physical methods have been developed for producing carbynoid materials (Kavan and Kastner,1994).

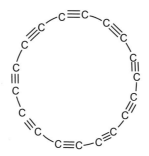

Fig. 6.3 A model of cyclo C-18 carbyne molecule.

6.5.4 Buckminster Fullerene

The C_{60} molecule was discovered in 1985 by Kroto, Smalley and coworkers. It was synthesized by laser-vapourization of graphite and the name Buckminster fullerene (C_{60}) was given after an architect-engineer, R. Buckminster Fuller who designed geodesic domes. Chemical reactivity resembles of strained electron deficient polyalkaene with localized double bonds. Fullerene is the most stable cage cluster structure, in which 20 hexagons and 12 pentagons are present as required by famous Euler's theorem and all the pentagons are separated by hexagons (the isolated pentagon rule). The bonding between carbon atoms in fullerene are sp^2 type with some sp^3 character due to its high curvature. Each carbon atom is bonded to three others by two longer bonds (length ~145pm) and one shorter bond (length ~140pm). Depending upon the solvent and method used for the crystallization, C_{60} forms fcc structure, hexagonal closed packed, hcp or orthorhombic structures. C_{60} behaves as 2π electron-deficient dienophiles and dipolarophiles, and therefore, cycloaddition reaction is normally used for its functionalization. Functionalized carbon atoms of fullerene change their sp^2 carbon to a less strained sp^3 configuration. Other types of fullerene C_{70}, C_{76}, C_{78}, C_{80} *etc.*, have also been

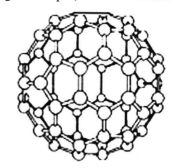

Fig. 6.4 A C_{60} Fullerene.

discovered. C_{60}-related materials (especially alkali metal doped K_3C_{60}) possesses superconductivity with Tc around 40K.

6.5.5 Amorphous Carbon

Amorphous carbon can be broadly classified as: (*i*) amorphous carbon a-C and (*ii*) hydrogenated carbon a-C:H. Both types of amorphous carbon contain different amounts of sp^2 and sp^3 carbon atoms. a-C carbon normally contain sp^2 carbon with varied amount of sp^3 in the range of 5–55%. Hardness of the a-C increases with increase in sp^3 carbon content, whereas, hardness of a-C:H is inversely related to its hydrogen content. Hard a-C:H is also known as diamond like carbon (DLC).

6.5.6 Carbon Nanotubes (CNTs)

A nanostructure of carbon atoms which appears as if a single layer of graphene sheet is rolled to form a cylindrical shape is designated as single walled carbon nanotubes (SWCNT). If more than one graphene sheets are rolled to form a hollow cylindrical shape, it is designated as multi walled carbon nanotubes (MWCNT). These graphene sheets are separated by a distance of 0.34 nm (Fig. 6.5).

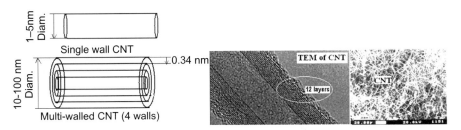

Fig. 6.5 Schematic sketches of single walled carbon nanotubes, multiwalled CNT and multiwalled carbon nanofibres.

They can be semiconducting or metallic, depending upon the diameter of the tube. SWCNTs are defect free, can be either semiconducting or metallic, have better mechanical strength *etc*. Their diameter ranges from 1–5nm. MWCNTs have very good mechanical character with high Young Modulus of around 1-terapascal and their diameter ranges from 10nm to 100 nm.

In other words, single wall carbon nanotubes can be visualized to be formed by bisecting a C_{60} molecule at its equator and joining the two resulting hemisphere with a cylindrical tube having the same diameter as the C_{60} molecule. The cylindrical tube consists of honeycomb structure of a single layer of graphite known as a graphene layer (Fig. 6.6).

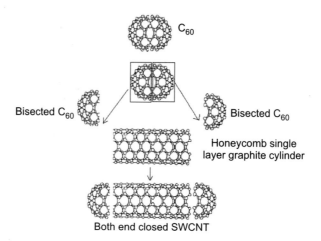

C_{60}

Bisected C_{60}

Bisected C_{60}

Honeycomb single layer graphite cylinder

Both end closed SWCNT

Fig. 6.6 Fullerene derived SWCNT. Each end capped with half fullerene.

6.5.7 Carbon Nanofibres

Carbon nanofibres (CNF) appear almost the same as CNT under the SEM. Both are like hollow cylinders. The difference is that the outer layer of CNT is made up of unbroken graphene sheet, whereas CNF is made up of broken graphene sheet (Fig. 6.7). Due to the high tensile strength and Young's modulus of some carbon fibres, and their low density, these materials are used as reinforcement in aerospace composites. Carbon fibres after activation by various processes, are being used as adsorbent, catalyst supports *etc*. Organic fibre precursors like PAN, poly (vinylidene chloride), poly (acrylonitrile), rayon fibres, phenolic resin, pitch is heated to around 300°C to render the fibres thermo set via cross-liking and then heated to around 1000°C under oxygen free atmosphere to convert it to carbon fibre. Recently, high performance rigid-rod polymeric fibres (commercially known as Kevlar) has been developed from precursors like poly p-phenylene terephthalamide or poly p-phenylene benzobisoxazole (PBQ).

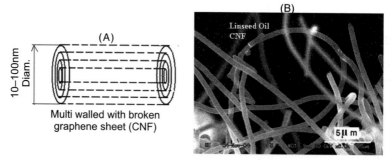

Fig. 6.7 (A) Schematic diagram of multi walled carbon nanofibres showing broken graphene sheet with dotted lines and **(B)** SEM of a Carbon Nanofibres.

Depending on the raw material and nanocatalyst used for synthesis of CNM different types of CNF has been reported *e.g.,*

(*a*) CNFs grown by CVD process from camphor using nickel powder catalyst (20–50nm size) have resulted in formation of *octopus shaped CNF* (Fig. 6.8A).

(*b*) *Cactus type CNF* has been fabricated by pyrolyzing camphor using nanosized nickel electroplated over copper as catalyst. When large size nickel particles were used as catalyst then vertically growing cylindrical CNF was formed and has been names as Cactus type carbon fibres (Fig. 6.8B).

(*c*) Camphor and turpentine on pyrolysis have produced *Y-junction* (Fig. 6.8C) and *bamboo-shaped branched CNT* (Fig. 6.8D), Spherical fluffy beads (Fig. 6.8E), porous beads (Fig. 6.8F).

(*d*) *Channel type fibre* has been synthesized by the pyrolysis of coconut fibre (Fig. 8G).

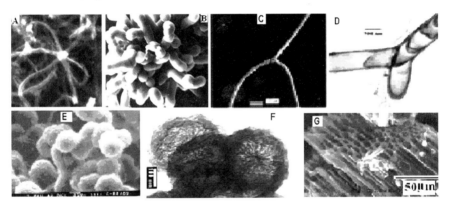

Fig. 6.8 Various types of CNF and CNB (A) Octopus like carbon fibre (B) Cactus type fibre (C) Spongy carbon beads (D) Bambo-shaped branched CNT (E) Cotton ball like CNB and (F) Porous bead (G) Channel type fibre from coconut fibre.

6.5.8 Active Carbon

Carbon produced from precursors like wood, cotton , bamboo *etc.*, are activated either by chemical or physical process. In physical process, carbon is exposed to an oxidizing atmosphere (usually steam or CO_2) at temperature in the range of $800 - 1000^\circ C$. By this process, carbon is gasified creating micropores (pores < 2 nm) and mesopores (pores 2–50 nm). Sometime it can also form macropores (>50 nm). Physical method is used with carbon produced from precursors which helps in leaving behind some structural defects in the carbon. Carbon produced from precursors like polyvinyl chloride are more order defect free. These carbons are activated by chemical process as they are difficult to activate by physical process. Chemical activation involves first mixing the precursor with compounds, such as zinc chloride, phosphoric acid or potassium hydroxide, followed by pyrolysis to temperatures in the range 400–850°C in oxygen free atmosphere, and final washing to remove activating agent. The mechanism of chemical activation is complicated. This method results in fine and high surface area powder.

6.5.9 Carbon Nanobeads

Recently, researchers have created a new form of spherical or beads of carbon: a spongy solid that is extremely lightweight and is attracted to magnets and has a size of 50–250 nm (Fig. 6.8E). This new structure was first created by Sharon, *et al.*, (1998) by CVD process. The spongy carbon nanobeads are hollow, interconnected and spherical having surface composed of broken graphene sheets. The inner surface is composed of amorphous carbon (Fig. 6.8F). The thickness of bead wall is around 80 nm and diameter of the beads is around 500 nm.

6.5.10 Nanofoam of Carbon

Recently, researchers have created a new form of carbon: a spongy solid that is extremely lightweight and is attracted to magnets and has a size of 1nm. This structure is created when a carbon target is illuminated with a high power laser beams (10,000 pulses a second). Illumination of target by laser causes a rise in temperature of target to around 10,000°C. At this temperature, carbon forms an intersecting web of carbon tubes, each just 1 nm in diameter. These spongy forms of carbon have been called as solid *nanofoam* (Giles, 2004).

6.6 Novel Properties of Nano Carbons

Though C_{60} was developed much earlier to Carbon nanotubes, but application wise, carbon nanotubes has surpass fullerene. Carbon nanotubes have many applications in field like chemical and biological separation, purification, energy storage, such as fuel cell, hydrogen storage and lithium battery. Some of the devices have also become a reality like sensors, transistors, nanoelectronics, field emission *etc*. These applications have been possible because of the unique structure and properties of carbon nanomaterials.

The structure and properties of nanotubes can be understood by examining the nature of carbon atoms present in them. As discussed earlier carbon atom can possess sp^3, sp^2 and sp hybridized orbitals as shown in Table 6.1. Carbon atoms in diamond possess sp^3 hybridized orbitals, which means it has four equivalent covalent bonds connected to four other carbon atoms via σ-type bonding. This provides four tetrahedral orientations giving three-dimensional interlocking structures. Such structure makes diamond hardest, electrically insulator material.

Graphite on the other hand possesses sp^2 type carbon atoms which help to form three in-plane bonds and one out of plane bond. As a result, graphite has parallel planar hexagonal net work with each plane separated by a spacing of 0.34 nm and is hold together by the Van der Waals forces. Interaction of the loose π-electrons with light causes graphite to appear black, while electron in diamond absorbs ultraviolet light making it transparent to our eyes.

Bonding in CNT is essentially sp^2 due to cylindrical nature which causes quantum confinement and three bonds are slightly out of plane. This makes nanotubes mechanically stronger, electrically and thermally a good conductor.

Unlike silicone, existence of sp^2 and sp^3 hybridization, and π-type bonding induces special properties to carbon nanomaterials. Depending upon the diameter of the CNT, it can show metallic and semiconducting properties. Tubular structure of hexagonal network makes the carbon nanotubes almost a defect free structure. Incorporation of pentagons and heptagons in hexagonal structure of CNT improve the chemical reactivity of the material. These types of incorporation help in capping, bending, branching *etc.*, of the tube. Due to cylindrical nature of the tube, electron confinement along the tube circumference helps in developing flat plate display unit. While silicon is only indirect band gap materials with a fixed band gap, carbon nanotubes can be synthesized of various band gaps (direct as well as indirect) by controlling the ratio of sp^3 and sp^2 carbon atoms, which indirectly control the diameter of the tube. Mechanical and electromechanical properties are very good because of its high Young's modulus and stress value. Cylindrical nature of tube provides electron to orbits in a circular motion induces many interesting magnetic and electromechanical properties such as quantum oscillation and metal-insulator transition. Its high thermal conductivity makes it as good materials in dissipating heat from electronic circuit.

Specific surface area of SWCNT is often larger than that of MWCNT. Typically, the total surface area of as grown SWCNT ranges between 400 – 900 m^2/g, whereas values ranging between 200–400 m^2/g for as grown MWCNT. It is observed that opening the end of the tube and cutting the size (length) as well as chemical treatment with KOH considerably increase the surface area of the CNT.

Tensile strength and Young's modules of graphite are the highest amongst the carbon materials. The reason is that a single crystal of graphite shows a maximum of anisotropy with a maximum of stiffness in the (001) plane due to the short C—C bond length of 0.142 nm (as compared to 0.154 nm for diamond) and easy (001) glides due to van der Waals spacing. Thus, Young's modulus in graphitic planes is 1036 GPa and the associated tensile strength is 100GPa. Hence, carbon materials like CNT, CNF *etc.*, will show intermediate of these. Tensile strength of SWCNT has been reported to be in the range of 45–63 GPa and that of MWCNT 1.72–52 GPa and Young's modulus 0.45 TPa. The variation in the measurements is because these properties depend upon the diameter of the tube, purity and structural defects *etc.* Nevertheless, these values are quite high as compared to even steel. Encouraged by their high tensile strength and Young's modulus, scientists have been trying to make composite of CNT/CNF with various types of polymers. In these experiments, due to variation in structure and diameters of the tube, contradicting results have been reported. Nevertheless, composite of CNT with iso-butyle ethylene and CNT with PVC have been found to increase the tensile strength by 20–30% giving interfacial shear stress 224 and 196 MPa, respectively. In selection of polymer for the composite, care has to be taken that thermal expansion of CNT must match with that of the polymer otherwise deformation of composite can take place. Efforts have also been made to prepare composite using carbon nanotubes as the fiber material and C_{60} crystal as the matrix. C_{60} crystal has unique mechanical properties owing to its highly symmetrical fcc structure. Some of these properties are discussed in detail here:

6.6.1 Electrical Properties

Carbon materials depending upon their structure cover a wide range of electrical properties from metallic to semiconductor to insulator (Fig. 6.9). The reasons for showing such wide behaviour can be understood by considering the electronic arrangement in carbon materials. Carbon materials exhibit either pure σ-bonds (when it contains only sp^3 carbon) or σ-bonds along with π-bonds (when it contains sp^2 carbon). The σ bonding and antibonding orbitals create a full valence band and an empty conduction band separated by a large energy gap. Without π-electrons *i.e.*, when it has only sp^3 carbon, the material behaves like insulator, as illustrated by diamond which has a large band gap of nearly 5.5 eV (Fig. 6.9). However, when π-electrons are present *i.e.*, when it contains sp^2 carbon, the valence and conduction bands, due to this new hybridization, fill the gap left by the σ-bands and materials show metallic behaviour which is the case with graphite with a band gap of ~0.25eV (Fig. 6.9). However, when carbon material contains both types carbon *i.e.*, mixture of sp^3 and sp^2, the material shows semiconducting behaviour with variable band gaps depending upon the ratio of sp^3 and sp^2 carbon (in the range of *i.e.*, >5.5eV and <0.25 eV), which is the case with carbon nanotubes. The band gap of carbon nanotubes depends upon its diameter which in turn depends upon the ratio of sp^3 and sp^2 carbon.

The electrical conductance of semiconductor SWCNT is highly sensitive to the change in the chemical composition of the surrounding atmosphere due to charge transfer between the nanotubes and the molecules from the gases adsorbed onto their surface. It is observed that there is a linear dependence between the concentration of the adsorbed gas and the difference in electrical properties.

The morphology and size of the carbon nanotubes plays a significant role in catalytic application due to their ability to disperse catalytic active metal particles.

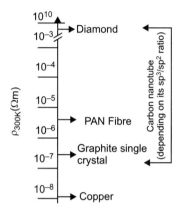

Fig. 6.9 Showing an approximate range of resistivity of various carbon materials and copper.

6.6.2 Properties of Carbon Fibres

Activated carbon can be used to adsorb SO_2, NO_2, NO coming out from combustion of coal and gasoline fuels. These materials convert SO_2 to sulfuric acid in moist air, NO/NO_2 to nitrogen. They have also been used to remove volatile organic compounds which are toxic *e.g.,* toluene, xylene, acetone, n-hexane. Granular active carbons are used for removing contaminants like trihalomethanes, pesticides *etc.,* from the drinking water. A second interesting application of ACF is in electrical double-layer capacitors because these materials are highly porous, high surface area and electrically conductor. Since some grades of mesophase pitch-based fibre have thermal conductivities three times that of copper, composites fabricated with these fibres are ideal for reducing thermal gradients. The ability to dissipate heat is an important factor in both structural composites and electronic systems.

6.6.3 Surface to Volume Ratio

Density of dangling bonds present on the surface of the nanomaterial is controlled by the total surface area available on the surface. Larger the surface area more would be the number of dangling bonds and hence more would be its reactivity. In a three-dimensional material, surface area is also related to the total volume of the materials. Therefore, it is important to get some idea about the surface to volume ratio because one would like to have this ratio as large as possible. Lets take a case of metal of radius 0.144 nm size crystallizing into a cubical structure and we calculate the number of atoms present on the surface as well as the total volume of the material of various sizes.

Number of metal atoms in a 1 metre cube

$$= [1m/(0.144 \times 2 \times 10^{-9}m) = 3.47 \times 10^9 m]$$

Volume of this cube

$$= (3.47 \times 10^9 m)^3 = 4.19 \times 10^{28} \text{ atoms of metal}$$

Surface area of this cube (has 6 faces)

$$= (3.47 \times 10^9 m)^2 \times 6 = 7.22 \times 10^{19} \text{ atoms}$$

Surface/volume ratio

$$= 7.22 \times 10^{19} \text{ atoms}/4.19 \times 10^{28} \text{ atoms}$$
$$= 1.72 \times 10^{-9}$$

If similar calculations are done for the various sizes of the cube upto a size 1.15 nm then it will be observed that the surface to volume ratio of a cube of size 2nm would be approximately 0.64, which is much larger than what one would get with larger size cube. A graph is plotted between the surface to volume ratio and the size of the particle in nm (Fig. 6.10). This graph suggests that when the size of the particle is more than 50nm, magnitude of surface is much smaller than what one would get with smaller size particles (less than 10 nm). Thus, number of dangling bonds present over surface of particle of 10 nm would be much more than larger size particles.

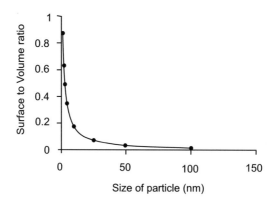

Fig. 6.10 A theoretical graph showing a relationship between the size of the particle (nm) and its corresponding surface to volume ratio.

6.7 Summary

In this chapter an overall view of carbon materials has been given. Discussion is made to explain why diamond without any doping shows p-type character. It is shown that almost all types of carbon can be classified by three forms of allotropes and there is no need to keep on increasing this number. Carbon nanomaterials have been synthesized to give various types of shape and size. Therefore, there is need to generalize the shape of these materials in mere two types: cylindrical and beads. Some of the popular carbons have been discussed and their special and specific properties are discussed.

Carbon Nanomaterials: Synthesis, Purification and Characterization

> I am among those who think that science has great beauty. A scientist in his laboratory is not only a technician: he is also a child placed before natural phenomena which impress him like a fairy tale.
>
> —Marie Curie

7.1 Introduction

A detailed account of fabrication or synthesis of nanomaterials is given in chapter 3. This chapter will be dealing briefly with some of the most common methods of synthesizing carbon nanomaterials only.

The fabrication of carbon nanotubes is not a difficult task, since they can be found also in common environments such as the flame of a candle. But is very difficult to control their size, orientation and structure, in order to be able to use them for technological tasks.

There are various methods used for the synthesis of carbon nanomaterials. Most of them are "Bottom-Up" approaches. The only "Top-Down" approach is by grinding the material using high energy "Ball-Milling", often using tungsten carbide balls. Since, the purity and uniformity of size by this method is low, it is not used for synthesis of carbon nanomaterials.

7.2 Various Methods of Synthesizing Carbon Nanomaterials

Though there are many methods tried by various workers, it would be difficult to discuss each of them in detail. However, some of the common methods which are popularly utilized for the synthesis of nano carbon are discussed in this section.

7.2.1 The Electric Arc Discharge Technique

An arc discharge is generated between two graphite electrodes placed face to face in the airtight chamber (Fig. 7.1) under a partial pressure of helium or argon (typically at 600 mbar). Carbon contained in the graphite sublimes into gaseous state. During sublimation, carbon atoms move towards the colder zone and get deposited on the cathode. The deposited carbon soot after purification contains carbon nanotubes. Soot collected on the reactor walls is found to contain fullerenes, amorphous carbon. The type of nanotubes formed depend upon the type of metal catalysts (*i.e.,* Fe, Co, Ni or Y) used in the graphite rod. Large scale synthesis of single walled carbon nanotubes is possible by this process. Kratschmer, *et al.,* (1990) were the first one to have used this technique to get large quantity of SWCNT.

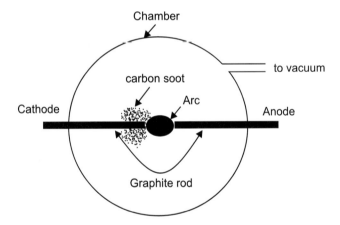

Fig. 7.1 A schematic diagram of electric arc discharge unit.

7.2.2 Thin Film by Evaporation Method

Carbon materials under vacuum can be heated by various sources like laser beam, microwave, electron beam, thermal heat *etc.,* to generate its vapour. The vapour is then allowed to get deposited on a substrate to get a thin film of the material. Except laser evaporation technique, almost all other method of evaporation does not ensure the composition of vapour to have similar stoichiometric composition as that of the target. As a result, laser evaporation technique is becoming popular for making thin film of carbon sample, this technique is discussed here.

7.2.2.1 *Laser Ablation Technique*

Laser is a monochromatic light source and hence its energy can be extremely high. Laser light is allowed to fall on a suitable target (which may be graphite or any solid organic precursors like camphor). Due to its high energy, carbon material present in the precursor gets vapourized and the vapour is allowed to condense on a suitable substrate to get a thin film of carbon. In order to allow the vapour (also known as plume) evaporated by the laser to travel towards the substrate without any hindrance, the entire process is done under vacuum (Fig. 7.2). If need be some desired environment can also be maintained in the chamber like oxidative or reducing environment. Two kinds of laser beams can be used for this purpose: continuous laser or pulsed laser. Nd-YAG pulse laser is one of common laser source used for the synthesis of carbon nanotubes. The main difference between both lasers is that the pulsed laser demands a much higher light intensity (100 kW/cm^2 compared with 12 W/cm^2).

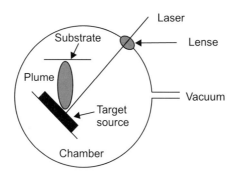

Fig. 7.2 A schematic diagram of PLD unit.

7.2.3 Vapourization Induced by a Solar Beam

This method uses a solar furnace (Fig. 7.3) to focus the sunlight on a graphite target which vapourizes carbon to give carbon soot. The target is made of a graphite crucible which is filled with a mixture of graphite powder and metallic catalysts. The soot is then condensed in a cold zone of the reactor. The target is placed at the centre of the chamber which is filled with suitable gas like argon. Concentrated solar light can generate a temperature around 3000K which is enough to vapourize graphite or decompose any precursor to give carbon soot. Depending on the catalysts and temperature, either MWNTs or SWNTs can be formed. The main problem with this method is that one needs to concentrate solar rays and hence tracking of solar rays is necessary for making the solar beam to fall at one point in the furnace.

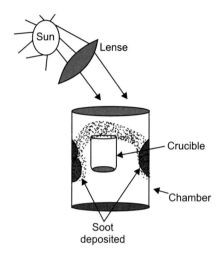

Fig. 7.3 A schematic diagram of solar unit used to prepare carbon nanomaterial.

7.2.4 Chemical Vapour Deposition (CVD)

CVD is a popular technique of growing thin films of various materials. The process involves passing a hydrocarbon vapour through a quartz tube kept in the furnace in which a catalyst material is present at sufficiently high temperature to decompose the hydrocarbon. CNTs grow over the catalyst in the furnace, which are collected upon cooling the system to room temperature. In the case of a liquid hydrocarbon (benzene, alcohol *etc.*), the liquid is heated in a flask and an inert gas purged through it carries its vapour into the (Fig. 7.4) reaction furnace; whereas, vaporization of a solid hydrocarbon (camphor, naphthalene *etc.*) is conveniently achieved in another furnace (low-temperature) before it reaches the main reaction furnace. Catalyst is either

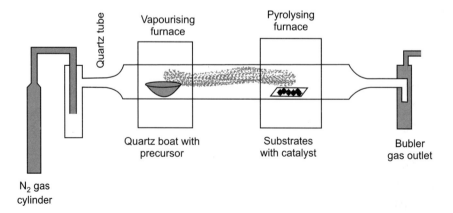

Fig. 7.4 A schematic diagram of a Chemical vapour deposition unit.

placed inside the furnace or fed from outside. Pyrolysis of the catalyst vapour at a suitable temperature liberates metal nanoparticles *in-situ* (such a process is known as floating catalyst method). Alternatively, catalyst-plated substrates can be placed in the hot zone of the furnace to catalyze CNT growth.

7.2.5 Flame Synthesis

Flame synthesis method is used for synthesizing single-walled carbon nano-tubes. A pyrolysis flame of CO/H_2 is established with introduction of the nanocatalyst precursor particles as an aerosol created by drying a nebulized solution of iron or iron colloid (in the form of ferrofluid).

7.2.6 Sputtering Technique

The material to be deposited as a thin film is bombarded by energetic ions, atoms, or molecules. This causes the ejection of material following their impact. The ejected material is then get deposited over a substrate giving a thin film. Sputter deposition is thus a physical vapour deposition (PVD) method of depositing thin films. In physical sputtering kinetic energy of the incident particle is transferred to the target material causing the ejection of atoms. In chemical sputtering incident particle reacts with the target material which leads to formation of volatile product. This product is thermally evaporated to get deposited on the substrate. These types of process are used to manufacture various types of semiconductor devices.

This process can remove certain unwanted impurities adhered to the surface of the thin film. A typical schematic diagram of a sputtering unit is shown in Fig. 7.5. In this there are two electrodes (anode and cathode). The material which is to be sputtered is kept over the cathode. Anode is attached with a substrate over which a thin film of the target material is to be deposited. Both electrodes are kept in a chamber which can be evacuated to a desired pressure. A dc potential is applied between the two electrodes and argon gas is allowed to enter the chamber at a very low pressure. Due to applied potential between the two electrodes, argon gas gets ionized and bombards the cathode causing ejection of ions from the target material. These ejected molecules/ions finally get deposited over the substrate giving a thin film of the target material. In this process, there are variable parameters like magnitude of dc potential, distance between the anode and cathode, and pressure of argon gas. These parameters are adjusted to get the desired type of thin film. This comes by practice only. Sometime a screen is kept in between the anode and cathode.

During the argon ion bombardment, any impurities adsorbed over the target material is allowed to deposit over the screen and after some time the screen is removed, and pure ejected ions are allowed to get deposited over the substrate.

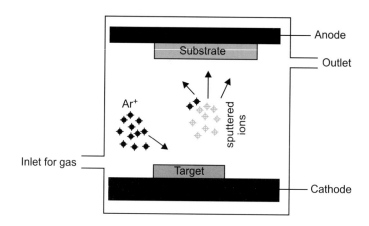

Fig. 7.5 A typical sketch of a dc sputtering unit.

7.2.7 Plasma Enhanced Chemical Vapour Deposition (PECVD) Method

Plasma enhanced chemical vapour deposition technique is very similar to sputtering technique, except that in latter technique, ions are bombarded on to the target to eject molecule/ions for the deposition process. In the Plasma enhanced CVD technique, the vapour of the materials is introduced into the chamber containing two electrodes. Vapour of the materials is introduced along with some carrier gas like argon. When the mixture of the gas is introduced into the chamber, argon gas under the dc potential applied between the two electrodes gets ionized and creates a plasma of ions. These energetic ions interacts with vapour of the materials causing it to decompose and gets deposited over the substrate. Like dc sputtering method, the substrate can be heated to maintain some desired temperature to facilitate the deposition of the decomposed products of the vapour. This method has been used to deposit thin film of carbon using vapour of some suitable precursors like camphor methane *etc*. A typical sketch of a plasma enhance CVD is shown in Fig. 7.6.

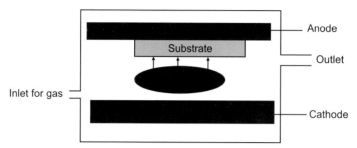

Fig. 7.6 A typical sketch of a plasma enhanced CVD unit.

7.3 Purification of CNM

The as-produced carbon nanomaterial soot contains impurities like graphite (wrapped up), amorphous carbon, metal catalyst (if it is used in synthesis) and some smaller fragments of carbon. These impurities are removed by oxidation and acid refluxing techniques. However, acidic treatments for purification may affect the structure of the nanotubes, pore size, particle size and surface area *etc*. Some of the common methods used for the purification are discussed here.

7.3.1 Oxidation

Carbon nanotubes, fullerenes and amorphous carbons have different degree of thermal oxidation. Most of carbon nanotubes are stable up to 500–600°C (Fig. 7.5), whereas amorphous carbon and graphite starts oxidizing even at 250–300°C. This difference in temperature being very large, it is possible to purify carbon soot to get pure carbon nanotubes by simple thermal oxidation at a controlled temperature. The main disadvantage of oxidation is that along with the oxidation of impurities; CNM may also get oxidized to some extent. Therefore, this process may cause creation of defects in CNM or make more open structures.

7.3.2 Acid Treatment

The acid treatment removes the metal catalyst. Diluted HNO_3 is preferred as it has an effect on the metal catalyst only and not on the SWNTs and other carbon particles. Whereas use of concentrated HCl is not favoured as it affects SWNTs and other carbon particles.

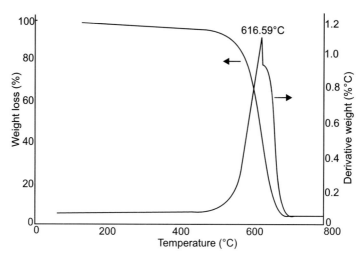

Fig. 7.7 Thermal gravimetric analysis graph of carbon nanotubes showing its stability at 616.59°C, suggesting that this material is stable up to this temperature. But there is a slow weight loss even at lower temperature also.

7.3.3 Ultra-Sonication

This technique separates particles, dispersed in a solvent, due to ultrasonic vibrations. Agglomerates of different nanoparticles are forced to vibrate and become more dispersed. When an acid is used during sonication, the exposure time plays an important role *e.g.*, exposure for a shorter period solvates the metal only, but a longer exposure time helps in chemically cutting the CNTs.

7.4 Activation of Carbon Material

Some time it is desirable to activate carbon nanotubes for some special purposes like adsorption of gases *etc*. Activation of carbon makes it extremely porous and increases surface area. Carbon material can be activated by exposing it to oxidizing/reducing atmospheres (*e.g.*, CO_2, oxygen, steam or their combination) at temperature in the range of 600–1200 °C.

In this method, the reaction is involved with carbon atom of carbon nanotubes and the oxidizing gas. Due to this, reaction pores are created. This method takes long time and also a large amount of internal carbon mass is eliminated to obtain well developed pore structure. Chemical activation on the other hand, is a single step process for the preparation of activated carbon. In this method activating agents like a strong acid, strong base, phosphoric acid, potassium hydroxide, sodium hydroxide, zinc chloride *etc*., are used.

7.5 Characterization of Carbon Nanomaterials

After the carbon nanomaterial is synthesized and purified, it is necessary to characterize the material to get an idea about its morphology, surface area, structure *etc.* Though there are many methods to characterize the carbon nanomaterials like NMR (which helps to find out the sp^2 and sp^3 carbon atoms), TGA/DTA (which is used to find out the temperature at which the materials starts decomposing in oxygen atmosphere, to confirm the presence/absence of amorphous carbon or presence of CNT), STM *etc.* But most common methods used for characterization of carbon materials are SEM, TEM, XRD and Raman analysis and these are briefly discussed here.

7.5.1 Electron Microscopy

Electron microscopy is a major technique for determining the nanoparticles size and their shape. This technique employs beam of accelerated electrons which on interaction with materials produces secondary electrons. These electrons on the analysis give the picture of the material being analyzed. Electron microscopy has two main units (*i*) TEM and (*ii*) SEM.

7.5.2 Transmission Electron Microscopy (TEM)

The sample under the observation is transilluminated by a beam of accelerated electrons with energy on 50–300KeV in very high vacuum. These electrons are deflected at small angle by the atoms of the sample and are allowed to pass through magnetic lenses to form a bright-field image of the internal structure of the sample on a screen. A resolution of 0.1 nm can be achieved

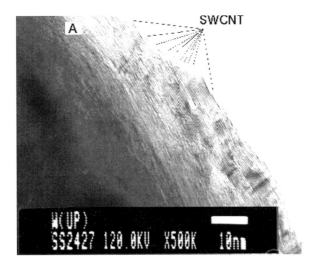

Fig. 7.8 TEM of SWCNT showing each graphine sheet.

which corresponds to a magnification of 10^6. Transmission microscope also makes it possible to obtain diffraction patterns, which provide information on the crystalline structure of the sample.

7.5.3 Scanning Electron Microscopy (SEM)

The scanning electron microscope (SEM) is a type of electron microscope capable of producing high-resolution images of a sample surface. Due to the manner in which the image is created, SEM images have a characteristic three-dimensional appearance and are useful for judging the surface structure of the sample. Scanning Electron Microscopy (SEM) is widely used for initial characterization of any nanostructure. It has contributed immensely in characterizing various forms of Carbon Nanomaterials *viz*. Carbon Nanofibres (CNF), Carbon Nanobeads (CNB), Carbon Nanotubes (CNT). However, it has limited application in characterizing SWCNT for which TEM is used. SEM can yield valuable information regarding the purity of the sample as well as an insight on the degree of aggregation of raw and purified CNT materials.

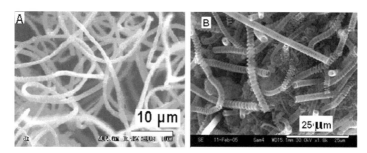

Fig. 7.9 SEM of carbonfibres obtained by the pyrolysis of (A) ethanol (B) acetylene.

7.5.4 X-ray Diffraction (XRD)

X-ray diffraction (XRD) is a versatile, non-destructive technique that reveals detailed information about the crystallographic structure of materials. Many nanomaterials are obtained only in powder form where crystals are randomly oriented. For a given wavelength, incident rays are reflected by a particular set of plane and are deviated by 2θ where θ is the Bragg angle. Therefore, for a system of randomly oriented crystals, X-rays reflected from corresponding set of plane will be deviated by 2θ from the direction of primary beam. The intensities of the reflected rays could be measured by a photomultiplier tube and plotting the number of photon per sec Vs 2θ gives information about various planes present in the crystal. From this graph, the possible structure of the material can be obtained (Fig. 7.8).

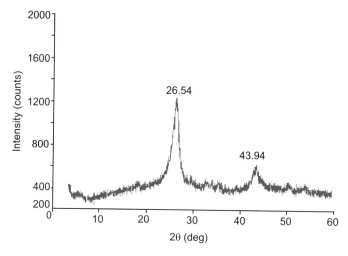

Fig. 7.10 A typical XRD of carbon nanotubes showing its two graphitic peaks at 26.54 and 43.94 two theta angles.

As the crystallite size reduces, the peak width increase and the intensity decrease. Peak broadening can also originate from variations in lattice spacing caused by lattice strain. From the broadening of the peak, it is possible to determine an average crystallite size by Debye-Scherrer formula:

$$D_{hkl} = \frac{k\lambda}{\beta_\theta \cos\theta} A^\circ$$

Where k varies in the range of 0.8 to 1.39 (usually close to unity),

λ is wavelength of the radiation,

β_θ is full width at half/maxima (FWHM)

D_{hkl} is the crystallite size in Å.

7.5.5 Raman Spectra

Raman spectroscopy is a spectroscopic technique used in condensed matter physics and chemistry to study vibrational, rotational, and other low-frequency modes in a system. It relies on inelastic scattering. The laser light interacts with phonons or other excitations in the system, resulting in the energy of the laser photons being shifted up or down.

Raman spectroscopy has opened an efficient technique to characterize materials. The biggest advantage of this technique is that it requires practically no sample preparation and with the advent of micro Raman techniques, very little amount of sample is needed.

Based on the symmetry of the carbon nanomaterials, A_{2u} and E_{1u} modes are IR active whereas, A_{1g}, E_{1g} and E_{2g} modes are Raman active. It is important to realize that number of IR and Raman active modes is independent of the diameter of nanotubes. However, the frequencies of these modes do vary with the nanotube's diameter. The strong lines between 1550 cm^{-1} and 1600 cm^{-1} may be assigned to three E_{1g}, E_{2g} and A_{1g} modes in carbon nanotubes with different diameters. It is important to note that the Raman intensity for graphite in the 1300 cm^{-1} – 1600 cm^{-1} region is sensitive to sample quality. Disordered graphite and carbons show a broad feature around 1350 cm^{-1} (Fig. 7.9). If we examine the Raman spectra of pure diamond and pure graphite, it will be noticed that while diamond shows one sharp peak around 1350 cm^{-1}, but graphite shows only one peak around 1550 cm^{-1}. Diamond is expected to contain 100% pure sp^3 configuration carbon, whereas graphite contains 100% pure sp^2 carbon. Hence, it can be concluded that Raman peak around 1350 cm^{-1} (known as D band) is an indication of sp^3 carbon and peak around 1550 cm^{-1} (known as G band) corresponds to sp^2 carbon. Raman peak around 1350 cm^{-1} in graphite is normally considered as disorder created in graphite. But the nature of disorder in graphite can be due to presence of sp^3 carbon. In other words, intensity of peak (or peak areas) of around 1350 cm^{-1} obtained with graphite may be considered as equivalent to amount of sp^3 carbon present in graphite. Similarly, presence of peak around 1550 cm^{-1} in diamond materials should be considered as the amount of sp^2 disordered present in the diamond. In diamond, like carbon which is assumed to contain both sp^2 and sp^3 carbon and Raman spectra does show the presence of both bands *i.e.,* presence of peak at 1350 cm^{-1} and 1550 cm^{-1}.

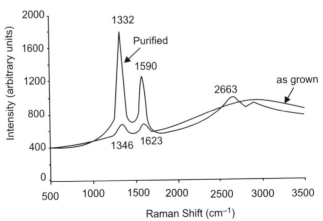

Fig. 7.11 Raman spectra of carbon nanotubes (purified and as grown). After purification there is sharp increase in the intensity of the peaks as well as the complementary peak at 2663 cm^{-1} is also seen.

At 186 cm^{-1} a strong A_{1g} breathing mode is found. This peak has strong dependence on nanotube diameter. Hence, the frequency of the A_{1g} breathing mode can be used as a marker for assigning the approximate diameter of the carbon nanotubes.

Among the several techniques used to characterize single walled carbon nanotubes, Raman Spectroscopy is perhaps the most powerful tool to get information on their vibrational and electronic structure. The most important observed Raman features of SWCNT are the following (*i*) the radial breathing mode, whose frequency varies according to the diameter of the tube; (*ii*) the tangential G band in the range of 1550–1605 cm^{-1}; and the disordered induced D band at about 1350 cm^{-1} and its second order harmonic (G band) at about 2700 cm^{-1}. Radial breathing mode has been found to be inversely proportional to the diameter of tube $\left(v = \dfrac{248}{d_t} \right)$ and is very sensitive to the charge transfer and tube tube interaction in a bundle. Radial breathing mode (RBM), thus makes a valuable probe to determine the structure and properties of SWCNT based materials.

7.5.6 Surface Area Measurement

In absence of BET method, to determine the surface area of powdered material, Methylene blue adsorption technique is an important alternative method. In this technique, a calibration graph is plotted between the amount of methylene blue and its optical density measured at its λ_{max}. Carbon nanomaterial is then kept in a solution of methylene blue of known concentration for 24 hrs. Methylene blue gets adsorbed at the surface of the carbon nanotube. By measuring the optical density of supernatant liquid, the amount of methylene blue adsorbed per g of carbon nanomaterial is calculated with the help of the calibration graph. Since size of one molecule of methylene blue is known, the total area on to which the methylene blue had been adsorbed is calculated. This value is taken as the surface area of the carbon nanotube.

7.7 Summary

In this chapter, we have learnt the various methods for the synthesis of carbon nanomaterials. Synthesis of carbon nanomaterials may contain some impurities like metal (which was used as catalysts) and some amorphous carbon. These impurities need to be removed. Different methods used for this purpose is also discussed. Sometime it is necessary to activate the carbon nanomaterials for some specific use like hydrogen adsorption. The method to activate carbon materials is discussed. It is also necessary to find out the surface area of the carbon nanamoaterials especially for applications like supercapacitor, hydrogen adsorption. There are two methods used for this purpose, BET method and the methylene blue technique. BET technique is not available in many laboratories. In absence of such facilities, a simple technique of methylene blue can be used to determine the surface area. This technique is briefly discussed. Finally it is required to characterize the carbon nanomaterials synthesized by the various techniques. Different techniques used for the characterization are briefly discussed.

8

Applications of Nano Carbon

> Imagine a medical device that travels through the human body to seek out and destroy small clusters of cancerous cells before they can spread. Or a box no larger than a sugar cube that contains the entire contents of the Library of Congress. Or materials much lighter than steel that possess ten times as much strength.
>
> —**U.S. National Science Foundation**

8.1 Introduction

Since, the discovery of fullerenes in 1985 and carbon nanotubes in 1992 there has been tremendous interest and efforts by scientists of various disciplines to look for its applications. Though variety of obstacles like its insolubility in water, possibility of cytotoxicity have been encountered during the course of investigation, it would not be an exaggeration to say that CNT has found application in almost all the branches of science.

As mentioned earlier, carbon nanoparticles are available in several geometries, including spherical fullerenes, cylindrical nanotubes and planar nanoplatelets. While these nanoparticles all share the same graphitic structure that imparts semiconductor properties, the diverse geometries afford a spectrum of unique chemical, electrical, magnetic, and optical properties; thus a plethora of applications. For the convenience of writing, we would describe some of the applications of nano carbon under two sub-headings (*i*) Applications of CNT in area other than Biological system, and (*ii*) Applications of CNT in Biological systems.

However, prior to looking for the various applications, it is important to know that for most of the applications, nano carbon needs to be functionalized *i.e.,* to increase its reactivity.

8.2 Dispersion and Functionalization of Carbon Nanotubes

CNT is usually dispersed by ultra-sonication. Addition of dispersion agents like polyvinylpyrrolidone (PVP) or sodium dodecyl benzene sulfonate, accelerates the dispersion effect.

Carbon nanotubes are insoluble in water, polymer resins, and most solvents. Thus, they are difficult to evenly disperse in a liquid matrix such as epoxies and other polymers. This complicates efforts to utilize the CNT's novel physical properties in the manufacture of composite materials, as well as in other applications. To make use of CNTs, it is necessary to physically or chemically attach certain molecules, or functional groups, to their smooth sidewalls without significantly changing their desired properties. This process is called functionalization.

Functionalization is done by chopping, oxidation or "wrapping" of the CNTs in certain polymers so that active bonding sites are created on the surface of the CNTs. For biological uses, CNTs can be functionalized by attaching biological molecules, such as lipids, proteins, biotins, etc. Then they can usefully mimic certain biological functions, such as protein adsorption, and bind to DNA and drug molecules. This would enable medically and commercially significant applications, such as gene therapy and drug delivery. In biochemical and chemical applications e.g., for biosensors, molecules such as carboxylic acid (COOH), poly amino benzoic sulfonic acid (PABS), polyimide, and polyvinyl alcohol (PVA) have been used to functionalize CNTs. Functionalized CNTs can become soluble in water and other highly polar, aqueous solvents.

8.2.1 Why do Carbon Nanomaterials and Fullerene are Easily Functionalized?

Functionalization involves oxidation, fluorination, amidation etc. Two paths are usually followed for the functionalization: attachment of organic moieties either to carboxylic groups that are formed by oxidation with strong acid, or by direct bonding to the surface double bonds.

The bonding between carbon atoms in fullerene are sp^2 type with some sp^3 character due to their high curvature. Thus, the pentagonal carbon atoms of fullerenes can be easily chemically or electrochemically oxidized. For example, efforts have been made to attach various types of functional groups like $R_1NHCH_2R_2$, porphyrines etc. (Hummelen, et al., 1999 and Satishkumar, et al., 1996).

Similarly, carbon nanotubes/fibres also contain five member rings at the end of the tubes. These pentagonal carbons, thus can be easily oxidized by chemical treatments with suitable agents. Refluxing CNTs in a H_2SO_4/HNO_3

mixture results in a clear, colourless solution, which on removal of solvent and excess acid gives a white solid containing functionalized CNTs. Oxidized carbon atoms can act as specific sites for adsorption of metal ions (Fig. 8.1).

Thus, these materials by simply treating with suitable agents can be attached with amid, amine, —COOH, —OH groups, making them an interesting material for many biological applications.

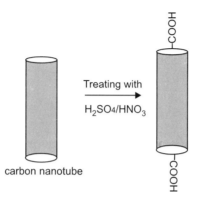

Fig. 8.1 A schematic diagram showing functionalization of carbon nanotubes (having both ends opened) with –COOH.

8.3 Applications of CNT in Areas Other than Biological Systems

8.3.1 Electron Field Emission

Field emission results from the tunneling of electrons from a metal into vacuum under application of a strong electric field. Emission from metal surfaces which leads to the well known Fowler–Nordheim equation:

$$I = \alpha E^2_{eff} \exp(-\beta/E_{eff}),$$

where α is a constant related to the geometry, E_{eff} is the effective field at the emitter tip, and β is a constant which is proportional to the work function. This type of field emission we normally observe with TV screens, where one has to apply very strong electrical field to eject electrons. The high electric field is needed by most metals because their work function is in the range of 4 to 5 eV, with the exception of Cs (1.81 eV). Carbon family, though also has work function in the same range as metal, but its Fermi level is different from various types of carbon allotropes. For example, Fermi level of carbon nanotubes, diamond in special is almost zero, which means energy required to remove electrons is almost zero. This special property of diamond and carbon nanotubes has attracted scientists to use them for electron emission purposes, because one would need almost zero energy for this purpose.

The electric field especially at the tip, nanotubes can be as much as 1200 times higher than E_0 (Fig. 8.2A) where E_0 (= V/d) is the potential difference between the anode to cathode voltage (V), divided by the anode to cathode distance(d) for metal emitter. As a result, field emission is readily achieved even for relatively low applied potentials (Fig. 8.2 B and C).

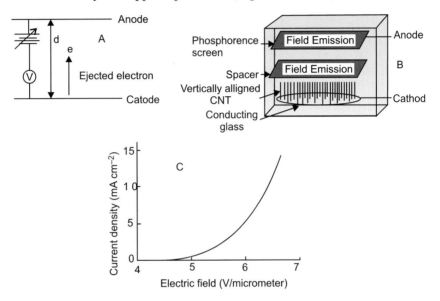

Fig. 8.2 (*A*) Schematic diagram showing the electric field (V/d) needed to emit electron from cathode and (*B*) A schematic diagram showing the formation of image by electron field emission process and (*C*) A typical graph of current versus the electric field obtained with CNT.

Field emission from individual MWCNTs was demonstrated by attaching a nanotube to a conducting wire and observing the current after applying a negative potential to the wire. It was concluded that carbon nanotube films could be used as field emission guns for technical applications, such as flat panel displays (Fig. 8.2B). Samsung Advanced Technology Institute is already manufacturing a prototype of a colour display, ready to commercialize this concept.

When studying the field emission properties of MWCNTs, it was noticed that together with electrons, light is emitted as well. This light emission occurred in the visible part of the spectrum, and could sometimes be seen with the naked eye. This light emission is not the luminescence due to resistive heating. The electron and light emission properties are very important features of carbon nanotubes.

Why do we need to use aligned tube for the emission? The major factors that determine the field emission properties of the tube are the tip radius and the position of these occupied levels with respect to the Fermi level, which depends primarily on the tip geometry (De Vita, *et al.*, 1999).

Theoretical calculations and STM measurements on SWCNTs and MWCNTs show that there is a distinctive difference between the electronic properties of the tip and the cylindrical part of the tube. For MWCNTs, the tube body is essentially graphitic. The local density of states at the tip exhibits sharp localized states that are correlated with the presence of pentagons, and most of the emitted current comes from occupied states just below the Fermi level. (Carroll, *et al.*, 1997)

8.3.2 Chemical Sensors

Chemical Sensors are those substances, which change their physical properties like surface conductivity when some specific materials get adsorbed at their surface. Surface conductivity depends upon the ease with which electrons can migrate from cathode to anode (Fig. 8.3).

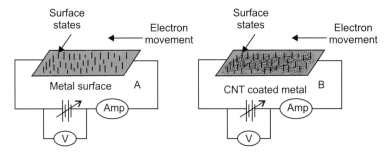

Fig. 8.3 (A) Dangling orbital shown by vertical lines present over a smooth metal surface, (B) Metal surface coated with carbon nanotubes. Each tube will possess several dangling orbitals, thus increasing the concentration of surface states.

Surface of any material will contain dangling orbitals which may be either positively or negatively charged. These sites are known as surface states. When electron migrates from cathode to anode, it encounters these surface states which hinder its transport to be arriving at anode. As a result, the conductivity of the surface depends upon these surface states. This metal when exposed to some other materials, adsorption of the materials occurs. The adsorbed materials would alter the percentage of dangling bonds present on the surface of metal. As a result of this alteration, electrons moving on the surface find a new situation, which alters the surface conductivity of the metal. The change in conductivity, thus becomes related to the concentration of material used for adsorption. A standard graph is established between the change in conductivity and the amount of material used for the adsorption. This graph is used for finding out unknown concentration of the same material. The efficiency of this method depends upon the magnitude of change in conductivity with concentration. The magnitude of change in conductivity depends upon the number of dangling bonds (*i.e.*, the density of surface states) which in turn depend upon the surface area of the substrate over which adsorption is carried out. Carbon nanotubes possessing large number of dangling bonds, thus are preferred for such purpose. Layers of carbon nanotubes are spread over a suitable conducting substrate (Fig. 8.3B). This surface is used for this purpose.

8.3.3 Hydrogen Storage, Fuel Cell, Supercapacitor

These three items are considered under one heading because these are inter related. Scientists are engaged globally to make electric driven vehicles. For this purpose, hydrogen is being considered as a suitable energy source. For making electric driven vehicles, one needs to transport hydrogen like petrol, use hydrogen either directly into the ignition cylinder or convert it into electrical power by using a fuel cell. In order to start an electrical motor or while driving the vehicle along an elevated road, at the start motor draws very heavy current. This power is needed for only few seconds. This is achieved by a supercapacitor. A scheme of hydrogen utilization is shown in Fig. 8.4.

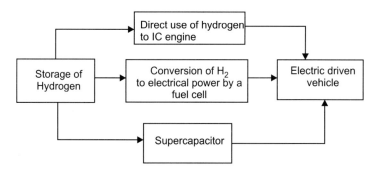

Fig. 8.4 A schematic diagram showing the path of utilization of hydrogen for making an electric driven vehicle.

8.3.3.1 Storage of Hydrogen

There are various methods to store hydrogen, but any system which can store hydrogen to the extent of 6.5wt%, is considered to be an economical system. Moreover, it is also necessary to reduce the weight of the container so that unnecessary weight of material is not added to the weight of the system. Considering the light weight of carbon, scientists are making efforts to utilize carbon nanomaterials for the storage of hydrogen.

There are two possibilities by which hydrogen can be adsorbed on carbon nanotubes. Hydrogen can enter the CNT through the hollow diameter

(Fig. 8.5 A) or it can be adsorbed on the surface of the CNT by the help of dangling (Fig. 8.5 B) bonds (*i.e.,* via the surface states). If it entered into the tube (Fig. 8.5 A), then desorption may be difficult. If it is adsorbed through the dangling bonds present at the surface, depending upon the nature of adsorption (*i.e.,* whether chemisorbed or physisorbed), desorption of hydrogen would take place easily. For the latter method, it would be desirable that surface of the CNT be as rough as possible.

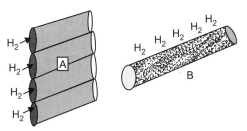

Fig. 8.5 (*A*) A schematic diagram showing the adsorption of hydrogen through the hollow space present in the CNT and (*B*) showing the adsorption of hydrogen on the surface of CNT via the dangling bonds.

This means it would be better to use carbon nanofibres rather than nano-tubes. Because carbon nanofibres surface is made of broken graphene sheet as compared to CNT, which consists of smooth unbroken graphene sheet as its surface.

8.3.3.2 Fuel Cell

Fuel cell is very much similar to a chemical secondary battery. The only difference is that the maximum electrical power which can be drawn from the battery though flexible, but once prepared it is fixed by the manufacturer. In fuel cell, it is possible to draw power so long the fuels are injected into the system. Fuel cell is an energy conversion device that theoretically has the capability of producing electrical energy for as long as fuel and oxidants are supplied to the electrodes. Fuel cell converts the chemical energy of a reaction directly into electrical energy. The basic physical structure, or building block, of a fuel cell consists of an electrolyte layer in contact with a porous anode and cathode on either side. A schematic representation of a fuel cell with the reactant/product gases and the ion conduction flow directions through the cell is shown in Fig. 8.6.

In this type of fuel cell, the following electrochemical reaction takes place:

At Anode $\qquad\qquad\qquad\qquad H_2 \rightarrow 2H^+ + 2e^-$

At cathodes $\qquad \frac{1}{2}O_2 + H_2O + 2e^- \rightarrow 2OH^-$

Overall reaction $\qquad\quad H_2 + \frac{1}{2}O_2 \rightarrow H_2O + Energy$

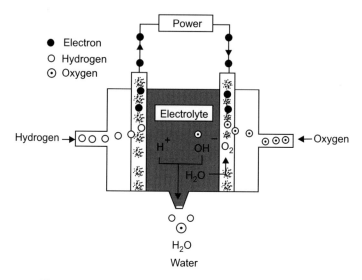

Fig. 8.6 A schematic diagram of hydrogen/oxygen fuel cell.

In such hydrogen/oxygen fuel cell, the by product is only water. Theoretically this type of cell can give a potential of 1.22V, but in reality due to polarization of electrodes, it is possible to get about 1.0V or slightly less than this value. Research is being carried out to develop a porous electrode (like pumice stone) over which non-noble metal (like Ni, Mg) is deposited as an electrocatalyst along with suitable form of carbon nanomaterials (Fig. 8.7).

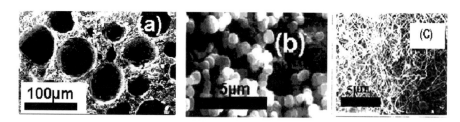

Fig. 8.7 (*a*) SEM of ceramic pumice stone showing its porosity, (*b*) carbon nanobeads or (*c*) carbon nanotubes which could be deposited over the ceramic pumice stone.

In a typical fuel cell, gaseous fuels are fed continuously to the anode (negative electrode) and an oxidant (*i.e.,* oxygen from air) is fed continuously to the cathode (positive electrode); the electrochemical reactions take place at the electrodes to produce an electric current.

Different types of fuel cells are available commercially, but due to their cost, they are not popular. The major cost lies in the type of electrodes being used for electrochemically converting fuels to get an electrical power. Hence, scientists are utilizing carbon nanomaterials to replace the expensive electrocatalyst (metals) so that the system could become affordable.

8.3.3.3 Supercapacitor

Electrochemical capacitors operate on principles similar to those of conventional electrostatic capacitors. Conventional capacitor stores energy in the form of electrical charge and a typical device consists of two conducting materials separated by a dielectric. This can store charge to the extent that there is no leakage of charge across the two plates through the dielectric media. The conventional capacitor can store charge to the extent of few microfarad per square cm area of electrode.

In electric double layer capacitor (EDLC) instead of high dielectric media, it uses electrolyte either of low pH (acidic media) or high pH (alkaline media) or organic liquid. It store electrical charge in a similar manner. The charge accumulates at the interface between the surface of a conductor and an electrolytic solution. The accumulated charge (known as an electric double-layer) is separated between the electrode and the electrolyte by a distance equivalent to diameter of water. A typical EDLC is shown in Fig. 8.8. This type of capacitor is named as supercapacitor, because its capacity is in the range of few hundreds of faraday per gram of material. Scientists are making efforts to increase the magnitude of capacitor to the tune of 300F/g.

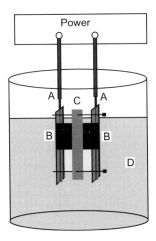

Fig. 8.8 A schematic diagram of a supercapacitor: (*A*) is conducting electrode over which (*B*) powder of carbon materials is sandwiched between a separator (*C*). The entire assembly is immersed in an electrolyte. The capacitance is developed between the two electrodes.

The actual capacitance of EDLC is about 12–14 μF/cm^2. But if the electrode material is selected which is porous and possess a surface area of 2000 m^2/g then its total capacitance becomes 12–14 μF/cm^2 times the surface area of the material. Scientists are making efforts to develop a carbon nanomaterial which is a good conductor of electricity and has surface area to the tune of 2000 m^2/g. Most of the inorganic materials are unstable in electrolyte and their surface area is not very large, whereas it has been possible to synthesize electrically conducting carbon nanomaterials with large surface area to the tune of 300–600 m^2/g. Efforts are on to improve their surface area to make high power supercapacitor.

8.3.4 Lithium Secondary Battery

The application of lithium battery as a portable energy source is increasing very fast especially because it can provide as high as 3 V cell potential. Moreover, cost of electrode is another consideration which needs to be lowered by developing cheaper materials. Lithium metal as anode, in conjunction with an organic electrolyte results in non-uniform formation of a passive film on the anode surface causing dendrite growth of lithium metal. Hence, scientists are looking for material like graphite *etc.*, which could be used as an anode in aqueous solution.

It has been observed that electrochemical intercalation of lithium in graphite generates considerable negative potential close to that of lithium and is less reactive and easily reversible. Therefore, graphite like material is expected to facilitate intercalation and deintercalation in lithium battery (Fig. 8.9) because they can either accept or donate electrons. As a result various attempts have been made to use carbon nanotubes (CNT) as the anode in lithium ion battery. The capacity of graphitized carbon is theoretically calculated to be 372 mAh/g with stoichiometric composition LiC$_6$.

The intercalation of Li ion with carbon depends upon various factors, such as pore size (either the size of the diameter of the CNT or pore size of channels present in fibrous carbon materials), density, surface area, and activation of CNM *etc.* One of the major disadvantage with carbon nanotubes in comparison with either graphite or coke as a raw material is the high cost of production.

In search for cheap material, it has been observed that carbon fibre obtained from bamboo shows reversibility in the charging/discharging of lithium battery. It has been reported that even after 100 cycles of charging/discharging process (Fig. 8.10), the capacity of the battery remains almost constant (150 mAh/g).

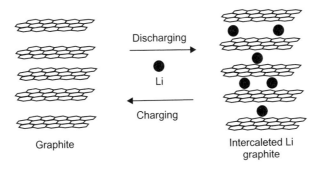

Fig. 8.9 A schematic diagram to show how the intercalation takes place in the host lattice (*e.g.*, graphite) and charging/discharging process in lithium battery.

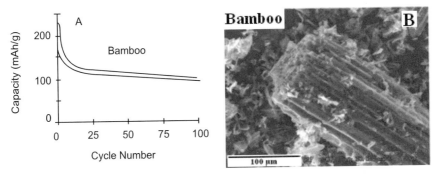

Fig. 8.10 (*A*) Graph showing the charging/discharging capacity of anode of lithium battery and (*B*) Carbon fibre obtained by the pyrolysis of bamboo fibre which showed highly reversible behaviour in lithium battery.

8.3.5 Microwave Absorption

Microwave absorbers have been widely used to prevent or minimize electromagnetic reflections from large structure such as aircrafts, ships and tanks. Magnetic materials such as ferrites, iron and Co-Ni alloys are used as absorbers. The main problem for the design of magnetic absorbers is related to the choice of the material. In addition, these materials require thick coating to meet practical demands. Recently scientists have started using carbon nanomaterials for studying microwave absorption properties. Like any optical absorption study, microwave of known frequency is allowed to fall on the sample and the intensity of the wave (R) which returns after getting reflected from the absorbing material is measured. In similar fashion intensity of microwave which gets transmitted (T) by the absorbing material is measured. From these two observations amount of microwave absorbed (A) by the material is calculated [*i.e.*, A = 1 − (R + T)]. Carbon nanomaterials have been used to study the microwave absorption from 8.0 GHz to 24 GHz (Fig. 8.11).

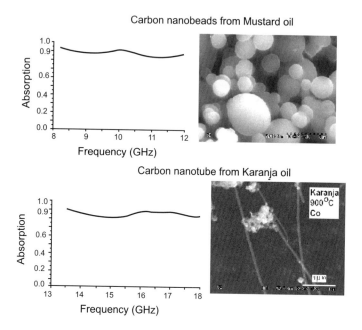

Fig. 8.11 Microwave absorption (almost 97%) from 8 GHz to 12GHz by carbon nanobeads obtained from mustard oil and from 13GHz to 18GHz by carbon nanotubes obtained from Karanja Oil.

8.3.6 Carbon Solar Cell

Photovoltaic cell is a device to convert light energy into electrical energy. If source of light is solar rays then the system is called solar cell. In the Table 8.1 band gaps of various materials are given. Each of them have specific band gap. Amongst them Si (1.1 eV) and GaAs (1.4 eV) have been exploited to make photovoltaic solar cell the most. But silicon based solar cell is still not very economical for terrestrial application and other cells like CdTe, suffers from the disadvantage as they are not environmentally friendly. These materials are toxic and hence extra precautions are needed for their disposal. Organic photovoltaic cells though are cheaper, but their life time needs to be ascertained for terrestrial application including their care in handling by common people.

In addition, to cover larger portion of solar spectrum, tandem type solar cell may be preferable. For such purpose, p:n junction should be made with materials possessing series of different band gap. This possess the problem of matching the Fermi levels such that flow of photogenerated carriers do not find any hurdle in travelling from one end of the cell to the back side for their collection with least possibility of losses due to trapping *etc*.

Carbon is the only material which can show a variation in the band gap from 0.25 eV to 5.5 eV. Hence, if system is perfected one could hope to make a graded junction with carbon without any fear of problem of mismatch of Fermi level. But the main hurdle with carbon materials is in finding the exact condition to get a material with a desired band gap. The other problem associated with carbon is in its low mobility of carriers. In addition, most of carbon materials show p-type character even without any external doping. Hence, preparation of an extrinsic semiconducting carbon is a challenge.

Table 8.1 Showing band gap of various materials

Material	Energy gap (eV) at 300 K
CdSe	1.74
CdTe	1.44
GaAs	1.43
GaSb	0.68
Gap	2.25
Ge	0.66
InAs	0.36
InP	1.27
InSb	0.17
Si	1.11
ZnO	3.2
ZnS	3.6

Sharon and his research group have selected to use Carbon because it has property very similar to silicon, if not better. Silicon has a fixed band gap of 1.1 eV (indirect), whereas carbon can possess band gap anything from 5.5 eV to 0.25 eV. Moreover, the work function of graphite (100% sp^2 carbon with band gap ~0.25 eV) and diamond (with 100% sp^3 carbon with band gap 5.5eV) is more or less same ~4.65eV. This makes the life simpler in making tandem type cells with band gap varying from 0.25 eV to 5.5 eV with almost least fear for matching the Fermi levels of each set of the junction. Such possibility does not exist with any inorganic material. The variation in the band gap can be achieved by controlling the ratio of sp^2 and sp^3 carbon atom in the film . Though to achieve the required ratio is not so easy, but at least there is a possibility of achieving this goal which is not available with any other inorganic material. Sharon and his group has been trying to develop homojunction carbon solar cell and they have achieved an efficiency of 2.28% so far (Fig. 8.12).

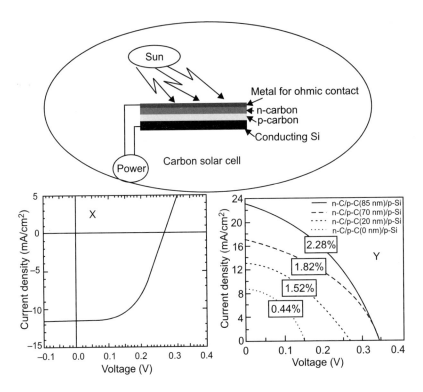

Fig. 8.12 (X) The Current-voltage (I-V under illuminated condition) characteristics of n-C/p-C deposited over conducting Si (shown as an inset) and (Y) I-V characteristics and corresponding efficiency of the cell with different thickness of n-C/p-C.

8.4 Applications of CNT in Biological Systems

Nanotechnology is the application of Nanoscience. Nanoscience is a rich branch of science which has inputs from all the disciplines of science *i.e.*, from physics, chemistry, electronics biology and engineering. Carbon nano-material (CNMs) has the potential of revolutionizing the Biosystem research as they can show superior performance because of their impressive structural, mechanical, and electronic properties such as small size and mass, high strength, higher electrical and thermal conductivity *etc*. The properties of CNMs which have influenced its application in Biosystem includes its surface morphology which makes it a suitable material for adsorption, absorption of gases and liquids, data storage, template for the various biological reactions, easier functionalization alters many of the characters of CNM *e.g.*, functionalized CNM have developed solubility in both polar and non-polar solvent-expanding its application in various biosystems *e.g.,* drug delivery, biosensors, biochips *etc*. Apart from surface morphology, electrical and electronic properties and conducting, semiconducitng properties make CNMs different from the other materials. Band gap of CNM as well as photochemical properties have found wide application in purification of water or degradation of organic and inorganic materials.

The unique physicochemical properties of nanomaterials (as mentioned in chapter 1) are not found in their bulk materials. In general, they have much higher reactivity, and because of their ultra small size and increased surface area, they can easily penetrate skin or cells, rapidly distribute in human body, and even directly interact with organelles within cells. Their huge surface area to volume ratio increases the chemical activities and therefore allows them to become efficient catalysts. These increased chemical and biological activities have resulted in many engineered nanoparticles, which are being designed for specific purposes, including diagnostic or therapeutic medical uses and environmental remediation.

As we have seen in chapter 6, the nano carbons are mixture of sp^3 and sp^2 carbons. The question arises

Why are structures containing mixture of sp^3 and sp^2 carbons interesting for Biological applications?

In most of biological applications, one is interested in using carbon nano-materials as carriers for delivering certain drugs into the body or for attaching to the body preferentially like in formation of bone tissue *etc*. This is perhaps due to two specific properties of carbon: (*i*) it is biologically acceptable materials (*ii*) it has very large surface to volume ratio and (*iii*) it contains large concentrations of dangling bonds which behaves either like Lewis base or Lewis acid.

So far biological acceptability is concerned, there is a dispute: one group of scientists feel that it is toxic and another group advocates that it is not chemically toxic, but because of its small size it may be toxic and hence depending upon the requirements, carbon can be safely used. Its large surface to volume ratio is useful because one can load larger concentration of material over smaller quantity of carbon. The advantage of large dangling bond is that one can temporarily load chemical to carbon nanomaterials and when need comes, it gets unloaded also fast. Existence of dangling bond is one of the important properties of carbon nanotubes, nanobeads. Dangling bonds help to form sort of a chemical bond which is not very strong like normal chemical bond (Fig. 8.13).

Carbon present on the surface will either be looking for electrons or may be ready to donate electrons, in such a manner that over all surface of the carbon is neutral. Thus, such carbon molecule when encounters any other material, depending upon whether the material is looking for electron or is prepared to donate electron, they will get attached to the sites of carbon molecule either by accepting electrons or donating electrons, accordingly. These types of bonding are known as *dangling bond.* It is the presence of these dangling orbitals which make carbon nanomaterials very useful in biological applications.

Therefore, CNTs, especially of those biologically compactable materials like titanium are promising for biomedical applications because one is able to tailor them for specific parts of the body.

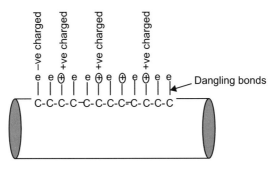

Fig. 8.13 Sketch of a cylindrical nano carbon showing presence of dangling bonds due to electronically unsatisfied carbon present at the surface of the structure. During the formation of the material, it may happen that the electron is lost from the surface creating a positive charge or it may contain unsatisfied electrons such that overall surface behaves like a neutral surface.

An understanding of Nanoscience and Bioscience reveals that NANO-TECHNOLOGY can be manifested in two ways either by mimicking the nature or by using biomaterial for development of Nanoscience and Nano-technology.

Some of such successful applications of carbon nanomaterials have been:

- Diagnostic Equipments *e.g., Biosensors, Nanoprobes* (DNA based Bio-chips) for cancer detection, Nanorobots and X-ray devices.
- Surgical Supplements *e.g.,* Nanomedicinal devices, Bioactive nanomaterial in bone grafting, Nanotweezers.
- CNT for tissue engineering.
- Gene Delivery using CNT instead of a vector.
- Pollution Control by CNT, by killing pathogenic microbes, filtration and physicochemical methods.
- Anticarcinogenic Activity of Carbon Nanomaterials.

- Drug Delivery using Carbon Nanomaterial.
- CNM for Neurodegenerative Disorder's Therapy.

Though some groups of scientists believe CNTs to be a biocompatible material and have shown interest in developing various biological applications. Other group believes it as toxic materials and hence do not advocate for its biological application. Nevertheless, considering our body being mainly built of carbon, it is difficult to assume that carbon in inert form can be toxic to human body. The physical property of carbon especially its nanometer size can cause some malfunction of the cells which may prove to be toxic. Hence, it is necessary to use CNT with caution for biological application. Though there are many applications suggested, but some of the popular applications that successful or at the verge of being implemented and some future possibilities are enumerated here:

(*i*)   CNT can be used for the fabrication of bioprobes and biosensors.

(*ii*) Transport of vaccines and drugs to a specific site of the human body by tagging them with CNT and suitable RNA directive.

(*iii*) As template to grow cells.

8.4.1 Diagnostic Equipments

8.4.1.1 Biosensors

Biosensors are defined as analytical devices incorporating biological material (*e.g.*, DNA, enzymes, antibodies, microorganisms *etc.*) associated with or integrated within a physico-chemical transducer. They act in the aqueous phase, particularly in physiological solutions. Sensors are based on the reaction between a bio-molecule immobilized on the sensor electrode with the analyte. The reaction produces an electric current, which is proportional to the analyte concentration. Such biosensors have been used in electrochemical oxidation of dopamine and to determine thyroxin, glucose in blood and urine *etc*. Glucose oxidase (flavine enzyme) is used to monitor blood glucose levels in diabetics. The enzyme adsorbed on CNT can be reversibly oxidized without loss of enzymatic activity. Bio-molecules, like proteins, enzymes, and DNAs, can also be attached to the CNTs. Biosensors can be applied for detecting diseases, particularly in cancer diagnostics. Detection of organo-phosphorus pesticides, nerve gases, detection of pathogens and toxins and direct determination of total cholesterol in blood by biosensors is also possible.

Biosensors can be made by using any inert metal also, but attachment of desired enzymes to the metal electrode is not very lasting. Thus, invariably one is needed to keep on attaching fresh enzyme to the electrode or it becomes one time use. Carbon nanotubes, on the other hand, due to the presence of dangling bonds (Fig. 8.9) adherence to metal electrode is relatively strong as well as the other end of the CNT attaches enzyme strongly through its dangling bonds which is almost rigid type. In addition, CNT can help in electron transfer from the analyte to enzyme to metal electrode efficiently. These qualities of CNT have made it superior material for developing biosensors. Ye, *et al.*, (2004) and Zhang, *et al.*, (2004) have shown that MWCNT functionalized with hemin are able to detect oxygen in solution, moreover they have shown that oxygen can get electroreduced by myoglobin on MWCNT modified glassy carbon electrode.

Some of the specific properties of CNM used for making Biosensors: A large surface-to-volume ratio and good electrical conductance make CNT an attractive option for making enzyme based electrodes in biosensors. The change in the chemical state of the surface because of the adsorption of chemicals or biological agents may result in a decline of electrons (holes) not only near the surface, but also in the entire volume of the nanostructure with an associated change in its Fermi level position within the band gap. By modifying CNM, self-assembly in sensor chips is achieved, so the sensors may be used numerous times.

Hydrophobic forces on the sidewalls of CNT have been found to be important for the insertion process. Along with Van der Waals forces, hydrophobic forces play a dominant role in the DNA-CNT interaction. CNM is uniform, well-structured, and chemically inert; hence, functionalization of CNM with appropriate proteins, such as antibodies help maintain the functioning conditions of the sensor as long as possible in working as well as storage conditions. CNTs serve both the functions (*i.e.,* as large immobilization matrices and as mediators) to improve electron transfer between the enzyme site and the electrochemical transducer.

Exposition of the majority of active sites to the analyte: An ordered molecular structure with a uniform orientation of the sensitive host biomolecules is favourable. CNM fits in very well here because it has a uniform structure, and its curve nature increases the sites for bonding of the analyte.

Functionalization of CNM helps the adhesion and provides ample space for adsorption of analytes. CNM can be deposited on various substrates, such as silicon wafers, ceramic plates, and quartz plates. These deposited films can act as sensitive membranes in a sensor.

Nature is our teacher to fabricate a nanosensor *e.g.,* highly sensitive smell perception in dogs, which is performed using nanoreceptors; certain plants also use nanosensors to detect sunlight; microbes and algae use nano-sensors to detect toxic molecules in aquatic conditions; fishes use nanosensors to detect minuscule vibrations in the surrounding water; many insects detect sex pheromones using nanosensors and ants through their nano sensing device detect sugar and other food grains. Mimicking the sensors used by ants can be extremely useful in detecting the presence of things like heroin, *etc.* Mimicking such biological sensors as a sensing material by functionalization of natural receptors with CNM, because CNT sidewalls are hydrophobic and chemically inert in nature is being tried. Moreover, bio-functionalization can reduce the problems of specific recognition and possibly biocompatibility. Hence, for biological sensing, immobilization of bio-molecules with specific functionalities on the sensing devices is necessary and can be targeted through antibody or enzymatic functionalization, nucleic acid hybridization and cellular interaction by biological barcodes such as DNA and RNA. Such functionalization needs opening up of end caps of CNT.

There are several kinds of biosensors, including enzyme, immune, DNA, and antibody-antigen sensors

Enzyme Sensors: Enzyme-based biosensors are based on electrochemical and amperometric principles. In most enzyme electrodes, Nafion is used along with CNT. Nafion is a perfluorinated sulfonic acid ionomer with a good biocompatibility and ion exchange properties. It has proven very effective as a protective coating for enzyme sensors. Based on molecules, sensors are known as glucose, pesticide and heavy metal, urea, triglyceride, butylcholine, and creatinine sensors.

DNA Sensors: DNA is a coding molecule of living organisms, which controls the functioning of the whole system. Therefore, malfunctioning body parts can be done by sensing DNA. Moreover, DNA can also be used in nano-sensors. CNM provides an excellent platform to attach DNA. Surface-confined MWCNTs have been shown useful to facilitate the adsorptive accumulation of the guanine a nucleo-base and greatly enhances its oxidation signal. SsDNA/SWCNT-FET sensors have been found to be very efficient and fast, and can detect a variety of odors within a fraction of second.

8.4.1.2 Nanoprobes

Probes are used to investigate and obtain information on an unknown region. Moloni, *et al.*, (2000) and Nguyen, *et al.*, (2004) have reported that CNTs can be used for making probes. One such example is use of CNT in AFM (Atomic Force Microscopy). AFM-generated image is dependent upon the shape of the tip of probe and surface structure of the sample to be studied. The probe tip should have radius of atomic dimensions. Conventional AFM tips are made of silicon or silicon nitride is pyramidical in shape and have a radius of curvature around 5 nm. Whereas, nanoprobes made of CNTs have high resolution, as their cylindrical shape and small tube diameter enable imaging in narrow and deep cavities. In addition, probe tips made of CNTs have mechanical robustness and low buckling force. Low buckling force lessens the imaging force exerted on the sample and therefore can be applied for imaging soft materials such as biological samples. Moreover, use of CNTs in AFM enhances the life of probes by minimizing sample damage during repeated hard crashes onto substrates (Baughman, *et al.*, 2002).

8.4.1.3 X-ray devices

Traditionally X-ray is generated by heating a metallic filament (cathode) that emits accelerated electron, which is bombarded on a metal target (anode) to generate X-rays (Fig. 8.14). There are several limitations to generation of X-ray by this method *e.g.*, it has slow response time, consumes high energy and has limited lifetime. Recent research has reported that field emission is a better mechanism of extracting electrons compared to thermionic emission (Romero, *et al.*, 2000). This is because electrons are emitted at room temperature and the output current is voltage controllable. In addition, giving the cathode the form of tips increases the local field at the tips and as a result, the voltage necessary for electron emission is lowered (Yeu, *et al.*, 2002). An optimal cathode material should have high melting point, low work function, and high thermal conductivity. CNTs meet all these requirements and hence are used as a cathode material for generating free flowing electrons.

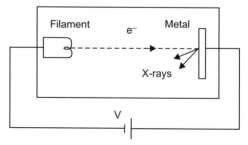

Fig. 8.14 Traditional method of generating X-rays.

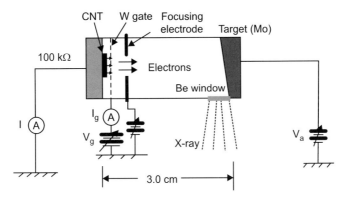

Fig. 8.15 Schematic drawing of the CNT-based microfocus X-ray tube.

Electrons are readily emitted from their tips either due to oxidized tips or because of curvature when a potential is applied between a CNT surface and an anode. Saito, *et al.*, (1997) and de Vita, *et al.*, (1999), generated continuous and pulsed X-rays using a CNT-based field emission cathode (Fig. 8.15). CNT-based X-ray devices has fast response time, fine focal spot, low power consumption, possible miniaturization (hence portability) longer life, it minimizes the need of cooling required by the conventional method (Sugie, *et al.*, 2001; Yue, *et al.*, 2002; Senda, *et al.*, 2004) and of low cost. The device can readily produce both continuous and pulsed X-ray (>100 KHz) with a programmable wave form and repetition rate. Moreover, miniaturized X-ray devices can be inserted into the body by endoscopy to deliver precise X-ray doses directly at a target area without damaging the surrounding healthy tissues.

8.4.2 Surgical Supplements

Surgical Aids: Surgery using macro instruments can be cumbersome for both the surgeon and the patient. On one hand, the patient experiences severe pain, scarring, and high healing time because of large cuts; on the other hand, the surgeon requires high concentration for a long period to perform the surgery accurately. Sometimes, it may lead to surgical error due to the surgeon's fatigue. In many cases, surgical error may result because of the limited view of the organs by the surgeon. In addition, macro surgical instruments are not suitable for certain delicate cases, such as surgeries related to heart, brain, eyes, and ears. One of the solutions is laparoscopic surgery, which uses a small entry port, long and narrow surgical instruments, and a rod-shaped telescope attached to a camera. However, laparoscopic surgery requires highly skilled surgeons for efficient surgery. Research needs to be carried out to investigate if smart instruments (such as forceps, scalpels, and grippers with embedded sensors to provide improved functionality and real-time information) using CNTs can be developed that can aid surgeons by providing specific properties of tissue to be cut and provide information about performance of their instruments during surgery. The usefulness of CNTs for optically guiding surgery should also be investigated. This can lead to easy removal of tumours and other diseased sites. Another option is the use of molecular nanotechnology (MNT) or nanorobotics in surgery. In nanorobotics, surgeons move joystick handles to manipulate robot arms containing miniature surgical instruments at the ports. Another robot arm contains a miniature camera for a broad view of the surgical site. It results in less stress for surgeons and less pain for patients; at the same time, high precision and safety is achieved. MNT allows *in vivo* surgery on individual human cells. Nanorobotics-based surgery can be used for gall bladder, cardiac, prostrate, bypass, colorectal, esophageal, and gynecological surgery. However, nanorobotic systems for performing surgery require the ability to build precise structures, actuators, and motors that operate at molecular level to enable manipulation and locomotion.

Nanotweezers (that can be used for manipulation and modification of biological systems such as structures within a cell) have already been created using CNTs, they have the potential to be used in medical nanorobotics. Nano tubes nanotweezers can be used for manipulation and modification of biological systems such as DNA and structures within a cell. Second, they can be used as nanoprobes for assembling structures. It will be helpful in increasing the value of measurement systems for characterization and manipulation at nanometer scale. Application of CNTs as nanoprobes for crossing the tumour, but not crossing into healthy brain tissue should also be investigated, as the presence of cancer in a brain tumour may result in weakening of the blood-brain barrier.

Besides, Cummings and Zettle (2000) have demonstrated that nested CNTs can make exceptionally low-friction nanobearings. These nanobearings can be used in many surgical tools. Therefore, there is a need for research to be extended to investigate the application of CNTs in other surgical tools.

8.4.3 CNT for Tissue Engineering

Nanocomposite for Tissue engineering: Tissue engineering involves replacement of anatomic structure of the damaged, injured or missing tissue or organs by agglomerating biomaterials, cells and biologically active molecules (Lanza, *et al.*, 2000; Ibarra, *et al.*, 2000 and Atala, 2000). It is clear till now that tissues engineering needs designing smart biomaterials which can support the cell growth without altering the biological functions required for growth of the tissue. Carbon nanotubes have been found to be a versatile scaffolds. Both SWCNT and MWCNT possess very interesting properties, such as ordered structures with high aspect ratio, ultra light weight, high mechanical strength, high electrical and thermal conductivity, metallic or semimetallic behaviour and high surface area (Sharon, *et al.*, 1995, 2007).These properties have made CNTs favourite material for nanofabrication of compatible biomaterials which can act as scaffold for engineering tissues (Martin, *et al.*, 2003).

8.4.4 Gene Delivery using CNT Instead of a Vector

To investigate the processes taking pace inside a cell or cell organelles, nano-sized devices are needed that can pass through the delicate cell membrane without damaging it. Carbon nanotubes (CNTs), with their needle-like geometry, high elasticity and strength, have recently shown such capacity. Chen, *et al.*, (2007) of University of California at Berkeley and Bertozzi from the and the Lawrence Berkeley National Laboratory have recently found that a CNT-based "nanoinjector" can penetrate a cell with no membrane damage, even after repeated use.

Cui's group (Pan Bi-feng, 2007) at Shanghai Jiao Tong University, China, have developed novel polyamidoamine dendrimer modified multi-walled carbon nanotubes (dMNTs) for gene delivery to improve the amount and delivery efficiency of genes taken by CNTs into human cancer cells. Compared with traditional gene-delivery carriers, polyamidoamine dendrimer modified multi-walled carbon nanotubes (dMNTs) exhibit maximal transfection efficiencies and inhibition effects on tumor cells.

Other successful trials of using CNT for delivery of gene were by: Gao, *et al.*, (2004), they found that Amino-functionalized MWCNTs interacted with negatively charged DNA and the cell membrane and delivered the GFP (green fluorescent protein) gene into mammalian cells without any cytotoxicity. Pantarotto, *et al.*, (2004) also found that functionalized, positively charged, water-soluble CNTs can penetrate into cells and can transport plasmid DNA by formation of non-covalent DNA-CNT complexes. Such CNTs can be used as novel non-viral delivery systems for gene transfer.

8.4.5 Pollution Control by CNT

CNT for photocatalytic killing of pathogenic microbes: Carbon nano-material can possess the property of semiconducting nature depending upon its internal diameter. In the living system, this property of CNM supports photocatalytic activity. Its photocatalytic behaviour has given an impetus to use it as an anti-microbial agent. Anti-microbial property or killing of microbes has been found to be caused by disrupting the cell membrane (Sharon, *et al.*, 2007). Moreover, CNTs have also been able to disrupt the plant cell wall and membrane (Gupta, 2008); hence, its various applications are being envisaged like killing the unwanted algae, as a sterilant *etc.*

Fighting microbes is commonly achieved by certain chemical agents. Two inorganic approaches can be used for sanitizing surfaces. The first is based on the photocatalytic activity. The second exploits the toxicity of certain metallic cations such as silver. Silver has long been known for its excellent antimicrobial effect due to the release of silver ions which are taken up by microbes and exert a toxic effect. Modern approaches increase its activity by dispersing silver in ultra fine particles. The extreme increase in surface area enhances silver's natural sanitizing ability. But when CNM is considered photocatalytic activity works the best and we will study that in detail. In photocatalytic process, particle of a semiconductor if in contact with another material, it can form a depletion region like one forms a p:n junction. When its interface is illuminated with a light of photon energy greater the band gap of the materials, formation of electron/hole pairs takes place in similar fashion as one gets when a p:n junction is illuminated. These photon generated electron/hole pairs are highly reactive and can initiate oxidation/reduction process with any organic material which comes in their physical contact. Considering this logic, anti-microbial effect of CNM on microbial cells have been studied; and efforts are on in many labs to use CNT for water borne microbe at large scale.

CNM is also being tried as filtration tool: Anti-microbial capacity and its use as a filter material can solve and improve the water purification and filtration. It is being thought as better alternative than existing water purification and filtration technology. Along with this the basic application of CNM for water purification, it can also be used as bandages and sterilant to clean surfaces that will be bacteria proof.

8.4.6 Anti-carcinogenic Activity of Carbon Nanomaterials

A vision to use CNT as a new substitute for metallic macro syringes to treat cancerous cell is being realized by several research groups who are also developing CNTs as targeted delivery vehicles for anticancer drugs. CNTs can ferry proteins as well as deliver anticancer drugs into cells and act as miniature thermal scalpels that can bake a cancer cell to death.

Prevention of precancerous cells by imaging agents and diagnostics from becoming malignant at early stages is a burning field where scientists are working on developing a method to detect cancer at very early stages so that its cure could be achievable. Two land mark achievements are: one by developing a highly sensitive immuno-detection of cancer biomarkers by amplifying CNTs (Yu, *et al.*, 2006); the other is by using protein coated Nano-cantilever (Gupta, *et al.*, 2006). In an attempt to increase the sensitivity of cancer biomarker detection and to decrease the need for large samples from which to detect those molecules, a "forest" of single-walled carbon nanotubes is used to detect lower levels of the prostate specific antigen (PSA) which is a protein produced by the cells of the prostate gland. When the prostate gland enlarges, PSA levels in the blood tend to rise. PSA levels can rise due to cancer or benign (not cancerous) conditions. As PSA is produced by the body and can be used to detect disease, it is sometimes called a biological marker or tumour marker. Moreover, this new system requires between 5 and 15 times less sample than does the other commercial detecting systems and can detect PSA at levels as low as 4 pg/ml in a sample size of 10 µl of serum. Whereas, the existing standard assay has detection limit of 10–100 pg/m and requires as much as 50–150 µl of sample. Bundles of SWCNT having chemically reactive groups on the ends of the SWCNTs to which enzymes or antibodies capable of reacting or binding to specific bio-molecules are attached was used for this purpose. An antibody that binds to PSA, which is used in detecting Prostrate as well as breast cancer was used. When to a serum sample, antibody-labeled CNTs is added then any PSA present in the serum sample gets bound to the antibody. Using electrical impulses, the number of PSA molecules bound to CNTs can be measured.

CNT has cylindrical and hollow tube shape and the number of walls enclosing a nanotube making it SWCNT or MWCNT and the arrangement of carbon atoms with respect to tube axis (there are many kinds of Chiral tubes) makes it a suitable material for preparing a cantilever. Mechanical (such as bending or bucking and high resilience to mechanical strains) and thermal (it shows thermal expansion under temperature) properties of CNT shows that cantilever can be fabricated from it and used for detecting cancer. Moreover, MWCNTs can adsorb light energy selectively depending on light polarity

To kill cancerous cells and identify biological targets attempt is being made to develop nanotube-based therapeutics and imaging agents. Since their excellent electrical conductivity, CNTs are poised to become the key component of ultra fast, miniaturized diagnostic gear that may soon be able to detect the earliest signs of cancer from a pinprick of blood immediately

Wang, *et al.*, (2004) have used *Electrodes modified with CNTs* to create highly sensitive devices capable of detecting specific sequences of DNA, including those associated with the breast cancer gene BRCA1. On the contrary, Sirdeshmukh, *et al.*, (2004) have attached antibodies to the surfaces of carbon nanotubes that can bind molecules shed by tumours into the bloodstream. In the both cases, the idea is to incorporate the tumour-sensing nanotubes into a small electrical circuit, which would signal the presence of the tumour marker with a change in electrical conductivity.

Nanoshell and SWCNT have been found to kill cancer cells by using infrared radiation or lasers with very high power (35 W/cm² for Nano-shell and 1–4 W/cm² for SWCNT) by heating the Nanoshell and SWCNT to 55–70°C. However, the level of temperature increases in SWCNT to an extraordinary level to create **Nanobombs** that explode at low laser intensities. By exploding SWCNT that are co-localized with cancer cells, one can selectively destroy the SWCNT as well as the cancer cells. This approach may be the only way of using SWCNT for therapeutics without causing any toxicity problems. Nanobombs are made by bundling the CNTs, and exposing them to light, thus the heat is generated. With a single CNT, the heat generated by the light is dissipated by surrounding air. But in bundles of CNTs, the heat cannot dissipate as quickly and results in explosion on the nanoscale. Nanobombs are created due to the optical thermal transitions in SWCNT and due to the thermal-energy confinement in SWCNT bundles and subsequent vapourization of liquids between SWCNT in bundles creating pressure that cause the eventual explosion. This phenomenon works in a host of different liquids such as alcohol, de-ionized (DI) water, and phosphate-buffered saline solutions. By hydrating SWCNT in sheets as well as by co-localizing them to cells, potent Nanobombs can be created to explode cancer cells completely. This process is highly localized with minimum collateral damage to neighbouring cells.

Efforts are on to use CNM as therapeutic agent for Cancer treatment because CNT can absorb near-infrared light waves, which can pass harmlessly through cells. When a beam of near-infrared light falls on a CNT, electrons in the CNT become excited and begin releasing excess energy in the form of heat (a solution of CNTs under a near-infrared laser beam heats up to about 70° C in two minutes). Dai and Flesher (2005) utilized this property of CNTs and used it as therapeutic agent. Nanotubes were placed inside cells and irradiated by the laser beam, the cells were quickly destroyed by the heat. However, cells without nanotubes showed no such effects when placed under near-infrared light.

8.4.7 Drug Delivery using Carbon Nanomaterials

The efficacy of a drug can be increased if it is delivered to its target selectively and its release profile is controlled. Moreover, targeting a drug to its site of action would not only improve the therapeutic efficacy, but also enable a reduction in total dose of the drug, thus minimizing unwanted toxic effects of the drugs. Nanoscale devices attached with antibodies and loaded with drugs can serve as a targeted drug delivery vehicle that can transport chemo-therapeutics or even therapeutic genes into diseased cells, while sparing the loading of healthy cells with drugs. CNTs have shown possibility of being used for drug delivery (Parihar, *et al.*, 2006). Diameters of SWCNTs are similar to the diameter of a molecule of DNA, hence can easily traverse through the cell. However, the length of CNM varies according to the method of production and it is important that they are tailored to right size for drug delivery. Filling of CNTs with DNA and proteins, through the opened end has already been done. This has given impetus to consider CNM for drug delivery. Depending on whether drug is to be adsorbed to the CNM surface or filled into the lumen of CNTs; the treatment of CNM for drug loading varies. Carbon Nanobeads (CNB) has recently attracted our attention to use them as drug carrying vehicle. Advantage of using carbon nanobeads would be that their smaller and desired sizes can be synthesized by controlling the CVD parameters and the precursor (Sharon, *et al.*, 1998). Since CNMs are insoluble in both polar as well as non-polar solvent, they exhibit very low process ability. Functionalization of CNT by —NH_2, —COOH, —OH and >C = O groups (Timea Kanyo, *et al.*, 2004). Overcomes the problem of process ability. The outer shell of MWCNT very often contains discontinuous spots and imperfections. These local vacancies are closed by functional groups mentioned above.

Since CNMs are prepared at very high temperature *i.e.,* around 750°C onwards, which may be damaging the drug molecules to be loaded, drug molecules can not be loaded during synthesis to CNT. Prior to loading the drug on to the CNM, detailed study of both drug and CNM characters is required. Drugs may be bound to the nanoparticle by:

(*a*) Incorporation in to the interiors of a nanocapsule (Soppimath, 2001).

(*b*) Covalent binding (Kopf, *et al.*, 1976 and Langer, *et al.*, 2000).

(*c*) Electrostatic binding (Langer, *et al.*, 1997 and Hoffmann, *et al.*, 1997).

(*d*) Surface adsorption (Berg, *et al.*, 1986 and Vora, *et al.*, 1993).

Various methods that have shown promises in loading metals and other atoms in CNM are:

Farajian, *et al.*, (1999) could fill endofullerenes with metals by collisions of accelerated atoms (With K.E. ≤ 150 eV for alkali metals) on SWCNTS. Hirahara, *et al.*, (2000) heated drug (beyond drug sublimation temperature) along with SWCNT under vacuum to fill the drug.

Monthiouxt, *et al.*, (2001) tried *In Situ Filling of open end* SWCNTs that are synthesized by the electric arc process. However, because of the high speed of the transient phenomena and the restricted volume, while SWNT formation occurs in the plasma, very low amount of filler entered the SWCNT before the closing of the tubule, while it grows.

Sloan, *et al.*, (1998) have filled SWCNTS with Ruthenium by soaking SWCNT in a solvent at room temperature.

Liquid phase filling of SWCNT along with the molten filler (molten salts) was tried by Govindraj, *et al.*, (2001), by putting them together in a sealed ampoule. Samarium oxide has been filled into MWCNT by this method.

Opening of MCWNTs in super critical water (SCW) medium creates alternative possibility for filling. However, for filling of MWCNTs, the very important criteria is the surface tension (v) threshold value of 100–200 mN/m, and must have route for escape of gas or air trapped in the MWCNTs. Thus, if the MWCNTs are opened inside a low surface tension liquid, the liquid should be pulled unhindered by capillarity (Ebbesen, 1996 and Ugarte, *et al.*, 1998). If opening were achieved in SCW it would be a useful mode to fill the desired compounds soluble in SCW. Since most of the modified compounds are soluble in super critical fluid it will be an efficient medium for drug delivery by filling the NTs with desired drugs. So far CNT has been filled with halides, oxides, C_{60}, heavy metals, gases like hydrogen, carbohydrates, DNA, enzymes and other proteins. But none of the drug has yet been loaded or filled in CNT.

Bianco, *et al.*, (2005) showed that CNTs are adept at entering the nuclei of cells, hence can be used as nanodelivery vehicle. They modified the Carbon nanotubes by heating them for several days in dimethyl formamide, which enabled short linking chains of tri-ethylene glycol (TEG) to be attached. Then a small peptide was bonded to the TEG molecule. When the modified CNTs were mixed with cultures of human fibroblast cells they rapidly entered and migrated towards the nucleus. At the low doses, the CNTs appeared to leave the cells unharmed, but as the concentration was increased cells began to die. Though the use of CNM is in its infancy in delivery of drugs; researchers have started to believe that one day they may be able to use CNT to deliver drug and vaccines.

8.4.8 CNM for Neurodegenerative Disorder Therapy

Neurodegenerative diseases are due to disorders of neurons. Neurons are the fundamental units of the central nervous system (CNS). It is made up of a cell body (soma), several dendrites, and a single axon (Fig. 8.16). More than one billion neurons exist in the CNS brain and spinal cord. Neurons are interconnected with sound precision; control our thoughts, emotions, movements, and sensation. Some of the Neurodegenerative Diseases are Parkinson's Disease, Friedreich's Ataxia, Amyotropic Lateral Sclerosis, Temporal Lobe Epilepsy and Alzheimer's Disease.

Since nanotechnology deals with materials of nanometric size, its applications in medicine are enormous. Nanomedicine has its impact in diagnosis, monitoring, and treatment of diseases. It is also useful in controlling and understanding biologic systems. Nanomedicine has especially contributed to the understanding of many neurologic diseases and has played a critical role in their diagnosis and therapy. Some of the special properties of CNTs that make them a suitable material for neurodegenerative disorder therapy are:

(*a*) Their very high mechanical strength and stiffness as well as high flexibility hence, CNTs can be used for building penetrating electrodes in neural prostheses.

(*b*) Their very high aspect ratio and small size allow making tiny electrodes, while maintaining a high electrical current density. Since they have good electrochemical stability, the possibility of damaging the electrodes and introducing abnormalities in neural function and cell structure is very less.

(*c*) CNTs of varied morphology (such as flat nanostructured continuous mats, sparse electrically conductive networks, localized three-dimensional nanoporous bushes, columnar closely packed forests, and spiked localized bundles or single fibres) can be grown on different surfaces. This allows tailoring of the neural interface morphology to better mimic the neural tissue microenvironment and to enhance electrical coupling.

(*d*) Chemical functionalization of CNT allows designing of appropriate functional electrode coupling down to the sub-cellular nanoscale.

(*e*) Neuroprotective properties are exhibited by many hydroxyfullerenes, CNTs. A neuroregenerative capacity is also seen in CNTs acting as a conduit, which can then be mixed with stem cells to repair neurons.

Biocompatibility of CNT has been favourably tested by Gabay, *et al.*, (2005); Hu, *et al.*, (2004); McKenzie, *et al.*, (2004) and Webster, *et al.*, (2004) for neural application. McKenzie, *et al.*, (2004) have further investigated the effects of varying diameter (60 to 200 nm) and surface energy of carbon nanofibre (CNF) on astrocyte (the cells largely responsible for the scar tissue formation seen with current implantable neural devices, proliferation and function. It was found that astrocytes preferentially adhered to and proliferated on CNFs with larger diameters and higher surface energies. The authors concluded that CNFs with diameters smaller than 100 nm show potential for neural applications because of a speculated reduction in neural scar tissue formation.

The authors concluded that these findings, coupled with the ability to tailor the electrical resistance of PU–CNF nanocomposites, warranted further investigation into their use in neural probe applications.

MWCNTs that were chemically functionalized with carboxylic acid, ethylenediamine, or poly-m-aminobenzene sulfonic acid (each holding a different ionic charge at physiologic pH) were each observed to provide a substrate for neurite extension by Hu, *et al.*, (2004). It was concluded that neurite extension was loosely based on the ionic charge, with positively charged ethylenediamine—MWCNT producing the most neurite extension. Neither the CNT patterned surfaces nor the functionalized MWCNTs demonstrated cytotoxicity toward neuronal cells. Recently Andrea, *et al.*, (2007) have developed a SWCNT-neuron hybrid system and demonstrated that CNTs can directly stimulate brain circuit activity. Mattson, *et al.*, (2004) used CNTs to served as substrates for neuronal growth.

It can be concluded that since CNTs possess a high tensile strength, are ultra lightweight, and have excellent chemical and electronic properties and thermal stability plus it also possesses semi-metallic and metallic conductive properties. Also, distinct electronic properties; it can be an excellent candidate that can be used as an artificial microdevice to replace the function of impaired nervous systems or sensory organs. Such neural interface implants could help to increase the independence of people with disabilities by allowing them to control various devices with their thoughts. Hybrid SWCNT-Neuron which can directly stimulate brain circuit activity, are found to be demonstrated as a future neuroprosthetic device for patients suffering from neurodegenerative disorders.

8.4.9 Some Others Biological Applications of CNT

Enzyme-containing polymer/SWNT composites have been used as unique biocatalytic materials by Rege, *et al.*, (2003).

Pantarotto, *et al.*, (2004) have functionalized CNT with bioactive peptides to make it act as a medium to transfer peptides across the cell membranes. This CNT grown on a sheet of paper can be used for transplanting cells into the retina. In this experiment, the substrate worked as a scaffold for the growth of the retinal cells taken from white rabbits, for implanting in other rabbits. The list is being added by efforts of scientists every day.

8.4.10 Possible Drawbacks in use of CNM in Biosystem

Use of carbon in therapeutic system has started with many apprehensions. Its suitability has been a big question mark in the mind of many scientists. Some of the apprehensions are that use of nanosized material in medicine poses unique problems; they should be cleared out of the body after the completion of their mission. Moreover, they also have large surface area as relative to their volume, which may allow unwanted friction. So far, as the breakdown products of nanocarbon are concerned, our knowledge at the moment is almost negligible and it needs further research. The brighter side is that looking at the possibilities and applicability of CNM in drug delivery; scientists have actively got involved into solving these inhibitions.

8.5 Summary

This chapter has dealt with various possibilities where CNM can be used to enhance the functioning wherever it is applied. Importance of functionalization of CNM has been stressed upon as it increases the reactivity of nano carbon. CNM has found its application in non-biological as well as biological systems e.g., for chemical sensors, hydrogen storage, as conducting material in fuel cell and supercapacitor, in Lithium secondary battery, solar cell, microwave absorption etc. Fullerenes, CNT, CNB, CNF due to presence of sp^3 and sp^2 carbon have been found to be very useful for biological systems such as, in diagnostic equipments, for tissue engineering, in surgical supplements, for gene delivery as vectors, pollution control due to its photocatalytic killing of microbes, anti-carcinogenic activity, drug delivery etc. However, CNM does present some limitations and drawbacks in its use.

□□□

9

Bio-Nanorobotics: Mimicking Life at the Nanoscale

> What would be the utility of such machines? Who knows? I cannot see exactly what would happen, but I can hardly doubt that when we have some control of the arrangement of things on a molecular scale we will get an enormously greater range of possible properties that substance can have, and of the different things we do.
>
> —Richard Feynman

9.1 Introduction

Life can be considered as the delicate and most magical sequestration of atoms. The multitalented crosstalks between these atoms via gigantic intermolecular and interatomic interactions generate living system, the highest altitude of complexity, perhaps, known to human being. With some level of uncertainty, the first person to decipher atoms was Indian philosopher Maharshi Kanad. He postulated that elements are formed due to specific arrangement of negatively charged bodies (later discovered as electrons). Behind activity ('karma') and characteristic properties of material, these fundaments ingredient of material play important role was one of the sensible comments made by him. He also credited electrons as smallest unit to measure time! You may be thinking this discussion extraneous, but understanding atomic behaviour is the only gate pass to enter into the realms of nanotechnology.

We have already discussed about the fundamental architecture and functioning of natural bio-nanomachines in restricted environment of biological cell. To repeat few statement from the chapter 2, *at nanoscale many properties are ambiguous in comparison with the macroscopic world of materials.* These problems become more extreme in case of biological system and hence mimicking biological system, called bio-nanorobotics, becomes an overwhelming task. In other words, this size related challenge is the ability to measure, manipulate and assemble matter with features on the scale of 1 to 100 nm. In order to attain cost-effectiveness in nanotechnology, it will be necessary to automate molecular manufacturing.

Due to granular nature *(molecular granularity)* of biological machines, it becomes extremely difficult to fabricate man made nanometric devices. This is mainly due to inability of gluing atoms into desired shape and size. *This also demands understanding the quantum mechanical properties of individual atoms and then of the resulting atomic network in order to make them functional in a specific environment.* The engineering of molecular products needs to be carried out by automated robotic devices which have been termed as *"nanorobots"*. A nanorobot is fundamentally a controllable machine at the nanometer or molecular scale that is fabricated of nanoscale components. Nanorobotics basically handles the design, construction, training or programming, and control of the nanoscale robots. Fortunately, we have nanomanipulators like atomic force microscope which work with nanometric precision (Requicha A., 2003; Stroscio and Eigler, 1991). It is due to the concerted efforts of scientists and engineers that finely tuned nanorobots have been constructed blending the principles of theoretical physics and engineering (Freitas R, 1998; Behkam B. and Sitti M., 2006; Cavalcanti A. and Freitas Jr R. A., 2005; M. Sitti, 2004).

Due to immense applications in medical, health care and environmental monitoring, which are attributable to the size of nanorobots comparable with biological entities, nanorobotics is at the apex molecular engineering. In this perspective, there is gluttonous development of in nanotechnological sciences which embraces nanofabrication, nanoelectronics and quantum computing, nanomaterials and nanomachines, nanophotonics and many more. The fruitful attempts of making nanojoints and nanoadhesives, and many nanostructure (Fukuda, *et al.*, 2003) using carbon nanotubes (CNTs), the single *"molecular vehicle"* (Shirai, *et al.*, 2005) with wheels as buckyballs, synthesis of *nanomotors fueled by chemical* (Kelly, *et al.*, 1999) or optical (Koumura, *et al.*, 1999) energy, fixing of ATPase to substrate used to rotate metallic or organic nanorods (Harada, *et al.*, 2002; Soong, *et al.*, 2000) as we have already seen in chapter 2, fabrication of nanoswimmers on the basis of supermagnetic elastic filaments (Dreyfus, *et al.*, 2005) are some of the initial leaps towards nanorobotics.

9.2 Small Size and Big Challenges

A very modest appreciation of nature, for idealizing biological machineries is required before entering in to the deeper intricacies of the subject. The feasibility of nanomachines or nanorobots is motivated by presence of organisms and biomolecules at same size scales performing very competently, in robust manner. In bringing nanorobotics from dreams to reality, there are several challenging problems. At nanoscale regime, as we already know, physical parameters, such as increased apparent viscosity of surrounding medium (Chil-ming Ho, 2001), wear and friction (Nosonovsky and Bhushan, 2005), considerable Brownian motion due to thermal agitation (Cavalcanti, *et al.*, 2006; Curtis, 2005) and non-rigid nature in nanosize regimes (Gauger and Stark, 2006) play very cardinal roles. *At low Reynolds number (refer in chapter 2), 'walking or crawling' are difficult in comparison to 'flying and swimming'.* This is one of the characteristic features at nanoscale which influence swimming of nanorobots. At this magical scale, nanorobots with flexible links are going to swim in fluids in *stokes regimes* (Purcell, 1977). Inertia becomes meaningless and pattern of movement is unaffected by time. *Time plays no role whether configuration is changed hastily or leisurely, the pattern of motion is exactly same. This makes swimming by reciprocal motion impossible.* [Concerned reader must refer luminous discussion by E.M. Purcell (Purcell, 1977) about swimming strategies of living system particularly bacteria. Of particular interest should be 'scallop theorem' to understand this discussion.]

The tremendous flexibility in *biological system is due to involvement of active polymerization of gel networks such as cytoskeletal proteins* (Kruse, *et al.*, 2004), molecular motors such as kinesins marching on tracks (chapter 2) formed by protein filaments (Julicher, *et al.*, 1997) and rotating or beating cilia or rotating flagella (Lighthill, 1976 and Brower, *et al.*, 2001). Another obstacle at nanoscale is *dominant thermal agitation and Brownian storms* (Feringa, 2001; Feringa, *et al.*, 2002, Purcell, 1997), subject already discussed in chapter 2. Already appreciated the majestic movement of bacteria assisted by molecular motors like cilia and flagella, and the tactics to exploit thermal fluctuations and Brownian motion to generate forces. As per the existing literatures, it is established that restricted stiffness or non-rigidity of beating filament, is essential in movement. The *non-rigidity of nanomechanisms,* due to low spring constant, has been supported by bead spring models (Gauger and Stark, 2006), slenderness theory, Kirchoff's-rod (Higdon, 1979; Kim and Powers, 2004) and many more. To understand more critical issue in nanodomain the reader is requested to read chapter 2 carefully.

9.3 Nature's Modernization to Shift in Nanosized Provinces

As we know, movement in biological system is a massive challenge and that's why, studied comprehensively. The nature in the form of miniature biological entities has developed innovative mechanism to tackle the problems related with nanometric domain. Living organisms prefer to swim rather crawl is very well known fact. Our cells are able to crawl on solid substratum by means of various attachments via specific proteins, such as fibronectin is some of the exceptions. *This motion involves, in general, three processes: the formation and projection of a thin lamellipod in front of the cell, the adhesion of the lamellipod to the substrate, and its withdrawal at the rear, pulling the cell forward. Other types of movement is achieved by biological motors* (F. Julicher, 1997) which convert chemical energy to effect stepwise linear or rotary motion. Both rotary and translatory motors are known to exist and march in a deterministic way along a cytoskeletal filament.

Bacteria use flagellar motors, spinning at the expense of ionic gradient, to swim in aqueous solution. These biologicals *cannot rely on drifting by inertia,* unlike macro-objects, such as submarines. They move by non-reciprocal periodic motion obtained by beating of flexible oars (cilia or flagella). The conformational shift over a period is asymmetric for elastic oars and assures a net motion. They are used by tiny cells like mammalian spermatozoa to swim and internal organs to pump liquids. Flagella of bacteria are helical rigid polymers set in movement at their base by a rotary motor. This generates non-reciprocating movement in cycles; swimming at low Reynold's with rigid oars and symmetric conformational changes is extremely thorny task. *Biological system therefore has approached the difficulty by having either flexible oars (cilia and flagella) or stiff helical shapes. Hydrodynamic friction converts the rotational motion of the helix into thrust along the helix axis.*

Another challenging aspect of motion in nanodomains is that the Brownian storms rage relentlessly and refuge from the random Brownian motion in solutions is found by resting on surface. Indeed, nature uses the concept of the *Brownian ratchet* to excel effects in the action of linear and rotary protein motors. Although biological motors are capable of complex and intricate functions, a key disadvantage of their applications *ex vivo* arises in their inherent instability and restrictions in the environmental conditions. *In vivo,* one can make nanoswimmers with asymmetric conformational changes by either using flexible oars or stiff helices, but even then low fuel efficiency issues predominates and Low Peclect number in nanodomains make the issue of movement more involved and interesting which is discussed next.

9.4 A Road Atlas to Construct Nanorobots: Birth of Thinking Machines

Thinking and intelligence are the features which discriminates us from inanimate objects. It is perhaps the most multifarious titles in science, to decipher. In addition, even more challenging task is to mimic this intelligence for constructing molecular machines, a branch called "Artificial intelligence". In order for machines to be capable of intelligently predigesting information, they should execute in a way the human mind thinks and acts accordingly. We must be able to enjoy talking with these man-made machines and give them all the commands to work according to our will. In other words, such "natural thinking machines" with their additional usual mathematical skills should influence our intellectual competence. The capabilities of man made nanorobots are still far from human intelligence is the fact of the time.

The definition of intelligence is rather multifarious. In biology, it is concerned with the survival and to further natural selection. It thus, in principle could be extended from humans to bacteria and plants. But in physics and chemistry it is anything which can be retraced to the skillful arrangement of molecules. In this respect the neurotransmission involving neurons can help us in extending research in nanorobotic intelligence. Current approaches are blended with new concepts and ideas promising significant progress within the next few years. It is urgent demand, in order to make more intelligent machines, to merge the principles of human thinking with the functional principles of biological cell system.

Nanorobots would represent any functional structure (nanoscale) which is having capacity of actuation, sensing, signalling, and processing of information, intelligent thinking and swarm behaviour at nano dimension. Some of the distinguishing skills that are desirable for nanorobots to function are:

- Decentralized and distributive intelligence *(Swarm intelligence)*
- Self-assembly and replication
- Nanoinformation processing and programmability
- Nano-to macro-world edge construction

Macro and nanoscale robots differ in many aspects, but the important being basic laws that regulate their dynamics. Macro scale robots are essentially in the *Newtonian mechanics domain where as the laws leading the nanorobots are in the regime of molecular quantum mechanics.*

Nanorobots fabricated with artificial components have not yet born. The energetic area in this field is focused more on molecular machines which are motivated by nature's way of functioning at nanoregime. Nature has perfected her own consortium of molecular machines during the course of evolution. The skill of maneuvering matter at nanometric dimensions is one of the hard core applications for which nanorobots could be the technological solution. For precise drug delivery to repairing cells and combating with cancerous cells, nanorobots are expected to play a cardinal role (Freitas, 1999 and 2003).

We have already observed the distinguishing features of natural bio-nanomotors in chapter 2; hence in this section, is given special emphasis on elementary principles behind fabrication of nanorobots.

9.5 Design Strategies of Nanorobotics System

For efficient designing of nanorobots contributions from quantum molecular dynamics is obligatory. Biological components, such as DNA and protein molecules can be exploited in fabricating nanorobots purely made up of biological spare parts. Though this idea is still in its gestation period, a frame work has been made for construction of such nanorobotics device. There are many complexities that are linked with the idea of using biological molecules such as self-assembly of protein in its native conformation in aqueous medium (protein folding) and aqueous medium itself, but the advantages are also quite substantial. Biological components offer enormous diversity and utilities at nanoscale where creating such components would be daunting task. As in case of ATP synthase, work efficiency is found to be 100%, using this as a molecular motor to navigate nanorobots can be very fertile idea from the point of view of thermodynamics.

Due to advancements in genetic engineering methodologies, it is now possible to get a mutated spare part of protein, which can be used in making nanorobots. It is certainly a prevailing application of proteins in nanorobotics. These mutations can consist of anything from simple amino acid side-chain swapping, amino acid insertions or deletions, amalgamation of non-natural amino acids, and even the combination of unrelated peptide domains into whole new architecture. In an exceptional experiment, zinc was used to control F1-ATPase, which could turn a nanopropeller at the expense of ATP. A computational algorithm (Hellinga and Richards, 1991) was used to resolve the mutations essential to construct an allosteric zinc binding site into the F1-ATPase using site-directed mutagenesis. *The mutant F1-ATPase would turn an actin filament in the cost of ATP generating an average torque of 34 pN nm. Zinc could control the movement and hence, the torque generation by ATP synthase. The rotation could be stopped with the addition of zinc, and re-established with the addition of a chelator to clear off the zinc from the allosteric binding site (Liu, et al., 2002).*

The Blueprint: There are several steps for construction of the nanorobots.

9.5.1 Bio-nanocomponents

This is the first endeavor towards designing nanorobots. But for that we must have a concrete understanding of how these biological components behave and about their efficient control. From the simple elements such as structural links to more sophisticated concepts like molecular motors, each component must be carefully observed and manipulated to meet the requirements. Pleomorphism in the shapes and sizes of DNA, carbon nanotubes opens many doors for fabrication of newer and complicated devices. Protein molecules can power nanorobots by collecting energy from the sun and converting it to mechanical and other form of energies just like solar cells. Light trapping proteins such as rhodopsins from eyes and similar bacteriorhodopsin from salt loving bacteria *Halobacterium salinarium* can be used for such valuable constructions. Like wise there are numerous biological molecules, such as heat shock proteins which can replace many parts of macroscopic robots.

9.5.2 Construction of Bio-nanorobots

In this step the preferred biological components are assembled into stable complexes.

The molecular arrangement defines the hierarchy of rules and special arrangements of various modules of the bio-nanorobots, such as the *inner core* (which act as energy source); *the actuation unit*, the *sensory unit* and the *signaling and information processing* unit. In this way, one will be able to construct and build such bionanosystems that will have enhanced mobile characteristics, and will be able to transport themselves as well as other objects to desired locations at nanoscale. Just to exemplify, is the robotic device which can sense oxygen scarcity and inspire other robotic components to produce oxygen, contributing the healthy physiological condition.

9.5.3 Molecular Collaboration (Distributive Intelligence)

Bio-nanorobots must be able to form a communication network among each other in order to work accurately and in perfect coordination. In other words, they must work in colonies, like bacteria, of similar nanorobots. This unique character could set a new altitude in "Bio-nanoswarms". In this way, these nanorobots can communicate with each other transmitting the signals to each other to work together like biological cells. To establish melodious relations with the macroscopic world, intelligent devices which can conjoin robots with an external machine such as computer must be made. In this way, we can control nanorobots proficiently.

9.5.4 Automation

Dedicated bionanorobotic flocks capable of going for tricky assignments like sensing, signaling and data storage. The next altitude in nanorobotics is automatic fabrication methodologies. The key consideration in this task is information processing. This will helps boiswarms to adjust according to fluctuations in environment, a term called homeostasis in biological system. Like microorganisms, these nanoswarms must be competent of using alternate energy sources in case if one is absent.

Another gigantic issue in fabricating nanorobots is *energy management, self-replication* and *self-repairing*. For having a better understanding of these issues and of course, their solution require a separate treatment of the topic.

9.6 Fueling Nanorobots: Nanomachines Competing with Nature

A giant leap towards nanorobotics possesses a serious problem in energy supply to nanoconstructs. This is like using petrol to run a nanocar. But certainly a different version of it! A solution to this problem can make many dreams or science fiction, come true. Microscopic robotic surgeons introduced in body to patch up blood capillaries, to locate and destroy senile plaques in case of Alzheimer's disease are some of the dreams we are striving to convert into reality. In recent years, chemists have created a range of incredible molecular constructs that can be used to make nanometric devices. Researchers at Rice University made nanocars with wheels made up of buckyballs, but without motors. So one can imagine running a nanocar powered by nanoengine or nanomotors in a desired direction unlike, one without motor. This goes randomly due to thermal fluctuations and Brownian motion.

Nature, during the course of evolution has perfected plethora of biological machines working within the realms of biological cell. *Biological nanomotors rely on impulsive reactions of energy rich biomolecules, such as hydrolysis of the biological fuel ATP and also using membrane potential and proton gradient to spin motors like flagella.* Thermodynamically competent way of converting chemical energy to mechanical force is secret behind many functional nanoworkhorses toiling for intra-cytoplasmic transport, energy production and transduction, muscle contraction and many more. Due to remarkable performance of biological motors, massive research efforts are currently being dedicated towards exploiting them in artificial micro scale assemblies (Van der Heuvel M.G. and Dekker C., 2002). The remarkable functioning of biomotors with some limitations, like limited life time *in vitro*, has attracted material scientists to construct nanorobots functioning on alternative source of energies (Ismagilov, *et al.*, 2002; Paxton, *et al.*, 2004, 2005 and 2006; Ozin, *et al.*, 2005). Such synthetic nanomachines are currently under intense inspection due its overwhelming applications in nanomachinery, nanomedicine, nanoscale transport and assembly, nanorobotics, fluidic transport and chemical sensing.

Catalytic nanomotors: Out of various attempts made to fuel a nanoscale device, synthetic nanomachine involving fuel-powered bimetal (Au/Pt) nanowires motor is most significant piece of efforts (Ismagilov, *et al.*, 2002; Paxton *et.al.*, 2004, 2005 and 2006; Ozin, *et al.*, 2005; Fournier-Bidoz, *et al.*, 2005; Kline, *et al.*, 2005). Such bi-segment nanowires are synthesized by *template-assisted elcetrodeposition* within the cylindrical nanopores followed by template removal. This method can produce multi-slice nanowires with variety of shape, size and composition. For generating directional forces, interestingly, *ordered deposition of gold and platinum can be framed into asymmetric nanowire* (Fig. 9.1). The resulting nanomotors are boosted by *electro-catalytic decomposition* of the hydrogen peroxide fuel (on both ends of the wire), with oxidation of the peroxide fuel taking place at the platinum anode and its reduction to water on gold cathode. *This is an intelligent solution to over come the random movement due to jittery thermal bangs and Brownian motion (in absence of the motor).* To be lucid, this leads to a random non-Brownian movement at 10–15 m/s towards platinum end (Paxton, *et al.*, 2004; Fournier-Bidoz, *et al.*, 2005; Kline, *et al.*, 2005). Another effort to generate directional movement is by administration of ferromagnetic nickel segment and aligning the magnetized nanowires vaguely using an external magnetic field (Kline *et al.*, 2005). Transforming between fluctuating magnetic field (weak and strong) can be exploited for instigating and halting the motion, with high spatial and temporal resolution (Laocharoensuk, *et al.*, 2008). Accurate navigation is accomplished by regulating the direction of the magnetic field. Catalytic nanowire motors present a chemotactic behaviour in response to a gradient of the fuel concentration; with a directed movement and increased speed toward higher peroxide concentrations (Hong Y., *et al.*, 2007). This is just like the chemotactic movement of bacterium towards a chemical stimulus such as carbon source. Controlling the concentration and level of fuel may be used for navigating nanomotors. This may also be done by using light energy.

Out of various mechanisms proposed, *self-propulsion of bimetallic catalytic nanomotors* is most preferred (Paxton, *et al.*, 2004 and 2005; Wang Y., 2006). The most acknowledged one being an electro kinetic self-electrophoresis (Wang Y., 2006). This mode of action instructs that in addition to the hydrogen peroxide reduction, the cathodic reaction on the gold segment involves also the four-electron reduction of oxygen to water. These cathodic reactions, along with the oxidation of the peroxide fuel at the platinum sector, effects in electron flux within the wire (toward the gold cathode) and generation of an electric field. These lead to electro immigration of protons in the electrical double layer (surrounding the nanowire) from the platinum end to the gold end and to self-electrophoresis and propulsion of the nanomotors.

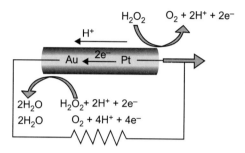

Fig. 9.1 Bipolar electrochemical mechanism for propulsion of catalytic nanowire motors in presence of hydrogen peroxide. The modus operandi is internal electron flow from one end to the other end of nanowire, along with the movement of protons in the double layer surrounding the double layer.

Ozin's group at Toronto also initiated bubble-driven mechanism and observed a rotary motion of Ni-Au nanowires with the gold end of the nanowire anchored to substrate (Fournier-Bidoz, *et al.*, 2005). The movement was credited to the evolution of oxygen bubbles at the nickel end. Gravitational forces also contribute in this self-propulsion phenomenon, a recent finding (Koutyukhova, *et al.*, 2008). Other mechanisms proposed for the self-propulsion of catalytic nanomotors comprise an *interfacial tension mechanism* and a *Brownian ratchet* mechanism in which the oxygen evolved on one segment decreases the local viscosity (Dhar, *et al.*, 2006).

Proficient energy conversion can lay the strongest ground to enhance the efficiency of catalytic nanorobots. Efforts should be taken to improve the velocity, force, lifetime, and versatility of man-made nanomotors by exploring new fuels to drive efficiently (Demirok, *et al.*, 2008; Laocharoensuk, *et al.*, 2008). Joseph Wang and his contemporaries verified that the integration of carbon nanotubes in to

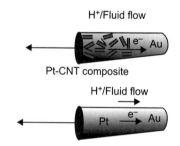

Fig. 9.2 High-speed catalytic nanomotors powered by CNT-Pt complex.

platinum slice of catalytic nanowire motors (Fig. 9.2) increase the speed and power (Laocharoensuk, *et al.*, 2008). Chemically driven catalytic nanomotors currently function within a very slender range of environmental condition, such as low ionic strength aqueous and limited fuels (hydrogen peroxide, hydrazine).

Extending the scope of man-made nanomachines to varied operations and wide range of environments would require the identification of new fuel sources and further modification in power and efficiency. To equip artificial nanomachines for doing more complex task, nanomachines should be automated so that it can choose the fuel as per need and also capable of communicating with each other and making judgments accordingly. The execution of such autonomous self-governing nanomachines and their amalgamation into functional micro-devices could lead to a new generation of laboratory-on-a chip system with unparalleled capabilities.

9.7 Medical Nanorobotics

Development in nanotechnology and nanorobotics helped medicine to take a giant dive towards molecular diagnosis, drug delivery and surgery. Intelligent molecular motors traversing in blood vessels in search of the errs and taking smart judgments to cure it, nanosurgeons equipped with camera rectifying the minute anatomical abnormalities in capillaries, are some of the thoughts which keep on fantasizing us. The inception of this innovative field was due to the brain child of famous personality and late Nobel physicists Richard Feynman, who worked on Manhattan project at Los Alamos during World War II and later taught at CalTech. In his extraordinarily prophetic talk *"There's plenty of Room at the Bottom"*, Feynman envisioned making tools at atomic dimensions using robotic devices which can tackle desired atomic arrangement (Feyman, 1960). He was highly influenced by the potent medical applications of nanorobotics and nanotechnology. He overwhelmingly acknowledged the idea of swallowing a nanosurgeon as a revolutionary method of surgery. With his friend, Albert Hibbs, he imagined tremendous potentialities of molecular machines saying that "you put mechanical surgeons inside the blood vessels and it goes into the heart and removes the blocked portion with a nanoknife". The serious outcome of Feynman's remark took more than two decades when K. Eric Drexel, while still a graduate student at MIT, wrote a paper (Drexel, 1981) signifying the prospect to construct nanometric machines that could scrutinize the cells of living human being and carry on repairs within them. This was followed a decade later by Drexel's outstanding technical book laying the fundamentals of molecular machine system and nanorobotics (Drexel, 1992).

9.7.1 Applications of Medical Nanorobot

In 1992, K. Eric Drexel of the Institute for Molecular Manufacturing hypothesized that a proficient nanochemical bearing could be made by bending two graphite sheets into cylinders of dissimilar diameters, then inserting the smaller one into larger one concentrically. John Cuming and Alex Zettl at University of California, Berkley, in 2000, experimentally confirmed the efficiency of nested carbon nanotubes as extremely *low-friction nanobearings* (J. Cumings, 2000).

Understanding details of molecular anatomy is beyond the scope of the book, however interested readers should refer brilliant coverage of medical nanorobots by Robert Freitas (R. A. Freitas, 2003).

The thought of surrendering our body to self-fueled and self-governing nanomachines might seem horrific to us, but in reality a living system is factory of such autonomous nanodevices (chapter 2) patrolling through out the body to do vital task. To give you an idea, more than 40 trillion single-celled microbes swim within our gut, many bacteria move by beating tiny flagellum, powered by 30 nm nanomotors which exploit electrochemical gradient and proton potential to run the flagellar filament. Apart from that, our body maintains trillions of different micro-devices including guard cells like neutrophils and lymphocytes. These beneficial nanorobots are continuously swarming around inside us, refurbishing injured tissues, destroying invading microbes, and transporting cargoes loaded with vital proteins. There are ongoing trials to construct *MEMS (Micro-Electro-Mechanical Systems) based microrobots.* For instance, the "MR-Sub" project of the nanorobotics Laboratory of Ecole Polytechnique in Montreal will employ a MRI system as a way of momentum for a microrobot in the blood vessels (Mathieu J, *et al.*, 2003). *In this approach, a magnetic force on robot made up of ferromagnetic particles, would be generated with variable MRI magnetic field. This will help in propulsion of the robot in desired direction in human body. This first generation micro-device might be exploited in targeted drug release, the dissolving the clot in the arteries, or taking tissue samples for pathological investigations.*

MEMS-based microrobots have many others vital functions to do which are covered in many literatures (Ishiyama K., 2003; Flitman S., 2000), including the magnetically controlled *"cytobots"* and *"karyobots"* (Chrusch D., 2002) for carrying out remote controlled intracellular surgery. There are preliminary ideas for fabricating hybrid Bionanorobots. ATPase motors have been modified to create a nanorobot capable of releasing active drug ingredients in the cell acting as *"molecular pharmacy"*. This magnificent nanorobot can also clutch the proteins which are not used by the cell and release them whenever needed.

The device would consist of a tiny nickel drum, attached to the ATP-powered biological motor, which is coated with antibodies that adsorb the target molecules, whereupon an electric field pulls the molecules to a storage chamber and holds them in place. Actual impetus in nanorobotics will be due to availability of molecular blue print to construct nanorobot form various components.

9.7.1.1 *Respirocytes*

These are the artificial red blood cells or "Respirocytes", patrolling in blood vessels and releasing oxygen (R. A. Freitas, 1998). Respirocytes (*see* Fig. 9.3) are blood borne spherical pressure vessel with vigorous oxygen pumping capacity fueled by glucose present in the serum, able to deliver much more oxygen than natural red blood cell and to tackle carbonic acidity. The nanorobot is fabricated of 18 billion atoms arranged in a diamondoid pressure tank that can be pumped full of up to 3 billion oxygen and carbon dioxide molecules. Same molecular pump can be used to release them in a very controlled mode.

The modus operandi is like natural *haemoglobin-filled red blood cells.* The concentrations of gases can be sensed by devices attached to Respirocytes, so that it can decide when to take the oxygen and unload carbon dioxide (Fig. 9.4). A nanocomputer and copious sensors facilitate complex programming of the device. Administration of roughly 1 liter of respirocyte saline suspension in human blood, (R. A. Freitas, 2003) could keep a patient with cardiac arrest, oxygenated up to 3 hours or it would permit a healthy person to sit calmly at the bottom of a swimming pool for 4 hrs., holding his breath, or to run at top speed for at least 15 min without breathing!. Respirocyte will be able to gear up with devastating medical complications like heart attack asphyxia cancer *etc.*

9.7.1.2 *Microbivores*

Microbivore (Fig. 9.5) is spheroidal nanodevice (diameter 3.4 micron), *containing* 610 *billion accurately arranged structural atoms and with dry mass of* 12.2 *pg.* This microdevice imitates white blood cells, dedicated to wrestle with the intruders such a pathogen (R.A. Freitas, 2001). Microbivore follows digest and discharge protocol in order to destroy pathogenic bacteria.

The device may use up to 200 picoW power while completely assimilating ensnared microbes at a maximum throughput of 2 μ^3 of organic material per 30 s cycle, which is sufficient to grab and chew a single microbe from any inimical pathogenic infection in a gulp. The nanorobots would several times

Fig. 9.3 An artificial red cell: the respirocyte.

Reprinted with permission from , Robert. A. Freitas, Jr.,
Artif. Cells, Blood Subst. Immobil. Biotech. 26, 411 (1998).
© 1998, *Forrest Bishop. http://www.rfreitas.com.*

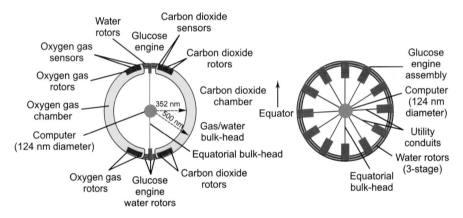

Fig. 9.4 Internal anatomy of respirocyte: Equatorial (left) and polar (right) view.

more competent in entrapping pathogens than the conventional biological machines called macrophages. Microbivore would fully eradicate septicemic infections within minutes unlike natural immune defenses which take several weeks to clear off the infection even after administration of the antibiotics.

Like a pitcher plant traps flies, bacterium is glued to the surface of microbivore via species specific binding sites (R.A. Freitas, 1999). After attachment to the surface via cell membrane, pathogens are transported to the ingestion port. After sufficient chopping up, the remains of the cell are flushed to a separate digestion chamber equipped with engineered enzymes.

These enzymes convert the remains into monoresidual amino acids, glycerol and simple sugars which are liberated back to the bloodstream without causing any harm.

Fig. 9.5 An artificial white cell: the microbivore.

Reprinted with permission from , Robert. A. Freitas, Jr. Designed by Robert A. Freitas, Jr., illustrator Forrest Bishop. © 2001, Zyvex Corp. http:// www.rfreitas.com.

9.7.1.3 *Clottocytes*

Blood clotting is an essential physiological concern in many ailments. Particularly during wound healing, role of platelets becomes cardinal and failing to do that will lead to hostile consequences such as haemophilia. The synthetic blood platelets or "Clottocytes" (R. A. Freitas, 2000) may facilitate complete process of blood clotting even in reasonably large abrasion, and the response time is several hundred times catalyzed than the natural system of blood clotting. The blueprint of Clottocytes can be imagined as *serum oxyglucose fueled spherical nanodevice packed with folded biodegradable fibre mesh.* Upon an external commend via sophisticated computers the device swiftly unfold the compact fibre mesh which acts as a patch to the wound to prevent bleeding. Soluble thin films covering certain ingredient of the mesh dissolves on contact with the blood plasma water content, revealing sticky sections analogous to surface antigens of red blood cells (and these sticky part play the role of fibrinogens). Blood cells are immediately trapped in the overlapping artificial netting released by many active Clottocytes and bleeding stops in a short while. This is done by initiating a progressive, carefully regulating mesh release cascade.

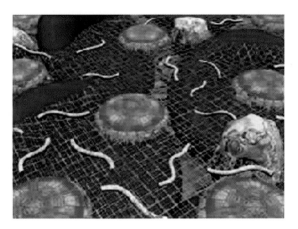

Fig. 9.6 Artificial blood clotting machine (Clottocytes).
Reprinted with permission from, Robert. A. Freitas, Jr
http://www.rfreitas.com.

9.7.1.4 *Vasculoid*

Blood is perhaps the most essential component of living organism. Deficiency of blood components can lead to dreadful calamities, such as hypovalamic shock and even blood cancer, jaundice and myocardial infarction. After molecular nanotechnology reaches to its youth, constructing a nanodevice which can replace blood becomes possible. This molecular machine is named as "vasculoid" or vascular like machine (R. A. Freitas, 2002). Such a nanorobot would replicate all vital thermo-chemical characteristics of the blood including exchange of respiratory gases, glucose, hormones, cytokines, waste products, and all necessary cellular ingredients components. This artificial blood vessel would mimic to the shape of natural blood vessels, replacing natural blood without perturbing physiological status of the body. The vasculoid apparatus would be tremendously complex, containing more than 500 trillion independent nanorobots!

9.8 Artificial Chemical Intelligence: Smart Chemical Machines

We have seen the versatile functioning of natural molecular motors inside the cell. A molecular machine is functional sequestration of discrete assembly of molecular components designed to carry out mechanical movement at the expense of external energy, as in case of natural molecular motors. The fundamental mode of operation, of a motor, is to convert the chemical energy to mechanical energy. A molecular machine is fabricated with such molecular motors in order to perform movement. Such miniature devices operate by means of electronic and/or nuclear rearrangements and exploit thermal fluctuations (R. D. Astumian, 2005). Like the macroscopic counterparts, they are differentiated by the kind of energy input supplied to make them work, the type of motion such as translation, rotation, or oscillation performed by their components, the way of scrutinizing their operation, the likelihood to repeat the actions in sequence, and the time scale required to complete a cycle.

Natural molecular motors are incredibly sophisticated systems, and it is open secrecy that the fabrication of systems of such structural and functional complexity by exploiting the bottom-up approach is still a daunting task. In late 1970, molecular electronic studies began to be healthy (Aviram, 1974; Carter, 1988; Metzger, 1991) under the nutritious nurturing of *supramolecular chemistry* (Pedersen, 1988; Cram, 1988; Lehn, 1988). Since then conception of the idea, those nanodevices can be fabricated from molecules as building blocks, rather than using atoms (much more difficult to handle) took place in few renowned laboratories (Lehn, 1988 and 1990; Balzani, *et al.*, 1987). The elegant idea of using molecules as building blocks is based on the following points:

- Molecules are stable species, whereas atoms are difficult to handle due to quantum mechanical obligations;
- Nature exploits molecules not atoms to fabricate the vast number and variety of nanodevices and biological nanomachines such as ATP synthase (Schliwa, 2003);
- Most laboratory chemical procedure follow protocols involving molecules, not atoms;
- Molecules are objects that already display discrete shapes and carry device-related properties (*e.g.*, properties that can be influenced by photochemical and electrochemical inputs); and
- Molecules can self-assemble or can be connected to make larger structures.

In subsequent years, supramolecular developed as indispensable tool for constructing synthetic molecular devices and functioning machines (Atwood and Steed, 2004). With further sophistication in understanding the molecular realms, it became more apparent that such intelligent methodologies could make exceptional contributions to peep into the enormously convoluted nanodevices and machines that catalyze biological process (Cramer, 1993; Schliwa, 2003).

9.8.1 Mechanically Intertwined Molecules as Nanoscale Machines

A molecular machine can be fabricated starting from several kinds of molecular and supramolecular systems involving DNA (Simmel, 2005; Sherman, 2004; Yin, *et al.*, 2004). Most of the other synthetic molecular machines constructed are based on interlock molecular species such, as *rotaxanes, catenins, and related species.* The molecular anatomy of rotaxanes is a dumbbell-shaped molecule bordered by a macrocyclic compound terminated by a massive group called "rings" and "stoppers", respectively (Fig. 9.7) that prevent disassembly and catenins are composed of intertwined macrocycles (Sauvage and Dietrich-Buchecker, 1999).

Fundamental features of rotaxanes and catenanes derive from non-covalent communications between components that contain complementary recognition site. Such interactions that catalyze proficient template-assisted synthesis (Schalley, 2004; Stoddart, 2002) of rotaxanes and catenanes (Fig. 9.8), include *charge-transfer (CT) ability, hydrogen bonding, hydrophobic-hydrophilic character, π-π piling, electrostatic interactions and, plethora of other strong bonds and metal-ligand bonding* (Balzani, 2007). Rotaxanes and catenanes are most attractive structure which can be exploited for fabrication of molecular machines.

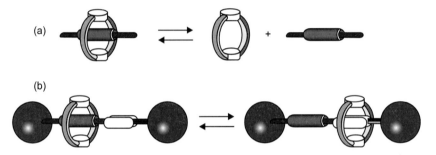

Fig. 9.7 (a and b)—Schematic representation of Rotaxanes.

Fig. 9.8 Structure of catenane.

The possible features which make them ideal candidates for the construction can be:

    (*i*) The structures are stable due to robust mechanical bonds which facilitate large range of mutual agreement of molecular components;

(*ii*) The intertwined design restricts the amplitude of the intercomponent motion in the three directions;

(*iii*) Such interactions can be modulated by external stimulation; and

(*iv*) The stability of a precise arrangement is determined by the strength of the intercomponent connections.

Movement in catenins and rotaxanes can be achieved in variety of the ways as shown in Figs. 9.8 and 9.9. The movement can be operated in two ways firstly, translations or shuttling of rings with respect to the axle; secondly, rotation of the rings around the axis. Hence, rotaxanes are good models for the assembly of both linear and rotary molecular motors.

9.9 Light Driven Molecular Machines: Rotaxanes and Catenanes

Dethreading/rethreading of the wire and ring component of a pseudorotaxane is analogous to spinning of a piston in a syringe. To accomplish a light-mediated dethreading in such piston/syringe system, pseudorotaxanes have been fabricated which incorporate a "light powered" molecular motor (Freemantle, 1998) in the wire (Ashton, 1998) (Fig. 9.10) or in a macrocyclic ring (Ashton, 1998) *see* Fig. 9.9 for better understanding. Oxygen free solution, excitation

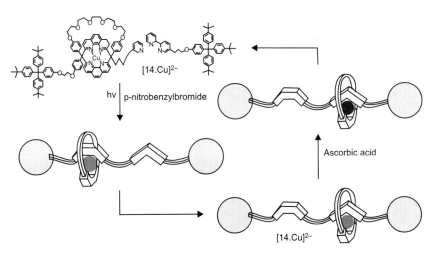

Fig. 9.9 Shuttling of the macrocyclic component of [14•Cu] [+] along its dumbbell-shaped component can be regulated photochemically by excitation of the metal Complex. Dark and light circles signify Cu (I) and Cu (II), correspondingly.

Reprinted from "From the photochemistry of coordination compounds to light-powered nanoscale devices and machines" Balzani, Bergamini & Ceroni, Coordination Chemistry Reviews 252: 2456–2469, (2008) © 2007 Elsevier with permission from Elsevier Ltd.

of the photosensitizer with visible light in presence of electron donor, catalyzes reduction of the electron-acceptor unit which further results in dethreading. Rethreading can be instigated by purging oxygen in the solution.

The ferry of macrocyclic component of a rotaxane along the linear portion of its dumbbell shaped component can be initiated photo-chemically (Armaroli, 1999). Irradiation (464 nm) of a CH_3CN solution of the rotaxane, in the presence of p-nitrobenzylbromide, guides the Cu (I) – based chromophoric unit to its metal-to-ligand charge-transfer excited state. Transport of electron from photo-excited rotaxane to p-nitrobenzylbromide pursues, producing a tetra-co-ordinated Cu (II) centre (Fig. 9.9). Since the preference of Cu (II) is penta-co-ordination geometry, the macrocycle shuttles away from the bidentate phenanthroline ligand of the dumbbell and surround the terdentate terpyridine ligand. In presence of ascorbic acid, the penta co-ordinated Cu (II) centre is reduced to a penta co-ordinated Cu (I) ion. In response to the preference of Cu (I) for tetra co-ordination geometry, as we have already discussed, the macrocycle shifts away from the terdentate terpyridine ligand and encircles the bidentate phenanthroline ligand to restore the original conformation. Reversible threading/dethreading of pseudorotaxane has also been obtained by exploiting the well-known reversible *trans-cis* photoisomerization of the azobenzene group (Balzani, 2001).

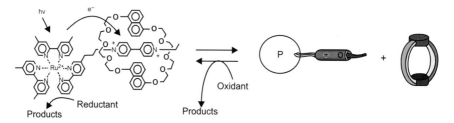

Fig. 9.10 Light-driven dethreading of a pseudorotaxane by excitation of a photosensitizer contained in the wire-type component.

Reprinted from "From the photochemistry of co-ordination compounds to light-powered nanoscale devices and machines" Balzani, Bergamini, Ceroni, Co-ordination Chemistry Reviews 252: 2456–2469, (2008) © 2007 Elsevier with permission from Elsevier Ltd.

By integrating the architectural features of the acid-base catalyzed switching of rotaxanes (Ashton, 1998) in to a triply threaded, two component supramolecular bundle, (Balzani, 2003) two-component nanoscale elevators (see Figs. 9.11 and 9.12) can be designed (Badjic, 2004). *The nanomachine, which is ~2.5 nm high and has a diameter of ~3.5 nm, consist of a tripod component equipped with two indentations—one ammonium center and one 4, 4'-bipyridinium unit—at different planes in each of its three legs. The latter are intertwined by a tritopic host, which plays the role of a platform that can be made to stop at the two different levels. The three legs of the tripod have bulky feet that prevent loss of the platform.*

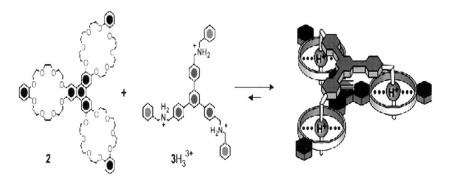

Fig. 9.11 Self-assembly of the tripodic host 2 and tripod component $3H_3^{3+}$ to afford a triply threaded supramolecular bundle. The adduct can be disassembled and reassembled in solution by addition of base and acid, respectively.

Reprinted with permission from Artificial nanomachines based on interlocked molecules, Alberto Credi, J. Phys. Condens. Matter 18 (2006) S1779–S1795, © 2006 IOP Publishing Ltd.

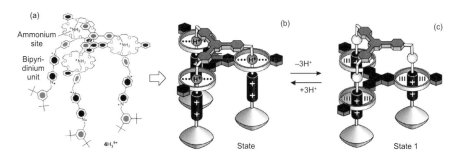

Fig. 9.12 Chemical formula (*a*) and operation scheme in solution ((*b*), (*c*)) of the molecular elevator $4H_3^{9+}$. The molecule is approximately 2.5 nm high and 3.5 nm wide.

Reprinted with permission from Artificial nanomachines based on interlocked molecules, Alberto Credi, J. Phys. Condens. Matter 18 (2006) S1779–S1795, © 2006 IOP Publishing Ltd.

9.9.1 Ring Toggling Processes in Catenanes

Light plays most cardinal role in many operations of molecular machines (Ballardini, 2001). In a catenane, structural alteration caused by rotation of one ring with respect to the other can be evidenced when one of the two rings contain non-equivalent units. Derivative of the $[\text{Ru (bpy) 3}]^{2+}$-type photosensitizer can be included in catenated structures (Ashton, 1998) and in tactically designed catenins of this type it is convenient to cause photochemical toggling between two conformations (Ballardini, 2001). In catenane, 16^{6+} (Fig. 9.10) excitation of the Ru-based compound should be followed by electron transfer to the bipyridinium unit. In the presence of an electron contributor, the oxidized

Ru-based complex would be promptly reduced and circumrotation of the cyclophane would occur to allow the formation of a more established configuration in which the *trans-bis* (pyridinium) ethylene unit inhabits the inside position. Oxidation of the reduced bipyridinium unit would then escort back to the original configuration.

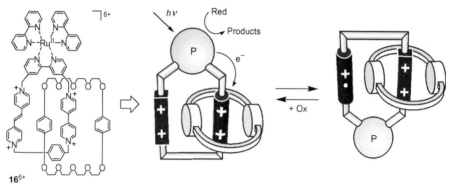

Fig. 9.13 A suggested system for achieving light-driven switching in catenanes.

The light initiated movement in Catenane 17^{2+*} (consists of a 63-membered ring which incorporates two phenanthroline units and a 42-membered ring which contains a bpy unit) is depicted in Fig. 2.11.

The proto-induced product 17′ consists of two detached rings, since the photo excitation of the Ru (II) moiety leads to decomplexation of the bpy unit contained in the 42-membered ring (Mobian, 2004).

Such synthetic molecular machines can act as tiny robots or can be used to navigate a part of nanorobotic device. Using variety of chemical rotors, shuttlers and other counterparts, nanobots can be fabricated and capable of movement in response to alteration in chemical composition. Such nanomachines can also be navigated in response to light.

9.10 Summary

With enhancing expertise to understand the atomic cross talk, scientists are developing new tools to formulate nanodevices capable of replacing many cellular types of machinery efficiently. Our inability to look at minute anatomical damages, such as loss of the receptor, vital proteins from cell membrane or a serious biochemical blunder in any part of the body will provide us a tangible to tackle molecular anomalies. And of course, mimicry of existing biological machine is the only way to fulfill our dreams. This helps us to overcome the obstacles seen when material steps into nanoworld. This utmost

---

*Refer the nomenclature of rotaxanes and catenanes.

need gave birth to "Bionanorobotics". Using the principles involved in atomic interactions, nanometric devices are made which are capable enough to compete with natural nanodevices such as flagella, ATP synthase, and most of the enzymes. Powering nanorobots, is another major apprehension needed to be solved. Initial efforts by moving nanorobots using reactions at gold and platinum rods can turn to be revolutionary idea. Using carbon nanotubes can also improve the fueling capacity. Nanorobotics has its major role to play in medicine and surgery. Nanorobotics devices marching in the body can give us abundant information for curing inimical physiological conditions, such as oxygen deficiency, blood related problems, cancer, minute surgeries *etc*. Chemical structures, such as rotaxanes and catenanes can be used to construct chemical based nanomachines, capable of moving in response to oxidation/ reduction or can be instigated with light flashes. Thanks to the concerted efforts of scientists to bring this fiction into reality.

Application of Quantum Dots in Medical Diagnostics

!!Sarve Bhavantu Sukhinah, Sarve Santu Niraamaya, Sarve Bhadraani Pashyantu Maa Kaschid Dukha Mapnuyat!!
Om Shanthi Shanthi Shanthi

"O Lord, In Thee May All Be Happy, May All Be Free From Misery, May All Realise Goodness, And May No One Suffer Pain."

—**Vedic mantra**

10.1 Introduction

Biological systems are charismatically orchestrated with functional nanometric devices, such as enzymes, motor proteins, and Nucleic acids, fabricated by intelligent assembly of macromolecules. These magical devices function with extreme accuracy in order to drive most perceptive bio-phenomena, such as protein folding, DNA replication, spatial and temporal expression of genes, cellular proliferation, and movement of the cell assisted by extracellular matrix are some among the never ending list of vital processes. Any incorrigible error happening due to faulty functioning of these biological nanowork-horses can lead to hostile consequences including death. Earliest detection of the disease can make a cardinal difference in treatment and mortality. Diagnosis of molecular signature markers of a particular disease (such as a tumour marker for a specific type of cancer), often remains a daunting task in molecular medicine mainly because the traceable stuffs are extremely low in concentration and also because of the unavailability of sensitive diagnostic tools capable of sensing such minute concentrations. Molecular diagnostics is of particular interest in detection of monstrous killers like Cancer and neurological complications where molecular detection, most of the time, is possible during autopsy. Others are viral diseases, such as Influenza, HIV, viral encephalitis and many bacterial infections. Thus, there is an urgent need of ultrasensitive diagnostic tool which can detect the molecular defects, be it at genomic or biochemical level, rapidly.

Most of the existing diagnostic tools are incapable because of the lack of interaction between the desired molecule and the one which is sensing molecules of interest. This is due to comparatively bigger size of the sensor substances used in detection and quantification. In other words, there should be one-one interaction between the sensing molecule and biomolecule of concern. Some of the widely exploited medical diagnostic tools are based on fluorescent and radio-labels to detect biological markers for disease. Such methods, apart from size related interactions, has major drawback of less photo-stability and undergo photo-bleaching. We are conversing about the fluorescence labels used in enzyme-substrate reactions.

The quest of finding the solutions landed us in the escalating monarchy of atoms and molecules. The controlled manipulation of atoms, keeping the laws of quantum mechanics in our brain capsule, materials with infinite applications can be explored. You must have speculated that we are talking about the mesmerizing field *"Nanotechnology"*, a brain child of late Nobel laureate Richard Feynman. Under the guidance of Nanoscience and nanotechnology, ultrasensitive bio-labels can be designed which can detect zeptomolar (10^{-21}M) quantities of proteins in samples! The heart of nanotechnology lies in the ability to compress the tools and devices to the nanometer range, and to accumulate atoms and molecules in to bulkier structures while the size remains very small.

With a very warm welcome to the world of Nanotechnology, let's try to walk around the possibilities and contribution of this field in medical nanodiagnostics and imaging.

As our acquaintance about understanding the jugglery of nanotechnology became more intense, we started apprehending the potential applications of nanoscale objects in detecting and curing dreadful ailments. This was the dawn of fruitful marriage between medicine and nanotechnology to give birth "Nanomedicine". Though this subject is at its apex in modern days, the use of nanoparticles is evident in ancient Ayurveda as "bhasmas", known for vigor and vitality. The principle used to make bhasmas, such as swarn (gold), rajat (silver), tamra (copper) and hirak (diamond) was top down approach starting from bulk and then crushing it to nanometric dimensions. The unique quantum mechanical properties and tunable optical band gaps make these metallic nanocrystals suitable for labeling and molecular imaging. Interestingly, or coincidently, whatever you say, biological system also exploits nanometer scale devices, such as enzymes, DNA and other biological macromolecules and cellular counter parts in order to sense battery of chemical stimulus, laying a very strong ground for nanodiagnostics.

In this chapter, we will make an attempt to explore properties of photonic nanocrystals and principles behind molecular imaging. This will be mainly in context with diagnosis and imaging in cancer and neurological disorders.

10.2 Why Metal Nanoparticles are used in Optical Detection and Imaging?

Nanotechnology got its initial impetus after the glorious birth of *"quantum mechanics"* after Neils Bohr unlocked the anonymity of discrete energy levels in atoms and proved the wave nature of electrons. This phenomenal innovation in modern physics helped in realizing us the power of miniature technology and finally nanotechnology emerged as a brain child of late Professor Richard Feynman during a burgeoning discussion about exploring the rooms at the bottom and finding out gigantic possibilities to manipulate atoms in to desired shapes and functional machineries. After several decades, the vision of Richard Feynman was translated in to reality. Nanoparticles are one such out come of controlled atomic manipulation. We will now roam around the medical applications of metal nanoparticles in diagnostics and imaging.

Nanoparticles are typically in the range of 1–100 nm (Liu, 2006), and can be tuned into various shapes and sizes, each having different quantum mechanical properties and hence different exploitations. Nanoparticles show sharp prejudice from their bulk in many respects which becomes bonus for medical diagnostics. Certain nanocrystals (crystalline nanoparticles) are attractive probes of biological markers because of:

- Small size (1–100 nm)
- Large surface to volume ratio (aspect ratio)
- Chemically alterable physical properties
- Change in the chemical and physical properties with respect to size and shape (Fig. 10.1)
- Strong affinity to target particularly proteins (in case of gold nanoparticles)
- Structural sturdiness in spite of atomic granularity
- Enhanced or delayed particles aggregation depending on the type of the surface modification, enhanced photoemission, high electrical and heat conductivity and improved surface catalytic activity (Liu, 2006;

Garg. *et al.*, 2008; McNeil, 2005; Rosi and Mirkin, 2005; Shrestha, *et al.*, 2006).

For robust target binding, size and shape of the nanoparticle plays very cardinal role. Tunable physical properties, as we have already conversed, are attractive feature of nanoparticle. This feature has been exploited in labeling biological material using nanoparticles from several decades (Baudhuin, *et al.*, 1989). Particularly the size, shape and chemical composition of the metallic nanoparticles and quantum dots can be tuned with specific emissive, absorptive, and light scattering properties (see Fig. 10.1), which make these particles ideal for biological detection (Liu, 2006; Jin, *et al.*, 2001, and 2003; Sun and Xia, 2002; Sau and Murphy, 2004).

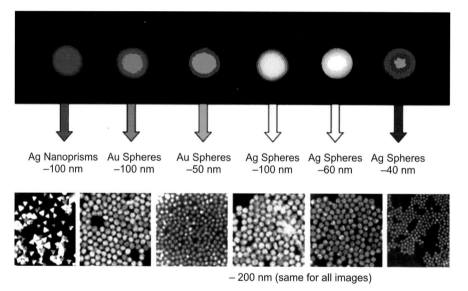

| Ag Nanoprisms −100 nm | Au Spheres −100 nm | Au Spheres −50 nm | Ag Spheres −100 nm | Ag Spheres −60 nm | Ag Spheres −40 nm |

− 200 nm (same for all images)

Fig. 10.1 Optical properties of the nanoparticles can be changed with different size, shapes and composition of metal nanoparticle.

Adapted with permission from Elsevier Ltd.

Most important aspect of nanoparticles which is atrributed for biological detection is one-on-one interaction between the nanoparticles and the desired biomolecule (Azzazy, *et al.*, 2006 and 2007). Among the plethora of metal photonic nanocrystals, quantum dots, gold nanoparticles and magnetic nanoparticles, owing to their unique optical detection and imagining characteristics, are widely used in medical diagnostics. A laconic outline of these nanocrystals is given in Table 10.1. We will treat each nanoparticle separately with respect to its optical properties and principle behind detection and imaging.

Table 10.1: Description and characteristics of QDs, AuNPs, Supramagnetic Nanoparticles (*Adapted with permission from Elsevier Limited*)

Nano-particles	Structure	Signal detection method (s)	Advantages	References
QDs	Nanocrystals typically composed of a core semiconductor *e.g.,* Cadmium selenide (CdSe) which is in a shell of another semiconductor with larger spectral band gap *e.g.,* zinc sulfide (ZnS) -A third silica shell can be added to the nanocrystals for water solubility	Fluorescence measurement using a fluorometer, fluorescence microscopy, or wide-field epifluorescence microscopy.	–Broad excitation range –narrow emission band –photo stability –optical tenability (control of emission wavelength by size control)	Azzazy, *et al.,* 2007, Alivisatos and Larabell, 2005; Goldman, 2006.
AuNPs	Either a dielectric core (normally gold sulfide or silica) enclosed within a thin gold shell (known as nanoshells) or simply a spherical gold nanoparticles.	Several detection methods can be used, such as colorimetric, scanometric, electronic and electrochemical, scattered light, surface-enhanced Raman scattering.	–Optical tunability –Strong optical signal	Azzazy, *et al.,* 2006; Baptista, *et al.,* 2008;
Super paramagnetic nanoparticles	Composed of magnetic metals like iron, or alloys of different metals.	Magnetometer	–High sensitivity due to detection of subtle modifications in magnetic character.	Jain, 2005

10.3 Quantum Dots: Semiconductors of Light

Instigation of these "artificial atomic structures" was due to unparalleled work on development of the novel semiconducting devices, in which the tunneling of the electrons could be systematically regulated, by Nobel laureate Leo Esaki in 1997. This was the adoption of idea of "Artificial solids" for the first time

(Esaki, 1992 and 1995). *The arousal of electronic and optical uniqueness of a semiconductor is due to quantum mechanical scattering of the valence electrons by atomic cores. In the first artificial solid, semiconductor atoms of varying composition were arranged serially in layers only a few atoms thick, so that the electrons were enforced to travel through an artificial potential, scattering as per the design.* In such quantum wells, electronic energy levels can be synchronized by adjusting the length over which the potential varies in comparison with the electron wavelength. The prior experiments in such "quantum confined" systems were extended from layers of atoms in sheet (quantum wells) to lines of atoms (quantum wires) and lastly to quantum dots (Duggan and Bitttner, 1999), miniature 3-D assemblages of atoms in which the electron motion is 'imprisoned' by potential barrier in all three dimensions.

In quantum dots (QDs) frequently christened as "Artificial atoms", there are distinct electronic energy levels, much as in an atom or molecule, but in this context, the spacing of electronic energy levels can be very accurately selected by the researcher. This was due to realization that colloidal quantum dots are the size of a protein, and that thus it can be injected in the biological cells. In 1998, the first use of QDs for biological detection and about its photochemical properties was reported (Bruchez, *et al.*, 1998; Chan and Nie, 1998). Medical diagnostics using colloidal QDs has got tremendous hoist from this milestone finding. QDs are robust and very stable light emitters and can be broadly tuned through size variations. In the past two years, wide range of protocols for bio-conjugating QDs (Tran, *et al.*, 2002; Wang, *et al.*, 2002; Guo, *et al.*, 2003) have been developed in diverse areas of applications: cell labeling (WU, *et al.*, 2003), cell tracking (Parak, *et al.*, 2002), *in vivo* imaging (Dubertret, *et al.*, 2002), DNA detection (Taylor, *et al.*, 200; Xu, *et al.*, 2003).

10.4 Quantum Dot: Quantum Confinement and Imaging Properties

These nanocrystals are composed of an interior of semiconductor, enclosed within a shell of other semiconductor having larger spectral band gap, see Fig. 10.2 for better transparency (Weiss, 1999; Jain, 2003; Salata, 2004; Michalet, *et al.*, 2005). The initial exploitation of QDs were for investigation of surface kinetics, where it was deciphered that the quantum yield of the photonic nanocrystals was very susceptible to the concentration of adsorbed species that endure reduction (Rossetti and Brus, 1982). Due to very small size (2–10 nm), one-on-one interaction is possible with biomolecules and quantum dots which forms strongest basis of biological diagnostics (West, *et al.*, 2003). Table 10.2 contains some jargons related with the QDs.

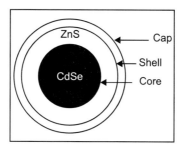

Fig. 10.2 Schematic representation of the quantum dot.

10.4.1 Fundamental Properties of Quantum Dots

The ability to tune the band gap of semiconductor (Brus, 1991) quantum dots such as CdS, once it enters to the nanometric dimension has revolutionized the field of optical detection. The alteration in the fundamental properties of such semiconductor can be exploited making light emitting diodes, and other optical and electrical devices. Semiconductor clusters or quantum dots are most ideal candidates for diagnosis because of several peculiar characteristics. For the sake of convenience, we will divide these properties in three parts:

1. There should be generous change in opto-electrical properties of quantum dots (or any other material) with the reduction of size: These transitions occur in semiconductors, for a given temperature, at a large size as compared to metals. This is because; the bands of a solid are centered about atomic energy levels, with the potency of the nearest neighbour interactions. These interactions are feeble in Van der Waals or molecular crystals (Alivisatos, 1996). The bands in the solids are very slender, and as a result not much size variation in optical or electrical properties is expected or observed in the nanocrystals regime. In metals where the *Fermi level lies in center of a band,* the relevant energy level spacing is small. At temperatures above a few Kelvin. The electrical and optical properties resemble those of a continuum, even at relatively small sizes (Cohen, *et al.*, 1987; Ellert, *et al.*, 1995). However in semiconductors, the Fermi level lies between two bands. Consequently optical excitations across the gap depend strongly on size for clusters as large as 10,000 atoms (Alivisatos, 1996, Fig. 10.3).

Table 10.2. Terms related with the QDs

Term	Definition
Spectral bandgap	The separation between electronic energy levels of a material.
Bohr radius	The natural separation distance between the positive and negative charges in the excited state of a material.

Term	Definition
Quantum yield	The ratio of photons absorbed to photons emitted by a fluorophore.
Blinking	The property of a fluorophore where it switches between fluorescent and non-fluorescent states. In the case of QDs, this occurs when they switch between the ionized and neutral staes.
Fluorescence resonance energy transfer (FRET)	A process in which energy is transferred from an excited donor molecule to an acceptor molecule through near-field dipole-dipole interaction. The process is very sensitive to the distance between the donor and acceptor molecules.
Stoke's shift	The energy (and thus wavelength) difference between absorption and emission spectra.

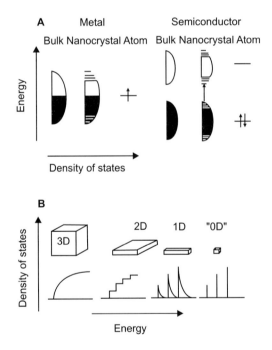

Fig. 10.3 (*A*) Schematic representation of the density of states in metals and semiconductor clusters. (*B*) Density of states in one band of a semiconductor as a function of dimension.

10.4.2 Photophysics of QDs and Its Relevance in Diagnosis

QDs link the gap between individual atom and bulk semiconductor solids. Due to this intermediate size, (2–8 nm) QDs acquire distinctive properties unavailable in either individual atoms or bulk material. In their biologically relevant form, QDs are colloids with equivalent dimension to bulkier proteins, dispersed in aqueous solvent and decorated with organic molecules in order to stabilize their dispersion.

The essential ingredient of QDs is inorganic semiconductors. They contain negatively electrons and positively charged holes (together called as *exciton*). Band gap energy *(minimum energy required to excite an electron to an energy level above its ground state)* of bulk semiconductors strongly depend upon the composition. Excitation can be initiated by the absorption of a photon of energy greater than the band gap energy, resulting in the generation of charge carriers. The exciton thus created can be return to the ground state accompanies with release of photon. Due to very small size of QDs, these spawned charge carriers are cramped to the space smaller than their bulk form. This is called the *quantum confinement of exciton* which makes these versatile semiconductors of light an ideal candidate in optical detection. The optical properties generated in these quantum dots are strictly size dependent (Alivisatos, 1996; Murphy and Coffer, 2002; Sapra and Sarma 2004). The degree of confinement increases with decrease in the size of the quantum dots. This results in production of an exciton of higher energy, thereby increasing the band gap energy. The most unique outcome of this property is that the bandgap and emission wavelength of QDs can be tuned by adjusting its size (Fig. 10.4).

By tuning their size and composition, QDs can now be prepared to emit fluorescent light from the ultraviolet (UV), though the visible, and into the infrared spectra (400–40000 nm) (Pietryga, *et al.*, 2004; Zhong, *et al.*, 2003; Qu and Peng, 2002). As an ultimate biological probe, QDs can absorb and emit light very efficiently, facilitating highly perceptive detection relative to conventionally exploited organic dyes and fluorescenct protein. In other words, QDs can have quantum efficiencies similar to that of organic dyes, individual quantum dots have been found to be much brighter than known organic dyes. This optical feature enables QDs to detect ultra low concentration of proteins such as cancer tumour marker. In addition to that QDs are several thousand times more stable against photo-bleaching than organic dyes (Chan and Nie, 1998; Bruchez, *et al.*, 1998).

Another noteworthy property of QDs is usage of multicolour QD probes to image and track multiple molecular targets concurrently. This important feature can be of particular importance in cancer marker detection because of involvement of multiple protein and genes in the progression of the diseases.

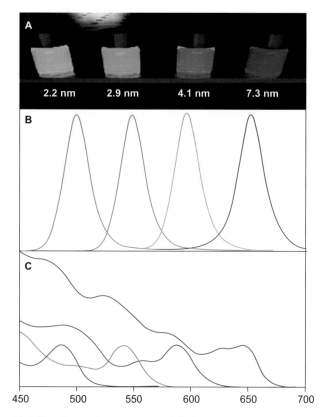

Fig. 10.4 Size dependent emission of CdSe quantum dots dispersed in chloroform, illustrating quantum confinement and size-tunable fluorescence emission. (*A*) Fluorescence image of four vials of mono-disperse QDs with sizes ranging from 2.2 nm to 7.3 nm in diameter. (*B*) Fluorescence spectra of the same four QD samples. Narrow emission bands (23–26 nm FWHM or full-width at half-maximum) indicate narrow particle size distributions. (*C*) Absorption spectra of the same four QD samples. Notice that the absorption spectra are very broad, allowing a broad wavelength range for excitation. Both the absorption and emission intensities are plotted in arbitrary units (AU).

Adapted with permission from Elsevier Ltd.

10.4.3 Modus operandi of Quantum Dots

Quantum confinement is most distinguishing property of quantum dots and hence the name QD (West and Halas, 2003). This peculiar effect is observed with the optical properties of semiconductors smaller than 10–20 nm (Bruchez, 2005). The size dependence of the QDs emission wavelength is accredited to quantum confinement effect with the following explanation:

When the size of the QD is smaller than the Bohr exciton radius, the energy levels of a photon are quantized. When light falls on such quantum dots, it absorbs a photon with a higher energy than that of the bandgap of the composing semiconductor. This initiates the formation of an exciton (a state of excited electron, also referred to as an electron-hole pair). As a result, the chance of absorption at higher energies will be increased creating a broad band absorption spectrum. When the exciton jumps back to a lower energy level, a sharp, energy band emission occurs. The florescence life time (the time the molecules remain in the excited state before emitting a photon) is quite long, about 10–40 ns (Alivisatos, *et al.*, 2005) which accounts for the stable and strong fluorescence. Even though QDs have established to be more photostable than conventional organic dyes, as we have already discussed, a significant loss of fluorescence has been noted upon injection into tissues and whole animals, and in ionic solutions (Akerman, *et al.*, 2000; Gao, *et al.*, 2004; Dyadyusha, *et al.*, 2005; Chen, *et al.*, 2002; Li, *et al.*, 2007). This signal weakening has been suggested to be due to slow deprivation of surface ligands and coating, and also due to ingredients absorbed to the surface, particularly when it is concerned with the body fluids, leading to surface defects and fluorescence quenching (Hess, *et al.*, 2001; Manna, *et al.*, 2002). For smart coating options, in order to avoid troubles concerned with the adsorption of various species on to the surface of QDs, there is an urgent demand of advancement in surface characterization technique. This type of sophistication will help in developing surface coatings, which provides minimal non-specific binding, whilst maintaining stability, avoiding oxidization and withstanding salt concentration in cells. It must also maintain strong fluorescence without bleaching, quenching, or blinking.

10.5 Synthesis and Surface Properties of Quantum Dots

Based on preferred usage, there are several tactics for synthesizing nanocrystals with plethora of properties. The most common way to synthesize monodisperse QDs is by injection of liquid precursor into hot (300°C) non-polar co-ordinating organic solvent, such as tri-n-octylphosphine oxide and hexadecylamine (Chan, *et al.*, 2002; Smith, *et al.*, 2006). Temperature and duration of the crystal growth play crucial role in size and shape of the desired quantum dots. The outer semiconductor shell of the quantum dots, such as zinc sulphide is epitaxially developed around the core (Michalet, *et al.*, 2005; Alivisatos, *et al.*, 2005). On account of their surface layer, ZnS-coated QDs synthesized by epitaxial growth method are by default hydrophobic and are thus inappropriate for use in biological media (Alivisatos, *et al.*, 2005; Ozkan, 2004; Jaiswal and Simon, 2004). Production of the water soluble quantum dots, the use of thiol as stabilizing agent in aqueous solution (Lin, *et al.*, 2004). A variety of surface alterations and orchestrations can be used to make biologically compatible QDs, appropriate for the particular target application (Fig. 10.5). As an attempt to improvise the quality of nanocrystal synthesis, Wang, *et al.*, (2005) urbanized a expedient policy for synthesis of QDs of various compositions and properties. To prepare QDs made of noble metals, the noble metal ions were reduced by ethanol at a temperature of 20 to 200 °C under hydrothermal or atmospheric conditions. This method relies on spontaneous phase transfer of the metal ions and a separation mechanism that occur at the liquid, solid, and solution interfaces during synthesis. This approach can be used for production of low disperse nanocrystals of different compositions, *e.g.*, semiconductors or rare earth metals, by tuning the reaction components.

For cardinal bilogical applications, solubilization of QDs is an obligation. Water insoluble QDs can be grown easily in hydrophobic inorganic solvents, but solubilisation needs stylish surface chemistry modification. Recent methodologies for solubilization without altering key properties are mostly based on exchange of the innovative hydrophobic surfactant layer with a hydrophilic one (Gerion, *et al.*, 2001; Bruchez, *et al.*, 1998; Kim and Bawendi, 2003), or the addition of a second layer, such as the amphiphilic molecule cyclodextrin (Pellegrino, *et al.*, 2004), which may also contain another functional group. Chitosan, a natural polymer with one amino group and two hydroxyl groups, has been used for intracellular delivery of specific molecules (Calvo, *et al.*, 1997; Miyazaki, *et al.*, 1990), and can be attached to the QD surface. Other methods for increasing solubility include encapsulation in phospholipid micelles (Dubertre, *et al.*, 2002), addition of dithiothreitol (Pathak, *et al.*, 2001), organic dendron (Guo, *et al.*, 2003; Wang, *et al.*, 2002), oligomeric ligands (Kim and Bawendi, 2003), and the addition of a second layer of poly (maleicanhydride alt-1-tetradecene) to the QD's surface. Silica and mercaptopropionic acid (MPA) are also commonly used (Gerion, *et al.*, 2001), and allow bioconjugation to ligands of interest. MPA accomplish this through carboxyl groups, and silica through the presence of thiol groups on its surface as depicted in Fig. 10.5 (Medintz, *et al.*, 2005). The boost in diameter brought about by such surface orchestrations, and conjugation with biomolecules, may substancially make intracellular drag delivery more difficult, and could increase toxicity (Weng, *et al.*, 2006). Another challenge is that there is no technique which consistently allows preparation of QDs with control over the ratio of biomolecules per QD and their orientation on the surface. Current strategy (based on modifying COOH groups on the QD surface for covalent attachment of amine groups) is limited by problems of reproducibility and aggregation (Mattoussi, *et al.*, 2000).

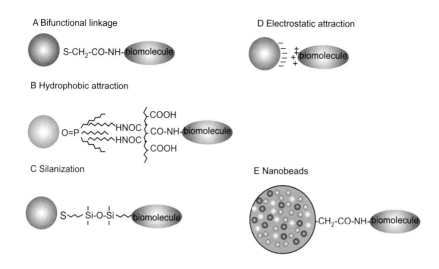

Fig. 10.5 Methods for functionalization for quantum dots.

10.6 Cytotoxic Aspect of QDs

One of the most important aspects in biological exploitation of quantum dots is its cytoxicity. Cytotoxicity of QDs has been investigated in a large number of *in vitro* studies (Medintz, *et al.*, 2005; Derfus, *et al.*, 2004) distressing cell growth and viability (Chen and Gerion, 2004). The extent of cytotoxicity has been found to be dependent upon a number of factors including size, capping materials, colour, dose of QDs, surface chemistry, coating bioactivity and processing parameters (Lovric, *et al.*, 2005; Shiohara, *et al.*, 2004). Vital mechanism which confer cytotoxity to QDS are desorption of free Cd (QD core degradation) (Medintz, *et al.*, 2005; Derfus, *et al.*, 2004) free radical formation, and interaction of QDs with intracellular components. Assessment of QD toxicity in a hepatocyte culture model showed that exposure of core CdSe to an oxidative environment causes decomposition and desorption of Cd ions. The formation fo free radicals, particularly reactive oxygen species has also been seen to contribute to toxicity (Clarke, *et al.*, 2006; Hoet, *et al.*, 2004; Oberdorster, *et al.*, 2005) in response to UV irradiation and even in dark, QDs lead to nicking of DNA. This is attributed to photo-generated and surface generated free radical exposure (Green and Howman, 2005). In a cytotoxicity assay by Chen, *et al.*, (2000) CdSe core QDs encouraged apoptosis in neuroblastoma cells by inaugurating a number of apoptotic pathways, and down regulation of survival signaling molecules (Chan, *et al.*, 2006). The composition of the core, and also the colour of the QD (a reflection of core size) appear to influence toxicity (Clarke, *et al.*, 2006). These studies also pragmated that adding ZnS shell was valuable, and reduced free radical production; however the DNA nicking observed was the result of incubation with CdSe/ZnS QDs with a biotin ligand. Whether or not, the generation of free radicals is dependent on Cd desorption is unclear, but is a possibility given that Cd has been shown to generate free radicals (Oh, *et al.*, 2006) and that a similar reduction in free radical generation as Cd desorption is seen on addition of a ZnS shell. In addition to the effects of the QD core, ligands added to render the probe biologically active may have toxic effects on cells. Mercaptopropionic acid (MPA) and mercaptoacetic acid, which are commonly used for solubilisation, have both been shown to be mildly cytotoxic (Kirchner, 2005).

10.7 Application of Quantum Dots in Medical Diagnosis

10.7.1 *In vitro* Diagnostics

In particular, the most crucial application of QDs is detection of the cancer biomarker. QDs and other nanoparticles probes can be exploited to quantify a battery of cancer biomarkers orchestrated on the surface of cancer cells, which act as diagnostics marker.

Due to fine tuned size, specific cross talk between single nanoparticle with multiple ligands, to improved binding affinity and superb specificity through a *"multivalency"* effect. These features are particularly significant towards the analysis of cancer biomarkers that are present at low concentrations or in small numbers of cells. As already conversed, traditional organic dyes used as biological labels are not photo-stable and undergo rapid photo-bleaching. Apart from that, these traditional proteins suffer spectral. Cross-talking, narrow excitation profiles and limited brightness/signal intensity. In contrast, the narrative optical properties of QDs defeat many of the problems and offer new applications which are either thorny or unfeasible with traditional fluorophores. For instance, due to their broad excitation profiles and narrow emission spectra, high-quality QDs are well matched for multiplexed tagging or encoding, in which multiple colours and intensities are combined to encode thousands of genes, proteins, or small-molecule compounds (Gao, *et al.*, 2002).

10.7.2 Cell Labeling

In an intensive study by Wu, *et al.*, (2003) it was for the first time demonstrated that QDs can be used to specifically and effectively label molecular targets at the sub-cellular level. In this study, QDs encapsulated within a polymer-shell were biofunctionalized with molecules and targeted to desired cell surface receptors or any other site such as actin or microtubules. QDs of two different colours (630 nm and 535 nm) were used concurrently and compared to an Alexa dye. QDs were more photostable in that study.

10.7.3 Quantum Dots in Neurosciences

Due to its unique optical sensing properties, QDs have occupied significant status in neuroscience research. These are proved to be very valuable tool for the studies of neurons as well as glial cells. Neuro-molecular phenomenon such as, signal transmission, which are ultrafast can also be tracked using suitable quantum dots. These properties are thorny to achieve using other techniques or approaches. Due to small size, quantum dots are useful for experiments that are limited by the constrained anatomy of neuronal and glial interactions, such as synaptic activity in neurons and possible molecular cross talks in order to achieve neurotransmission and other complicated interactions like an astrocyte and neuron. Small size and optical resolution helps in tracking for a longer time the molecular dynamics of intracellular and/or intercellular molecular processes. However, it should be cherished that the hydrodynamic radius of functionalized quantum dots is larger (15–20 nm) than their actual size of 5–8 nm (Larson, *et al.*, 2003). Antibody functionalized quantum dots help in tracking lateral diffusion of glycine receptors in cultures of primary spinal cord neurons (Dahan, *et al.*, 2003). Using the same technique, it became possible to trace the roadways of individual glycine receptors for tens of minutes at spatial resolutions of 5–10 nm, demonstrating that the diffusion dynamics varied depending on whether the receptors were synaptic, perisynaptic, or extrasynaptic. Using quantum dots, neuronal differentiation can be studied by tagging quantum dots with nerve growth factor (Vu, *et al.*, 2005). To tackle with biological systems, quantum dots should be functionalized to make it biocompatible which should retain its optical properties. Brinker and colleagues (Fan, *et al.*, 2005) developed a technique to produce biocompatible water-soluble quantum dot micelles that retain the optical properties of individual quantum dots. These micelles showed uptake and intracellular dispersion in cultured hippocampal neurons.

Eventually, quantum dot nanotechnologies will require easy-to-use approaches that can be straightforwardly replicated in a typical neurobiology lab.

10.7.4 Quantum Dots for Cancer Diagnosis and Therapy

Perhaps the most important usage of the quantum dots is diagnosis and therapy of cancer. Due to high photo-stability QDs allows real-time monitoring or tracking of intracellular processes *in vivo* over comprehensive periods. As the labeling of individual molecules or cell structures in living cells or tissues is becoming an increasingly important tool in diagnostics, QDs, because of their many advantages over organic dyes, have a large potential for new and improved diagnostic tests in medicine. Supermagnetic nanoparticles contain a metal core (*e.g.,* iron, cobalt, or nickel) that is magnetically active, and are used as contrast enhancement agents to improve the sensitivity of MRI. Magnetic particles, when coated with an organic outer layer, can also be conjugated to biomolecules and used as site-specific drug-delivery agents for cancer treatment. Iron-oxide-based magnetic materials have been used widely in clinical practice as magnetic resonance agents and in studies of gene expression, angiogenesis imaging, and cellular trafficking. Metal nanoparticles in combination with fluorescent active molecules can be used for combined optical and magnetic imaging (Perez, *et al.*, 2002).

10.8 Summary

Diagnosis of a disease in its very early stage can play important role in treatment.

Due to phenomenal advancement in nanotechnology, quantum dots have emerged as pivotal tool for detection of a particular biological marker with extreme accuracy. Quantum dots being very photo-stable and optically sensitive can be used as labeling and can be easily traced with ordinary equipment. Early detection of tumour markers using quantum dots is proved to be boon for cancer diagnosis. Use of QDs has also helped in unlocking complex neurological phenomenon, such as molecular activities at synapse during neurotransmission. QDs also give important information about receptor movement if tagged with suitable antibodies. In short, optical stability and easy to handle properties have made QDs to remain at the apex of medical diagnostics.

11

Nanomedicine: A Monstrous Entry of Multifunctional Nanoclinics into Realm of Healthcare

> *"Om Namo Bhagavate Maha Sudharshana Vasudevaya Dhanvantaraye; Amrutha Kalasa Hasthaaya Sarva Bhaya Vinasaya Sarva Roka Nivaranaya Tri Lokya Pathaye Tri Lokya Nithaye Sri Maha Vishnu Swarupa Sri Dharivantri Swarupa Sri Sri Sri Aoushata Chakra Narayana Swaha"* .
>
> (We pray to the God, who is known as Sudarshana Vasudev Dhanvantri. He holds the Kalasa full of nectar of immortality. Lord Dhanvantri removes all fears and removes all diseases. He is the well wisher and the preserver of the three worlds. Dhanvantri is like Lord Vishnu, empowered to heal the souls. We bow to the Lord of Ayurveda.)
>
> —**Prayer of Lord Dhanvantri (An Abode of Medicinal Knowledge)**

11.1 Introduction: A Paradigm Shift in the World of Medicine

Lord Dhanvantri is regarded as the God of *Ayurveda (The Science of Life),* overarching all the remedial mysteries existing in this world. The edifice of Ayurveda was built on the foundations of basic concepts which were impervious to change. Since every embodied individual is composed of a body, mind and soul, the ancient teachings were based on Ayurveda which is most concerned with the physical basis of life concentrating on its harmony of mind and spirit. *Charak Samhita* (a compendium of internal medicine) and *Sushruta Samhita* (a compendium of surgery) are the gifts from Lord Dhanvantri to mankind and the transmitter of this knowledge were Acharya Charak and Acharya Sushruta, respectively. One of the most important concerns of Ayurveda is *Noble Metal Bhasma* as well as *decoctions* and *extractions* of various medicinal herbs. It is a matter of adoration that the ancient teachers were not oblivious of *Nanoparticles in the form of Bhasma*. This ancient Indian wisdom was then disseminated to the whole world, giving rise to different branches of medicine like Allopathy, Homeopathy, Unani and Siddha. As time passed on, Greek philosophers like Hippocrates started using methods, such as bleeding, leeching and blistering from patients suffering from the disease. The disease was attributed to imbalance of four Humors (*i.e.,* blood, phlegm, black and yellow bile). Hippocratic doctrine involved balance of all the four humors using various drugs. Traditional herbal and medicinal therapies were also carried out by Tibbetan, Chinese and Japanese medical researchers after learning from Ayurveda. European scientist Descartes after referring to Charak and Sushruta Samhitas then gave an eloquent report in the form of Materia Medica. Harvey, Galen and Vesalius all gave a detailed account of anatomical and physiological survey of the human body. Edward Jenner's epoch-making demonstration of vaccination led to an inception of Preventive medicine. With Alexander Fleming, an era of antibiotics was started. There was a global resurgence in Allopathy which focused on a very prominent window of opportunity. It involved purified chemicals from plant extracts and antibiotics secreted from bacteria, fungi and Actinomycetes.

But epidemiological transition has led to the realization that the elimination of infectious diseases by antibiotics was hindered by the development of bacterial resistance and more importantly, non-communicable diseases, such as atherosclerosis, cancer and mental disorders which were multifactorial in origin and not amenable to "one antibiotic-one microbe approach". Furthermore, these multifaceted diseases need pertinent approaches for targeted delivery of drugs as they can have inimical effects on normal cells also. Against this background, several tenets and practices need exploration which can bid adieu to all the traditional practices of drug delivery. This has led to the global emergence of *Nanomedicine* which has a targeted approach like a "silver-bullet". This mode of drug-delivery is referred to as synaphic (docking-based) targeting. Engineered Nanopolymers can mimic a macrophage, possessing antibacterial granules within and exude it out specifically on the targeted bacteria in the body. Nanotechnology has also got a Theranostic approach which constitutes both diagnosis and Therapeutic measures. The nanostructural design involves conjugating Luminescent Quantum dots with nanomaterials and then linking this coupled complex to drugs and other drug-delivery functionalities for *in vivo* detection and targeting of diseased cells especially in the case of cancer.

What did defy the ravages of time was the knowledge of metallic Bhasma and herbal wealth, which has triumphantly reappeared again in the form of a more modernized buzzword known as *"Nanomedicine"*. It is a sobering thought that the knock of this technology on the doors of science is growing louder by the day.

11.2 Nanomedicine: Small World Big Hopes

Nanomedicine was first coined by Institute for Scientific Information, Thompson, Philadelphia, USA in the year 2000. Though Nanomedicine has become something more than a semantic fashion, giving a precise definition for it is really a difficult task with its blurred borderlines encompassing biotech and Microsystems Technology. The most acceptable definition is monitoring, repair, construction, and control of human biological systems at the molecular level, using engineered nanodevices and nanostructures. It involves nanoscale manipulation and assembly to applications at the clinical level of medical sciences. Although Nanotechnology has strengthened its roots and is an established discipline, nanomedicine is still at its infant stage of development. There is an utmost requirement of nanomedicinal products which can add to the armamentarium of therapeutics. Nanomedicine has got critical advantages over other drug delivery systems which are as follows:

(*i*) Poorly water-soluble drugs can be delivered effectively.

(*ii*) Targeted drug delivery is possible in a cell or a tissue specific manner and transcytosis of drugs across epithelial and endothelial barrier can occur with precision.

(*iii*) Intracellular drug delivery in a particular site of action is the biggest achievement.

(*iv*) Combinatorial therapies are also possible if required and their visualization can also be done by using imaging modalities.

(*v*) Real-time analysis is also possible for the comprehension of a therapeutic agent.

The biggest hurdle of many critical medications, hampering them from getting into the pipelines of pharmaceutical companies, is their insolubility as well as their complexity of regulatory processes. Many cardinal drugs fail to get into the market just because they are insoluble in nature. Nanomedicine is a vehicle with the help of which such insoluble drugs will get a way out to become soluble as well as be pumped into the pharmaceutical pipeline. Nanotechnology has now been embraced by various pharmaceutical companies which can surmount inherent solubility problems, targeted delivery to specific sites and bestow upon enhanced bioavailability.

11.2.1 Targeting Drugs Using a Nanovehicle: Is it a Necessity or Luxury?

There are certain important parameters for the development of a successful targeted drug delivery vehicle. (Gu, *et al.*, 2008) These include:

(*a*) The vehicle must be bio-compatible.

(*b*) The biophysicochemical properties of the vehicle must be favourable enough for effective bio-distribution and possible cell uptake.

(*c*) Scale-up operations of targeted drug delivery systems must be possible for clinical translation.

The biophysicochemical properties (such as size, charge, surface hydrophilicity) of the drug delivery vehicle and the nature of the ligand attached on the surface of the vehicle can have their impact on the circulating half-life of the particles as well as their biodistribution (Gu, *et al.*, 2008). The drugs attached to drug-delivery systems will lead to their interaction with a subset of cells; potentially enhancing cellular uptake by receptor-mediated endocytosis. The type of drug needs a critical consideration before choosing a targeted drug delivery system. The delivery of drugs whose site of action is somewhere in the intracellular region, will pertinently need an efficient targeted drug delivery system. In such cases, it is very important that the therapeutic drug bound to a vehicle may require homogeneous tissue penetration and optimum cellular uptake. In many cases, it is believed that the delivery system gets anchored to certain tissues thus decreasing the efficiency of diffusion and uniform tissue distribution. Thus, it is required that the ligand must be optimally bound to the vehicle facilitating balance between tissue penetration and cellular uptake, consequently leading to optimal therapeutic efficacy. The synaphic targeting of drug delivery systems allows receptor-mediated endocytosis of the ligand associated vehicle.

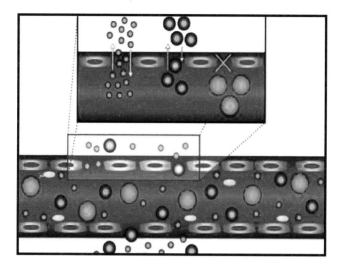

Fig. 11.1 Efficacy of nanoparticles as delivery vehicles is highly size and shape-dependent. The size of the nanoparticles affects their movement in and out of the vasculature, whereas the margination of particles to vessel wall is impacted by their shape.

Reprint from Impact of Nanotechnology on Drug delivery, Omid C. Farokhzad and Robert Langer, ACS Nano (Perspective), 3(1), 16–20, 2009 © American Chemical Society with permission from American Chemical Society (Courtesy of Omid C. Farokhzad and Robert Langer).

Though there are plethora of disease associated biomarkers and their ligands for targeted drug delivery applications, nanotechnology has first captured the most dreadful areas in Healthcare *i.e.*, Cancer and Neurological disorders. In fact, currently many nanotechnology therapeutic products have been approved for cancer therapy. There are many nanobased drug delivery systems for the treatment of cancer which piggyback many anti-cancerous drugs, accumulating through Enhanced Permeability and Retention effect (EPR effect) thus releasing their payload in the extra vascular tumour tissue for antitumour efficacy. Anti-cancerous drugs like methotrexate, Doxorubicin are cytotoxic to normal cells and hence, there is a need of targeted drug delivery system to surmount their inimical effects. Tumour tissue accumulation is a passive process requiring a long circulating half-life to facilitate time dependent extravasations of drug delivery systems through the leaky tumour microvasculature and accumulation of drugs in the tumour tissue. This process is largely mediated by the biophysicochemical properties of the nanoparticles and not by active targeting. Therefore, even in the absence of targeting ligands, drug delivery systems can be engineered to better target a particular tissue, or non-specifically absorbed by cells, by optimizing their biophysicochemical properties. However, once particles extravasate out of the vasculature into the tumour tissue, their retention and specific uptake by cancer cells is facilitated by active targeting and receptor-mediated endocytosis (Fig. 11.1). This process can result in higher intracellular drug concentration and increased cellular Cytotoxicity. While there is relatively modest improvement in tumour tissue accumulation of targeted drug delivery systems relative to non-targeted drug delivery systems, the difference in cellular Cytotoxicity is more pronounced. In the case of vascular endothelial targeting for oncology or cardiovascular indications, ligand-mediated targeting is of critical importance as tissue accumulation is not a function of EPR.

11.2.2 Nanoapproaches to Pharmaceuticals

Nanomedicine conventionally can be classified into two broad classes; "nanocarriers" and "nanoparticle engineering". Nanocarriers are those delivery systems (such as liposomes, dendrimers, polymeric nanoparticles) which can encapsulate, or allow an interaction with the active pharmaceutical ingredient (API)so that there is effective transfer of drugs in a particular target. Apart from polymer/dendrimer conjugation, these often require self-assembly of drug and carrier. Nanoparticle engineering aims to generate particles of API in the sub-micron range, often in the form of redispersible solids or dispersions in water. Successful micron-range particle engineering has produced many commercial oral medicines with enhanced bioavailability using attrition techniques, such as air-jet micronisation and milling. The current focus is to move even smaller to produce particles with diameters <500 nm. Nanomilling is widely considered as the "gold-standard" because it has led to successful new commercial products, which demonstrate increased dissolution kinetics, increased bioavailability, slow release and reduced dose.

This chapter focuses on applications in the cellular and intracellular delivery of therapeutic agents. We explore various types of nanoparticles ranging from polymers to liposomes, as well as current methodologies to develop inorganic nanoparticles. A brief discussion of the pharmacokinetic parameters and specific targeting strategies of these nanoparticles follows, presenting suggestions for the mechanisms of cellular and intracellular uptake. Since the remarkable drug delivery challenges in cancer and central nervous system's blood-brain barrier, illustrative examples of nanoparticles in the treatment of cancer, neurovascular disorders, and neurodegenerative diseases are provided.

11.3 Nanoparticles and its Amalgamation with Drugs for Drug-Delivery

11.3.1 Polymeric Nanoparticle

Nanomedicine formulation depends on the choice of suitable polymeric system having maximum encapsulation (higher encapsulation efficiency), improvement of bioavailability and retention time. Nanoformulations comprising of polymeric Nanoparticles and drugs are superior to traditional medicine with respect to controlled release, targeted delivery and therapeutic impact. These targeting capabilities of nanomedicines are influenced by particle size, surface charge, surface modification, and hydrophobicity. Among these, the size and size distributions of nanoparticles are important to determine their interaction with the cell membrane and their penetration across the physiological drug barriers. The utmost requirement of any drug delivery system is its size which can traverse through biological barriers and reach towards their target. Surface charge also plays critical role since they can conglomerate in blood and affect the Hemodynamic shear forces of blood or they can interact with oppositely charged cell membrane and deliver the cargo via internalization. Cationic surface charge is desirable as it promotes interaction of the nanoparticles with the cells and hence increases the rate and extent of internalization. For targeted delivery, persistence of nanoparticles is required in systemic circulation of the body. But conventional nanoparticles with hydrophobic surface are rapidly opsonized and massively cleared by the fixed macrophages of the mononuclear phagocytic system (MPS) organs. For increasing circulation time and persistence in the blood, surface of conventional nanoparticles are modified with different molecules. Coating of hydrophilic polymers can create a cloud of chains at the particle surface which will repel plasma proteins. Finally, the performance of nanoparticles *in vivo* is influenced by morphological characteristics, surface chemistry, and molecular weight. Surface modified nanoparticles have anti-adhesive properties by virtue of the extended configuration on the particle surface which acts as steric barrier reducing the extent of clearance by circulating macrophages of the liver and promoting the possibility of undergoing enhanced permeation process.

Release mechanism can be modulated by the molecular weight of the polymer used. Higher the molecular weight of polymer slower will be the *in vitro* release of drugs. Careful design of these delivery systems with respect to target and route of administration may solve some of the problems faced by new classes of active molecules (Kumari, *et al.*, 2010).

Polymeric nanoparticles have been synthesized using various methods (Pinto Reis, *et al.*, 2006) according to needs of its application and type of drugs to be encapsulated. These nanoparticles are extensively used for the nanoencapsulation of various useful bioactive molecules and medicinal drugs to develop nanomedicine are highly preferred because they show promise in drug delivery system. Biodegradable polymeric nanoparticles provide controlled/ sustained release property, sub-cellular size and biocompatibility with tissue and cells (Panyam, *et al.*, 2003). Apart from this, these nanomedicines are stable in blood, non-toxic, non-thrombogenic, non-immunogenic, non-inflammatory, biodegradable, avoid reticuloendothelial system and applicable to various molecules, such as drugs, proteins, peptides, or nucleic acids (Rieux, *et al.*, 2006). The general synthesis and encapsulation of biodegradable nanomedicines are represented in Fig. 11.2.

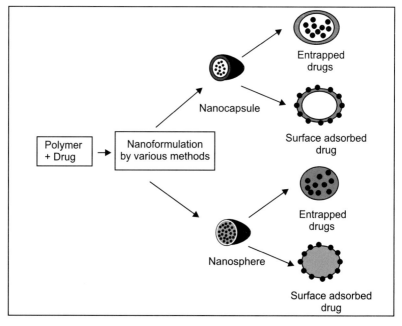

Fig. 11.2 Type of biodegradable nanoparticles: According to the structural organization, biodegradable nanoparticles are classified as nanocapsule, and nanosphere. The drug molecules are either entrapped inside or adsorbed on the surface.

Originally adapted from Tiyaboonchai, Chitosan nanoparticles: a promising system for drug delivery, Naresuan Univ. J. 11 (2003) 51.

The drug molecules either bound to surface as nanosphere or get encapsulated inside as nanocapsule. For the past two decades, countless work has been conducted to develop most effective nanomedicines from biocompatible and biodegradable nanopolymers (Kreuter and Kreuter, 1994). The administration, activity and therapeutic importance of some medicinal drugs on different nanosystems are different, for example taxol (anticancer drug) nanomedicine have 100% and 20% encapsulation efficiency on PLGA

(Mu, *et al.*, 2003) and PCL (Kim and Lee, 2001) nanodevices, respectively. However, the therapeutic activity and stability of PCL nanomedicines are reasonably higher than PLGA nanomedicine (Kim and Lee, 2001). The most commonly and extensively used polymeric nanoparticles are poly-dl-lactide-co-glycolide, polylactic acid, poly-caprolactone, poly-alkyl-cyanoacrylates, chitosan and gelatin.

Polymeric nanoparticles exhibit plethora of properties and proves its excellent candidature due to its biodegradability, biocompatibility, high entrapment efficiency and enhanced surface modifiable potency.

Furthermore, polymer-based coatings may be functionalized onto the other types of nanoparticles to change and improve their biodistribution properties. The biologically inert polymer polyethylene glycol (PEG) has been covalently linked onto the surface of nanoparticles. This polymeric coating is thought to reduce immunogenicity, and limit the phagocytosis of nanoparticles by the reticuloendothelial system, resulting in increased blood levels of drug in organs such as the brain, intestines, and kidneys. The US Food and Drug Administration (FDA) have approved biodegradable polymeric nanoparticles, such as PLA and PLGA, for human use. They may be formulated to encapsulate several classes of therapeutic agents including, but not limited to, low molecular weight compounds. (Panyam, *et al.*, 2004). Moreover, polymeric nanoparticles have been applied in gene therapy to breast cancer cells, resulting in antiproliferative effects. The polymer matrix prevents drug degradation and may also provide management of drug release from these nanoparticles. (Faraji, *et al.*, 2009).

11.3.2 Liposomes

The two legendary institutes *viz.*, Christian Dior Laboratories and Pasteur Institute in the year 1986 prepared a formulation using Liposomes, in collaboration with each other. Since Liposomes were considered to possess limited stability in blood due to reticuloendothelial clearance, they were further conjugated with Polyethylene Glycol to augment residence time in the blood. The liposome bilayer can be composed of either synthetic or natural phospholipids. The predominant physical and chemical properties of a liposome are based on the net properties of the constituent phospholipids including permeability, charge density and steric hindrance. Hydrophobic phosphate groups of phospholipids interact with water molecules leading to the spontaneous formation of a lipid bilayer known as liposome. There is plethora of drug-loading mechanisms, as follows 1) Liposomes can be formed in a drug saturated aqueous solution and encapsulation of lipophilic drugs is also possible for the same.

Table 11.1. Structure and Synthesis of Biodegradable Polymeric Nanoparticles

Polymer	Structure	Encapsulant & Encapsulation Efficiency	Synthesis Methods	Release Mechanisms	References
PLGA	x - Number of units of Lactic Acid y - Number of units of Glycolic Acid	Taxol and 100% EE	Solvent evaporation/solvent extraction technique	Diffusion, matrix swelling and polymer erosion	Mu and Feng, 2003 Interfacial deposition method
		Paclitaxel and > 90% EE	Interfacial deposition method	Dissolution and Diffusion	Fonseca, et al., 2002
		9-Nitrocamptothecin and 33%	Nanoprecipitation	Diffusion	Derakhshandeh, et al., 2007
PLA		Haloperidol and 30% EE	Solvent evaporative method	Diffusion	Budhian A, et al., 2005
		Zidovudine and 55% EE	Solvent evaporation method		Mainardes, et al., 2009
		Neurotoxin-1 and 35.5%	Double emulsion method	Brain delivery of NT-1 enhanced due to diffusion	Cheng, et al., 2008

Polymer	Structure	Encapsulant & Encapsulation Efficiency	Synthesis Methods	Release Mechanisms	References
Poly-ε-Caprolactone (PCL)		Tomoxifen and 90% EE	Solvent displacement	Preferential Tumour targeting and circulating drug reservoir	Shenoy, et al., 2005
Gelatin		Taxol and 20.79% EE	Micelles	Higher taxol loading and diffusion	Kim, et al., 2001
		Paclitaxel and 33–78% EE	Desolvation method	Diffusion	Lu, et al., 2004
Chitosan		Cyclosporin A and 73% EE	Ionic gelation method	Dissolution method	De Campos, et al., 2001

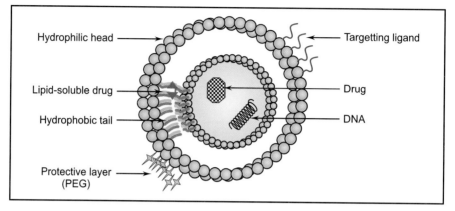

Hydrophilic head

Lipid-soluble drug

Hydrophobic tail

Protective layer (PEG)

Targetting ligand

Drug

DNA

Fig. 11.3 Diagram of a bilaminar liposome. The hydrophobic region traps drugs in the central core when the liposomes are prepared. The outer surface can be functionalized with ligands for active targeting or PEGylated. Liposomes can vary in the number of lipid bilayers they possess and can be classified into three categories: (*i*) multilamellar vesicles, (*ii*) large unilamellar vesicles and (*iii*) small unilamellar vesicles.

Liposomes generally reach their site of action by extravasation into the interstitial space from the bloodstream. Liposomes can target specific tissues through both active and passive targeting strategies (Figure 11.3). This is because liposomes can easily be manipulated by adding additional molecules to the outer surface of the lipid bilayer. Since liposomes are of the order of 400 nm in size, they are rapidly cleared by the mononuclear phagocytic system (MPS). Reducing opsonization of liposomes by PEGylation, therefore reduces clearance by the MPS, increasing the circulation half-life. Liposomal formulations of anticancer drugs have already been approved for human use. Doxill is a liposomal formulation of the anthracycline drug doxorubicin used to treat cancer in AIDS-related Kaposi sarcoma and multiple myeloma (Ning, *et al.*, 2007). Its advantages over free doxorubicin are greater efficacy and lower cardiotoxicity. These advantages are attributed to passive targeting of tumours, due to leaky tumour vasculature (Ogawara, K., *et al.*, 2008) and the EPR effect, and to lower concentrations of free doxorubicin at healthy tissue sites.

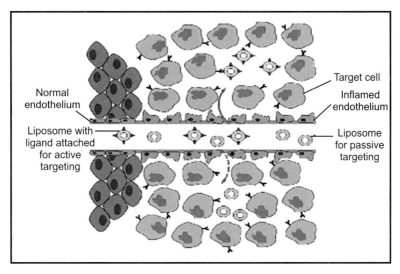

Fig. 11.4 Active and passive targeting of cells for drug targeting using liposomes. At sites of pathology where the endothelium layer is inflamed, mediators such as bradykinin, vascular endothelium growth factor and prostaglandins increase the endothelial permeability. Underlying pathology includes cancer, rheumatoid arthritis and infection. Liposomes extravasate through the gaps between cells and enter the interstitial fluid. Active targeting is achieved by conjugating ligands to the liposome that bind to a specific target cell receptor, leading to internalization or release of the drug. Passive targeting can be mediated by internalization or local high-concentration release of the drug.

Adapted from Partha Ghosha, Gang Hana, Mrinmoy Dea, Chae Kyu Kima and Vincent M. Rotello, Gold nanoparticles in delivery applications, (2008) Adv. Drug Deliv. Rev. 60, 1307–1315, 2008 © Elsevier Science Ltd. with permission from Elsevier Limited.

There is evidence that liposomal Doxill is metabolized by leukemia cells via a different mechanism than that for free doxorubicin, which might explain the improved efficacy and lower toxicity. Furthermore, Doxill is under clinical trial for the treatment of breast cancer. One of the most interesting developments in this field is the potential of liposomes to combat the increasing problem of multidrug resistance (MDR) acquired by cancers, which drastically reduces chemotherapeutic efficacy. Proposed mechanisms underlying MDR at the cellular level include: (*i*) increased metabolism of drugs due to increased enzyme expression, especially of glutathione S-transferase; (*ii*) drug transporters and efflux proteins and (*iii*) point mutations in proteins that are therapeutic or drug targets. Ogawara, *et al.*, (2009) investigated the effect of PEG liposomal doxorubicin (Doxill) in a male mouse tumour model inoculated with either colon cancer (C26) cells or their doxorubicin-resistant (MDR) subclone, which overexpresses P-gp efflux pumps. The results showed that PEG liposomal doxorubicin had anti-tumour effects on both doxorubicin-resistant and non-doxorubicin-resistant C26 cells. With increasing incidence of resistance to chemotherapy, the use of liposomes offers effective treatment without the need for the costly discovery of new chemotherapeutic drugs because current drugs can be reformulated. To date, no specific *in vivo* study has compared the efficacy of liposomes to that of other nanoparticle delivery systems; therefore, we cannot comment on the relative efficacy of liposomes. Liposomes are firmly established with the success of Doxill, and new liposomal formulations of other anticancer drugs are now being intensively explored to improve chemotherapy outcomes and reduce toxicity (Malam, *et al.*, 2009).

11.3.3 Solid Lipid Nanoparticles

Solid Lipid Nanoparticles (SLNs)are colloidal nanocarriers which have been proved to be an efficient pharmaceutical alternative to liposomes and emulsions. SLNs comprises of a rigid core of hydrophobic lipids such as mono, di, and triglycerides, fatty acids, waxes, which are more stable at room and body temperatures. They are comparatively even more stable than Liposomes. They are biodegradable and biocompatible with least toxicity, but needs stabilization by surfactants to synthesize stable administrable emulsions. They also have got a strong lipophilic scaffold allowing proper loading of drugs for efficient drug release.

The principal factors affecting drug loading into the SLN matrix are:

(*i*) The solubility of the drug in lipid (the drug must be lipophilic).

(*ii*) The chemical and physical properties of the lipid or mixture of lipids.

(*iii*) The crystalline characteristics of the lipid(s) at biological temperature.

(*iv*) The polymorphic form of the lipid(s) used. (Malam, *et al.*, 2009).

They have controllable pharmacokinetic parameters and can be engineered with three types of hydrophobic core designs: a homogenous matrix, a drug-enriched shell, or a drug-enriched core. Two primary production methods exist, including a high-pressure homogenization technique devised by Müller and Lucks, (1996) and a microemulsion technique pioneered by Gasco (1993). It has been demonstrated that the compound payload exits the hydrophobic core at warmer temperatures; conversely, the compound payload enters the hydrophobic core at cooler temperatures. These principles are used to load and unload solid lipid nanoparticles for the delivery of therapeutic agents, taking advantage of recent techniques to selectively produce hypo- and hyperthermia.

Use of a heterogeneous lipid mixture promotes an imperfect crystalline structure with larger gaps for superior drug loading. SLNs have been investigated for the delivery of various anticancer drugs with promising results in preclinical mouse trials specifically showing that SLNs might help to overcome MDR in cancers. Serpe, *et al.*, using colon cancer cells *in vitro*, demonstrated the benefits of SLNs in the delivery of doxorubicin, cholesteryl butyrate and paclitaxel.

In an exciting development, mitoxantrone, a topoisomerase inhibitor that blocks DNA replication, was loaded into SLNs and used *in vivo* as a local injection to treat breast cancer and lymph node metastases in mice. The results revealed a nearly three-fold reduction in lymph node size compared to free mitoxantrone, which is a significant improvement on the existing treatment. SLNs offer an alternative platform for drug delivery in cancer. However, more *in vivo* studies are required before they can be translated to human treatment.

11.3.4 Nanocrystals

Drug nanocrystals are crystals with a size in the nanometer range, which means they are nanoparticles of size 20–100nm with a crystalline character. A further characteristic is that drug nanocrystals are composed of 100% drug; there is no carrier material as in polymeric nanoparticles. Dispersion of drug nanocrystals in liquid media leads to so called "nanosuspensions" (in contrast to "microsuspensions" or "macrosuspensions"). In general, the dispersed particles need to be stabilized, such as by surfactants or polymeric stabilizers. Dispersion media can be water, aqueous solutions or non-aqueous media (*e.g.*, liquid polyethylene glycol [PEG], oils). A nanocrystalline species may be prepared from a hydrophobic compound and coated with a thin hydrophilic layer like surfactants. These surfactants allow a favorable biological interaction between nanocrystal and cells. This hydrophilic layer also exhibits high bioavailability and biodistribution preventing any agglomerates to be formed in blood, thus enhancing drug delivery efficiency. High dosages can be achieved with this formulation, and poorly soluble drugs can be formulated to increase bioavailability via treatment with an appropriate coating layer. Both oral and parenteral deliveries are possible, and the limited carrier, consisting of primarily the thin coating of surfactant, may reduce potential toxicity. A drawback, however, is that the stability of nanocrystals is limited. Moreover, this technique requires crystallization; some therapeutic compounds may not be easily crystallized.

The main reasons for the increased dissolution velocity and thus increased bioavailability are:

(*i*) Increase of dissolution velocity by surface area enlargement.

(*ii*) Increase in saturation solubility.

Nanocrystals have been proved to be an efficient candidate for water insoluble drugs. The key factors responsible for enhanced drug delivery and bioavailability of drugs are augmented particle surface, particle curvature, saturation solubility and enhanced dissolution velocity.

Nanocrystal technology offers great benefits and is ideally suited for drugs with solubility problems. Particle size diminution and the resulting increase in particle surface, curvature, saturation solubility, and consequently the increased dissolution velocity, are important factors. Solubility enhancement alone is not the only important factor. It becomes even more important when a drug has a narrow therapeutic window where it can be absorbed. In these cases, the increased solubility and dissolution velocity lead to an acceptable bioavailability. In addition, the nanocrystal technology enables formulations to be developed without the need of problematic surfactants (*e.g.*, Cremophor EL) which may cause enhanced side effects or adverse reactions. Furthermore, nanocrystals allow for a fast action onset, as the drug is absorbed quickly due to the fast dissolution of the nanoparticles. This is an advantage, especially for drugs which need to work fast (*e.g.*, naproxen for headache relief). As discussed the enhanced solubility also leads to an identical or very similar absorption in fed and fasted conditions. Drugs which normally require food to become soluble will be bioequivalent as nanocrystals in fed and fasted states. If it is necessary to give a large dosage in order to achieve reasonable blood levels for poorly soluble drugs resulting in increased side effects, the nanocrystal technology allows for smaller doses and thus decreased side effects. In addition, improved technologies are required to be able to produce tablets with high drug nanocrystal loads to formulate high dose drugs in–preferentially–one single tablet. In future, more drugs will be poorly soluble and thus require smart formulation technologies to make them soluble and bioavailable. An increased awareness in patients not willing to suffer from unnecessary side effects will lead to an increased number of products using nanocrystals to reduce these risks. By modifying the nanocrystal surface it is possible to achieve a prolonged or a targeted release.

11.3.5 Metallic Nanoparticles

(*a*) **Gold nanoparticles:** Metal nanoparticles have fascinated material scientists from ages, since they have unique size- and shape-dependent optical properties. The most ancient use of colloidal gold was done by Alchemists in Egypt, who skillfully exploited the brilliant colours of nanosized colloidal nanoparticles of Ag, Au and Cu. Faraday was the first to study bright colours of colloidal gold and Mie explained the origin of the phenomenon by using Maxwells' Electromagnetic equation when a photon of light interacted with a spherical nanoparticles. An electromagnetic field at a certain frequency induces a resonant, coherent oscillation of the metal free electrons across the nanoparticle, for all the spherical nanoparticles which are much smaller than the wavelength of light (diameter $d << \lambda$). This oscillation is known as the surface Plasmon resonance. The resonance lies at visible frequencies for the noble metals Au, Ag, Cu. The Surface Plasmon oscillation of the metal electrons results in a strong enhancement of absorption and scattering of electromagnetic radiation in resonance with SPR frequency of the noble metal nanoparticles, giving them intense colours and interesting optical properties. The frequency and cross-section of the SPR absorption and scattering is dependent on the metal composition, nanoparticle size and shape, dielectrics of the surrounding medium/substrate and presence of inter-particle interactions. Au and Ag are the plasmonic metals of choice due to their stability. Gold nanoparticles can be synthesized in a wide range of sizes and shapes (4–80 nm) by facile chemistry involving the reduction of Au ions in solution.

Gold nanoparticle is well-capable paraphernalia, useful in therapeutics as well as diagnostics. There are certain advantages of gold NPs to be used as contrast agents in Biomedicine:

1. Immuno-gold nanoparticles conjugated to antibodies and fab fragments have been used for biological labeling and staining in electron microscopy based on their charged properties. Gold nanoparticles are not susceptible to photobleaching and they appear biocompatible and non-cytotoxic. (Connor, 2005).

2. The facile conjugation and modification of Gold nanoparticle surface can be done by thiols, disulfides and amines due to their strong binding affinities towards the nanoparticles. Thus DNA, proteins and various peptides conjugate either by a naturally available thiol/amine group or

Table 11.2. Overview of current state of development of drugs using the Nanocrystal technology or others

Trade Name	Drug	Indication	Applied Technology	Company	Status
Rapammune	Rapamycin	Immunosuppressive	Nanocrystal elan	Wyeth	Marketed
Emend	Aprepitant	Anti-emetic	Nanocrystal elan	Merck	Marketed
Tricor	Fenofibrate	Hypercholesterolemia	Nanocrystal elan	Abbott	Marketed
Megace ES	Megestrol	Anti-anorexic	Nanocrystal elan	Par Pharmaceutical companies	Marketed
Triglide	Fenofibrate	Hypercholesterolemia	IDD-P Skyepharma	Scielepharma Inc.	Marketed
Nucryst	Silver	Anti-bacterial	Patented	Nucryst Pharmaceuticals	Phase II

synthetically incorporated thiol group. Surface modification with thiolated polyethylene glycol passivates nanoparticles and masks them from the intravascular immune system, allowing increased blood circulation times.

3. The phenomenon of Surface Plasmon Resonance makes gold nanoparticles significantly superior to the absorbing and fluorescing dyes conventionally used in biological and biomedical imaging.

The preparation of gold nanoparticles commonly involves the chemical reduction of gold salts in aqueous, organic, or mixed solvent systems. However, the gold surface is extremely reactive, and under these conditions aggregation occurs. To circumvent this issue, gold nanoparticles are regularly reduced in the presence of a stabilizer, which binds to the surface and precludes aggregation via favourable cross-linking and charge properties. Several stabilizers exist for passivation of the gold nanoparticle surface, including citrate, thiol-containing organic groups, encapsulation within microemulsions, and polymeric coatings. In particular, gold nanoparticles may be encrusted with biomolecules, with exciting prospects in biological sensing and imaging. Several synthetic strategies exist, such as the two phase liquid-liquid method initially described to create metal colloidal suspensions by Faraday in 1857. Faraday reduced an aqueous gold salt with phosphorous in carbon disulfide to obtain a ruby-coloured aqueous suspension of colloidal gold particles. The Brust–Schiffrin method further optimized this two phase liquid-liquid system with gold salts being transferred from water to toluene using tetraoctylammonium bromide as the phase transfer reagent, with reduction by aqueous sodium borohydride in the presence of dodecanethiol (Brust, *et al.*, 1994). Using modifications of this method, gold nanoparticles have been synthesized with numerous biomolecular coatings (Fabris, *et al.*, 2006; *Higashi*, 2005). The resulting gold nanoparticles have biological applications; for instance in the detection of polynucleotides via hybridization to oligonucleotides appended on the nanoparticle surface (Elghanian, *et al.*, 1997).

(*b*) **Iron oxide nanoparticles:** Paul Ehrlich's famous movie Dr. Ehrlich's magic bullet mentioned about a saga of a bacterium killed by a toxin attached bullet. This scientific fiction was brought into reality in 1960s when magnetic microparticles were exploited as contrast agents for localized radiation therapy and to induce vascular occlusion of the tumours (antiangiogenic therapy). In 1976, magnetic erythrocytes were used for the delivery of cytotoxic drugs. Magnetic albumin microspheres were also exorbitantly used to encapsulate an anticancer drug (doxorubicin) in animal models. Later biodegradable poly-lactic acid microspheres were used which incorporated magnetite and the beta-emitter 90Y for targeted radiotherapy and successfully applied to subcutaneous tumours. But all the above approaches were microsized. Later it was proposed that magnetic nanoparticles could be transported through the vascular system and concentrated in a specific part of the body with the aid of a magnetic field. Magnetic nanoparticles were also used for the first time in animal models by Lubbe, *et al.*, (1996) and the first Phase I trial was also carried out by the same group in patients with advanced and unsuccessfully pretreated cancers using magnetic NPs loaded with epirubicin. Chemicell GmbH currently commercializes Target MAG-doxorubicin NPs involving a multidomain magnetic core and a cross-linked starch matrix with terminal cations that can be reversibly exchanged by the positively charged doxorubicin. The particles have a hydrodynamic diameter of 50 nm and are coated with 3mg/ml doxorubicin. These magnetic NPs loaded with mitoxantrone have already been used in animal models with successful results.

The main advantage of magnetic (organic or inorganic) NPs is that they can be:

(*i*)   Visualized (super magnetic NPs are used in MRI).

(*ii*) Guided or held in place by means of a magnetic field.

(*iii*) Heated in a magnetic field to trigger drug release or to produce hyperthermia / ablation of a tissue.

Based on a synthetic procedure magnetic nanoparticles or nanocapsules can be obtained. Magnetic Nanoparticles are mentioned to be covalently bound to drug via entrapment or are adsorbed within the pores of the magnetic carrier (polymer, mesoporous silica *etc.*). Nanocapsules (reservoirs) designate magnetic vesicular systems where the drug is confined to an aqueous or oily cavity, usually prepared by reverse micelle procedure and surrounded by magnetoliposome.

The very critical and pertinent parameter of magnetic nanoparticles are related to surface chemistry, size, magnetic code, hydrodyanamic volumes, size distribution and magnetic properties (magnetic moment, remanance, coercivity). The surface chemistry of magnetic nanoparticles is pertinent enough so that it can avoid the reticuloendothelial system evading immune surveillance mechanism and augment the half life in the blood stream. Surface coating of nanoparticles with a neutral and hydrophilic compound *i.e.*, polyethylene glycol, polysaccharides, dysoponins (HSA) *etc.*, helps in augmenting the circulatory half-life from minutes to days. The most effective drug-delivery system can be designed using super paramagnetic iron oxide Nanoparticles (SPION) in conjunction with magnetic fields.

Magnetic nanoparticles for biomedical applications must be endowed with the specific characteristics required and the first requirement is often super-paramagnetism. Super-paramagnetism occurs in magnetic materials composed of very small crystallites (threshold size depends on the nature of the material for instance, Fe based Nanoparticles become super-paramagnetic at size < 25 nm (Lee, *et al.*, 1996). In a paramagnetic material, the thermal energy overcomes the coupling forces between neighbouring atoms above the Curie temperature, causing random fluctuations in the magnetization direction that result in a null overall magnetic moment. However in super-paramagnetic materials, the fluctuations affect the direction of magnetization of entire crystallites. The magnetic moments of individual crystallites compensate for each other and the overall magnetic moment becomes null.

When an external magnetic field is applied, the behaviour is similar to paramagnetism except that, instead of each individual atom being independently influenced by an external magnetic field, the magnetic moment of entire crystallites aligns with the magnetic field. In large nanoparticles energetic considerations favour the formation of domain walls. However, when the particle size decreases below a certain value, the formation of domain walls becomes unfavourable and each particle comprises a single domain. This is the case for super-paramagnetic nanoparticles. Super-paramagnetism in drug delivery is necessary because once the external magnetic field is removed, magnetization disappears (negligible remanence and coercivity) and thus agglomeration and the possible embolization of capillary vessel is avoided.

Another limitation relates to the small size of Nanoparticles, a requisite for super-paramagnetism, which is in turn needed to avoid magnetic agglomeration once the magnetic field is removed. A small size implies a magnetic response of reduced strength, making it difficult to direct particles and keep them in the proximity of the target, while withstanding the drag of blood flow. Targeting is likely to be more effective in regions of slower blood velocity, and particularly when the magnetic field source is close to the target site.

In addition, coatings play an essential role in retarding clearance by the RES. Depending on their size, surface functionalization and hydrophilicity, a rapid uptake of uncoated nanoparticles by the mononuclear phagocytic system (MPS) is likely after systemic administration, followed by clearance to the liver, spleen and bone marrow. Different proteins (antibodies) of the blood serum (opsonins) bind to the surface of foreign bodies, accelerating phagocitation of the particles. To avoid this, biodegradable (dextran) and non-biodegradable organic and inorganic coatings can be used as a means to retard detection and uptake by macrophages of the RES.

11.4 Pharmacokinetics of Nanocargoes

11.4.1 Distribution and Clearance

Trojan Nanoparticles conjugated with drug and stealth endowing hydrophilic PEG need proficient natural clearance and excretion mechanism. As soon as a pharmaceutical agent is administered intravenously, it is distributed systemically via the vascular and lymphatic systems. The distribution of a drug in any tissue is correlated with the relative amount of cardiac output passing through that tissue. Hence tissues or organs having very high blood flow (brain, Liver, Heart, intestines, lungs, kidneys, spleen *etc.*) may possess high concentrations of a drug provided that the drug is able to permeate from the vasculature. Cardiac output can therefore act as a filter to nanomaterial distribution. Similarly, passive targeting mechanism involves altering the size of the nanoparticle carrier, thus altering the biodistribution profile of the nanomaterials. The 1–20 nm sized nanomaterials have long circulatory residence times and slower extravasation from the vasculature into the interstitial spaces, (Winter, *et al.*, 2003) causing slower attainment of the maximal volume of distribution or even an altered volume of distribution when administered intravenously. The optimum size of nanoparticles is 30–100 nm is which would be sufficient to avoid leakage into capillaries, but also small enough to avoid reticuloendthelial clearance. Surface manipulating agents can further control the extent of localization at interstitial sites and limit clearance.

Hydrophilic pegylated nanomaterials, endow stealth properties to it, circumventing the problem of fast clearance from the vascular system. This augments the residence time of the nanomaterial in vascular system as well as in a particular tissue, thus emerging to be circulating depots of drug. The biodistribution properties of the drug ultimately depend on the kinetic of payload movement from the nanoparticle carrier; fast loss of the drug payload before the nanoparticle reaches its target may result in decreased drug efficacy. In spherical nanoparticles, the surface area to volume ratio is inversely proportional to the radius. Thus if internally loaded nanoparticles become smaller, a greater proportion of the drug payload will be located on the surface and have access to the exterior aqueous phase, thus leading to significant alterations in the pharmacokinetic parameters.

The most dominant and proficient mechanism of biodistribution of nanoparticles is endothelial disruption or alteration. Inflammation, solid tumours and disrupted endothelia lead to an increased leakiness that provides vascular contents greater access to extravascular targets. Rapid angiogenesis due to outgrowth of tumour thus leads to leaky and highly fenestrated endothelial cells and enhanced permeation and retention (EPR) of the drug.

Blood brain barrier may be weakened by solid tumours such as *Glioblastoma multiforme*, thus providing better distribution of therapeutic agents to the CNS and tumour. Nanoparticles of size 100–150 nm in diameter will tend to accumulate in tumours due to their poor extravasation from normal vasculature. The presence of disturbed porous vascular beds at the tumour allow for selective targeting by this passive mechanism.

After a long discussion on passive targeting mechanism for effective biodistribution of nanoparticle conjugated drug, let's discuss certain examples of active targeting. Comprehension of specific cell markers, ligands, linkers and various surface proteins are required for the effective exploitation of active targeting mechanism. The surface of nanomaterials can be ligated to a peptide, an antibody or an aptamer for proper targeting of a nanoparticle. All the above Trojan horse strategies are useful for proper penetration and distribution of nanoparticle ligated drugs surrounding a target. Larger polymeric microparticles of 60 μm size get collated at a particular site, while 100–200 nm sized nanoparticles get biodistributed efficiently. When microparticles of size range 25–250 μm are injected peritoneally in mice, they remain therefore at least 2 weeks. In contrast, nanoparticles of the same material exhibited almost complete clearance from the peritoneum in the same time frame (Kokahe, *et al.*, 2002 and 2006). Further, mice treated with nanoparticles exhibited enlarged spleen and foamy macrophages presumably resulting from accumulation of a large amount of polymeric material, thus obviating the role of reticuloendothelial system in removing foreign objects from biological milieu. As part of the immune responses, monocytes and macrophages readily absorb circulating nanomaterials and then accumulate in lymph nodes and spleen for further processing.

11.4.2 Toxicity

Nanoparticles architecture and composition influence systemic toxicity enormously. The size of nanoparticles should be such that they should not easily traverse the normal vasculature, but permeates only the damaged/disrupted vascular system as in the case of tumour. Moreover, particles > 5 μm in diameter may cause embolization of blood vessels, similarly < 100nm particle size may lead to aggregation, thus forming a cluster that can embolize and occlude blood flow. This property has been exploited to intentionally occlude the vasculature of tumour, such as in the case of transarterial (hemoembolization of hepatocellular carcinoma). But alternatively, undesired consequences may also result including lodging of these aggregates in various organs, causing pulmonary embolism, strokes, myocardial infarctions and other microinfarctions at distant sites and organs. Thus, it becomes very pertinent that nanoparticle administration should not result in aggregation and consequently leading to embolic phenomena.

Nanoparticles when injected in the neurons may directly stimulate alterations in the central nervous system and cardiovascular autonomic function.

De Lorenzo, *et al.*, (1970) administered intranasally colloidal gold nanoparticles (approximately 50 nm) in squirrel monkeys which were translocated anterogradely in the axons of the olfactory nerves to the olfactory bulbs (Lorenzo, *et al.*, 1970). The movements of these 50nm gold nanoparticles were traced and was found to be traversing synapses to the olfactory glomerulus within 1 hr. of intranasal administration with a calculated neuronal transport velocity of 2.5 mm/hr. It was later found that the gold nanoparticles in the olfactory bulb were not freely distributed in the cytoplasm; instead they were preferentially located in mitochondria and got accrued over there. But this pathway appears to circumvent the BBB and may be exploited as a delivery alternative for drugs and nanoparticles that are otherwise unable to breach the Blood Brain Barrier. It is likely that nanoparticles when administered cause inflammation of the olfactory mucosa, olfactory bulb, cortical and subcortical regions of the brain.

In general, nanoparticles may trigger an inflammatory process resulting in the release of cytokines and chemokines such as IL-6, IL-1β, TNF-α, reactive Oxygen species, C-reactive protein and transcription factors. This cascade results in the activation of mitogen-activating protein kinase (MAPK), redox sensitive transcription factors, nuclear factor kappa-B (NF-κ B) and activating protein-1. As a result, we can give a concluding remark that nanoparticles may promote, if not trigger, low level systemic inflammation at distant organs and tissues depending on nanoparticles access to the vasculature via penetration of small blood vessels and capillaries.

11.5 Fundamentals of Targeting a Diseased Cell

A pharmaceutical agent has to face multilayered obstacles from the site of administration to the targeted intracellular structures. Many a times the agent is lost due to ineffective partitioning across the biological membranes. Polar compounds tend to distribute irregularly, while non-polar or lipophilic compounds easily bypass this obstacle with greater membrane permeation generally via diffusion. The condition becomes more complicated when myriad of other cellular processes like endocytosis mechanisms, intracellular trafficking, release of the therapeutic agent into the cytoplasm, diffusion, translocation of the agent to its susceptible target and thus partitioning into nucleus or other organelles alter and mask the activity of the therapeutic agent from its biological environment.

But nanoparticles pose an ambrosial opportunity, masking the drug and protecting it from the influence of biological milieu so that it can reach to the intracellular target. The properties and surface characteristics of the nanoparticle play a greater role in compound delivery at the intracellular sites.

Endocytosis is the only modus operandi operating for a biological cell for the ingestion of nanoparticles. They are further of 3 types:

(*a*) Phagocytosis is a process in which ingestion of materials up to 10 μm in diameter can be accomplished by few cells, such as macrophages, neutrophils and dendritic cells.

(*b*) Pinocytosis is an uptake mechanism which is virtually all cell types. Hence, larger micro particles can be engulfed by phagocytic cells and smaller nanoparticles are engulfed by all cells.

(*c*) Receptor mediated endocytosis is the most efficient mechanism for selective cellular targeting. The cell-surface is crowded by myriads of receptors, which when bound to extracellular ligands then collate other accessory molecules on the surface which then transduce the extracellular signals to the intracellular space. This signal actuates plethora of biochemical pathways inside the cell, but it may also cause internalization of the ligand and its conjugated nanoparticles via endocytosis. Internalization is mediated by clathrin coats which generate a membrane indentation with a radius of curvature as small as approximately 50nm and invaginate further upon binding of the ligand. Interaction of receptors via ligands attached to nanoparticles result in a more pronounced membrane crater with subsequent enfolding and reunification of the cellular membrane forming an endosome. Nanoparticles of size range between 25 and 50nm are a prerequisite for optical endocytosis and intracellular localization (Aoyama, *et al.*, 2003; Nakai, *et al.*, 2003; Chithrani, *et al.*, 2006; and Jiang, *et al.*, 2008). Furthermore, selective active targeting of nanoparticles to specific tissues may take advantage of the differential expression of receptors between cellular types.

11.6 Nanoparticle Drug Conjugates Delivery for Human Therapeutics

Jeopardy at the molecular level of a cell cannot be prevented, since it gives rise to many complexities and affects many pathways multifactorially. Drugs having molecular actions need to be efficiently targeted as they act like a double-edged sword. They can affect normal cells also which can cause mayhem in the normal functioning of the body, thus leading to many side-effects. Nanotechnology is a weapon which can hammer specifically on the target due to its size and synaphic qualities. This makes it efficient paraphernalia for the treatment of most dreadful and devastating diseases like Cancer and Neurological disorders which will be in priori of therapeutics in near future.

11.6.1 Nanobullets and Cancer

Cancer research has sprung with rich and complex body of information with its revelation of dynamic changes in the blueprint of life *i.e.,* genome. It has now become a justified fact that cancer is caused due to mutations that produce oncogenes with dominant gain of function and tumour suppressor genes with recessive loss of function. Cancer cell genome is a manifestation of six essential alterations in cell physiology that collectively dictate malignant growth:

1. Self-sufficiency in growth signals
2. Insensitivity to growth inhibitory signals
3. Evasion of programmed cell death (apoptosis)
4. Limitless replicative potential
5. Sustained angiogenesis
6. Tissue invasion and metastasis.

All the above manifestations breach, all the immune surveillance mechanism hardwired into cells and tissues.

Normal cells require growth signals before they can move from a quiescent state (G0) into an active proliferative state. Such signals are transduced into the cell by transmembrane receptors which bind specific signaling ligands, such as diffusible growth factors, extracellular matrix components and cell-to-cell adhesion/ interaction molecules. Most of the Oncogenes in the cancer catalog act by mimicking normal growth signaling in one way or another. Normal cells are completely dependent on extraneous signals for proliferation, whereas in cancer cells, it has been found that they exhibit reduced dependence on the exogenous growth signals. This liberation from dependence on exogenously derived signals disrupts a critically important homeostatic mechanism that normally exists ensuring a behaviour of the various cell types within a tissue.

Similarly, all normal cells proliferate and expand in number in a tightly regulated fashion but if the number exceeds then the cell possesses another mechanism of cell attrition via Programmed cell death. A body of knowledge explains that Apoptotic program is in its latent form in all cells, but once actuated by a variety of physiological signals, this program unfurls a precisely choreographed series of steps. The internal cellular organelles are blebbed and fragmented, cytoplasmic and nuclear skeletons are destroyed, all in a span of 30–120 min. Such shriveled cell corpse is then chewed upon by phagocytic cells typically within 24 hrs. Apoptotic machinery can be divided into two classes of components–sensors and effectors. The sensors are responsible for policing the intra and extracellular milieu for detection of any abnormality which then transduce the signal to certain effectors of apoptotic death. This can be exemplified by the binding of TNF-alpha to TNF receptor or binding of Fas Ligand on Fas receptors on the cell surface which act as sensors and then the signals are transduced to the effectors of the cell, such as caspase 8 and caspase 9, and many other proapoptotic factors (Bax, Bid and Bim). There are many drugs which induce Apoptosis in normal cells as well as Cancerous Cell, but a cancerous cell exhibits unique and distinct identity by over-expression of folate receptors, or certain tumour specific cell surface marker. Antibodies against such sentinels can lead to targeted delivery of certain drug delivery systems which transport drugs to such tumorigenic sites and induce apoptosis. Nanoparticles of size 100–150 nm can be engineered so that they can reach to such sites either passively or actively.

Another mechanism of survival of a cancerous cell is sustained angiogenesis. For a normal cellular functioning and survival, oxygen and nutrients supplied by the vasculature are crucial, obligating virtually all cells in a tissue to reside within 100 um of a capillary blood vessel. During organogenesis, this closeness is ensured by coordinated growth of vessels and parenchyma. Once a tissue is formed, the growth of new blood vessels—the process of angiogenesis—is transitory and carefully regulated. Due to this dependence on nearby capillaries, it would seem plausible that proliferating

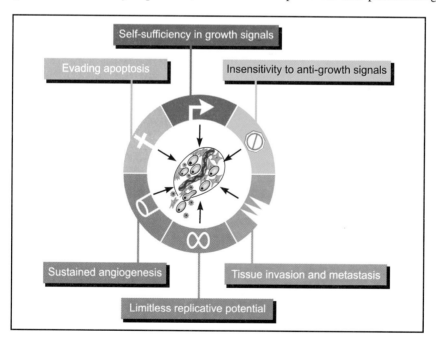

Fig. 11.5 Acquired Capabilities of Cancer.

Reprinted from The Hallmarks of cancer, D. Hanahan and R. Weinberg, Cell, 100, 57, 2000 © 2000 Elsevier Science Ltd. with permission from Elsevier Limited (Courtesy of Robert Weinberg).

cells within a tissue would have an intrinsic ability to encourage blood vessel growth. But in the case of a tumorigenic tissue, the whole story is different. Induction of angiogenesis for the fulfilment of augmented oxygen and nutrient requirement becomes imperative in such a defective and aberrant tissue. Such incipient neoplasias secrete angiogenesis-initiating signals like VEGF (Vascular Endothelial Growth Factor). Hence, to inhibit angiogenesis, the most critical weapon is anti-VEGF which can again be targeted using Nanocontainers (Hanahan and Weinberg, 2000).

11.6.1.1 *Nanoparticles Hitting the Culprit: Maiming the Cancerous Cell*

The most important thing to be noted is that all anti-cancer drugs after administration must be able to reach their target *i.e.,* tumour tissues through the penetration of barriers in the body with minimal loss of their concentration and activity in the blood circulation. Secondly, they must specifically maim the target tumour cells and unload its toxic content in it without affecting the normal cells. Both these characteristics can be accomplished with the help of nanoparticles due to their *Size and Surface Characteristics* (Moghimi, *et al.,* 2001) which can be explained in details as follows:

(*a*) **Size:** One of the advantages of nanoparticles is that their size is tunable. The size of nanoparticles used in a drug delivery system should be large enough to prevent their rapid leakage into blood capillaries, but small enough to escape capture by fixed macrophages that are lodged in the reticuloendothelial system, such as the liver and spleen. The size of the sinusoid in the spleen and fenestra of the Kupffer cells in the liver varies from 150 to 200 nm and the size of gap junction between endothelial cells of the leaky tumour vasculature may vary from 100 to 600 nm. Consequently, the size of nanoparticles should be up to 100 nm to reach tumour tissues by passing through these two particular vascular structures.

(*b*) **Surface characteristics:** In addition to their size, the surface characteristics of nanoparticles are also an important factor determining their life span and fate during circulation relating to their capture by macrophages. Nanoparticles should ideally have a hydrophilic surface to escape macrophage capture. This can be achieved in two ways: coating the surface of nanoparticles with a hydrophilic polymer, such as PEG, protects them from opsonization by repelling plasma proteins; alternatively, nanoparticles can be formed from block copolymers with hydrophilic and hydrophobic domains.

11.6.1.2 *Passive Targeting by Nanoparticles*

Smaller the better, is the *mantra* for effective drug delivery and this endeavour is fulfiled by nanoparticles which can easily extravasate out through the leaky vasculature of tumours. The vascular system of tumour microenvironment has a gap of 100–600 nm between endothelial cells. This has also resulted in enhanced permeability and retention of nanoparticles at such tumour environments. The passive targeting of nanoparticles is thus possible and the mechanism of which can be elucidated as follows:

(*a*) **Enhanced permeability and retention effect:** Nanoparticles of size 50–200nm have got the ability to escape the reticuloendothelial system capture and can stay in blood circulation for longer time, thus augmenting the chance of reaching the targeted tumour tissues. Similarly, all the macromolecules like Nanoparticles get accumulated in tumour tissues using the same mechanism. Unregulated growth of cancerous cells increases their population in a particular area which causes deficiency of oxygen and nutrients. This leads to the release of angiogenic factors, such as VEGF (Vascular Endothelial Growth factor) by cancerous cells. This release is caused by an imbalance of angiogenic regulators, such as growth factors and matrix metalloproteinases which makes tumour vessels highly disorganized and dilated with numerous pores showing enlarged gap junctions between endothelial cells and compromised lymphatic drainage. These features are called the enhanced permeability and retention effect, which constitutes an important mechanism by which macromolecules, including nanoparticles, with a molecular weight above 50 kDa, can selectively accumulate in the tumour interstitium.

(*b*) **Tumour micro-environment:** The inimitable micro-environment surrounding the tumour cells also play a pertinent role in passive targeting of nanoparticles. Fast-growing, hyper proliferative cancer cells show a high metabolic rate, and the supply of oxygen and nutrients is usually not sufficient for them to maintain this. Therefore, tumour cells use glycolysis to obtain extra energy, resulting in acidic environment. There is a need for the development and design of pH-sensitive nanoparticles which are stable at a physiologic pH of 7.4, but are degraded to release active drug in target tissues in which the pH is less than physiologic values, such as in the acidic environment of tumour cells. In general, the tumour interstitial compartment is characterized by large interstitial space, high collagen concentration, low proteoglycan and hyaluronate concentrations, high interstitial fluid pressure and flow, absence of anatomically well-defined functioning lymphatic network, high effective interstitial diffusion coefficient of macromolecules, as well as large hydraulic conductivity and interstitial convection compared to most normal tissues (Jain, 1987).

11.6.1.3 *Active Targeting by Nanoparticles*

Passive targeting of nanoparticles has certain limitations in the sense that they can affect normal cells also in and around a tumour tissue. Secondly, they cannot distinguish between a normal cell and transformed cell. This makes the whole procedure completely non-specific. Thus, such a drug-delivery system is required which can be conjugated to antibodies and then targeted specifically to the cancerous cell. Earlier researchers used myriad of mechanisms for drug targeting. One of them was conjugating drugs with antibodies, but preservation of immune recognition was limited. The recent development and introduction of a wide variety of liposomes and polymers as drug delivery carriers increases the potential number of drugs that can be conjugated to targeted nanoparticles without compromising their targeting affinity relative to earlier antibody-drug conjugates. Taking advantage of this, many recently developed active targeting drug conjugates use a ternary structure composed of a ligand or antibody as a targeting moiety, a polymer or lipid as a carrier, and an active chemotherapeutic drug. When constructing ternary structure of nanoparticles, some factors must be considered to create more efficient delivery systems:

(*a*) **Antigen or receptor expression:** Ideally, cell-surface antigens and receptors should have several properties that render them particularly suitable tumour-specific targets. *First*, they should be expressed exclusively on tumour cells and not expressed on normal cells. *Second*, they should be expressed homogeneously on all targeted tumour cells. Last, cell-surface antigens and receptors should not be shed into the blood circulation.

(*b*) **Internalization of targeted conjugates:** Whether targeted conjugates can be internalized after binding to target cells is an important criterion in the selection of proper targeting ligands. Internalization usually occurs via receptor-mediated endocytosis. Using the example of the folate receptor, when a folate-targeted conjugate binds with folate receptor on the cell surface, the invaginating plasma membrane envelopes the complex of the receptor and ligand to form an endosome. Newly formed endosomes are transferred to target organelles. As the pH value in the interior of the endosome becomes acidic and lysosomes are activated, the drug is released from the conjugate and enters the cytoplasm, provided the drug has the proper physico-chemical properties to cross the endosomal membrane. Released drugs are then trafficked by their target organelle depending on the drug. Meanwhile, the folate receptor released from the conjugate returns to the cell membrane to start a second round of transport by binding with new folate-targeted conjugates.

11.6.1.4 *Therapeutic Application of Ligand-Targeted Nanoparticles*

Conventional chemotherapy has its own limitation in the treatment of cancer since many drugs have poor specificity in reaching tumour tissue, and are often restricted by dose-limiting toxicity. A combo of controlled release technology

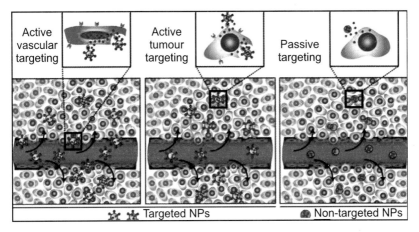

Fig. 11.6 Passive *vs.* active targeting (Right) Particles tend to passively extravasate through the leaky vasculature, which is characteristic of solid tumours and inflamed tissue, and preferentially accumulate through the EPR effect. In this case, the drug may be released in the extracellular matrix and diffuse throughout the tissue for bioactivity. (Middle) Once particles have extravasated in the target tissue, the presence of ligands on the particle surface can result in active targeting of particles to receptors that are present on target cell or tissue resulting in enhanced accumulation and cell uptake through receptor-mediated endocytosis. This process, referred to as "active targeting", can enhance the therapeutic efficacy of drugs, especially those which do not readily permeate the cell membrane and require an intracellular site of action for bioactivity. (Left) The particles can be engineered for vascular targeting by incorporating ligands that bind to endothelial cell-surface receptors. While, the presence of leaky vasculature is not required for vascular targeting, when present as is the case in tumours and inflamed, this strategy may potentially work synergistically for drug delivery to target both the vascular tissue and target cells within the diseased tissue for enhanced therapeutic.

Reprint from Impact of Nanotechnology on Drug delivery, Omid C. Farokhzad and Robert Langer, ACS Nano (Perspective), 3(1), 16–20, 2009©American Chemical Society with permission from American Chemical Society (Courtesy of Omid C. Farokhzad and Robert Langer).

and targeted drug delivery is a humble solution to conventional chemotherapy. There is enormous interest in developing nanodrug delivery system which can store chemotherapeutic agents and can directly unload its piggyback into a cancerous cell after maiming it with the help of monoclonal antibodies or Aptamers (Aptamers are DNA or RNA oligonucleotides or modified DNA or RNA oligonucleotides that fold by intramolecular interaction into unique conformations with ligand-binding characteristics) (Wilson and Szostak, 1999) or peptides or simply folic acid. Controlled drug release occurs when polymeric nanoparticles can encapsulate the drug and release it as and when the optimal concentrations of the drug decrease in the tumour micro-environment. The release of the drug may be constant or may be triggered by the environment or other external events, such as targeting of the ligand and the receptor, thus increasing the efficacy of the drug and maximizing patient compliance. Several classes of nanomaterials have been used for controlled release and targeting the drug effectively against the tumour tissue like Polymeric nanoparticles and Liposomes. Metallic nanoparticles like Fe_3O_4 and Gold nanoparticles have been used for the lysis of cancerous cells due to their thermal activity.

(*a*) **Polymeric nanoparticles:** Natural polymers such as albumin, chitosan and heparin can be effectively used as nanoscale drug-delivery systems. The treatment of metastatic breast cancer involves Nanometer-sized serum albumin conjugated with Paclitaxel (Abraxane). The conjugate has arrived in the clinics for the treatment (Gradishar, *et al.*, 2005). Among synthetic polymers, such as N-(2-hydroxypropyl)-methacrylamide copolymer (HPMA), polystyrene-maleic anhydride copolymer, polyethylene glycol (PEG), and poly-L-glutamic acid (PGA). PGA was the first biodegradable polymer to be used for conjugate synthesis. Several representative chemotherapeutics that are used widely in the clinic have been tested as conjugates with PGA *in vitro* and *in vivo* and showed encouraging abilities to circumvent the shortcomings of their free drug counterparts (Li, 2002). Among them, Xyotax (PGA-Paclitaxel) (Sabbatini, *et al.*, 2004) and CT-2106 (PGA-camptothecin) (Bhatt, *et al.*, 2003) are now in clinical trials.

In another instance, for effective treatment of cancer, a combinational therapy of conventional chemotherapeutic drug and anti-angiogenesis agent was modelled. But this faced two obstacles. The first one was the long-term shut down of tumour vasculature by the anti-angiogenic agent, thus preventing the tumour from receiving a therapeutic concentration of the drug. The second hurdle was that inhibiting blood supply drives the intra-tumoural accumulation and over expression of hypoxia inducible factor-1a (HIF1a), thus causing increased tumour invasiveness and resistance to chemotherapy. (Sengupta, *et al.*, 2005). This led to the engineering of a drug-delivery system, also known as a nanocell which circumvents both these barriers exclusively found in solid tumours. This nanocell is composed of a nanoscale pegylated-phospholipid block-copolymer conjugated to an anti-angiogenic agent combretastatin-A4 envelope coating a nuclear Poly Lactic-co-Glycolic acid nanoparticle conjugated to doxorubicin (*see* Fig. 11.7). Such a complex strategic nanocell was engineered due to sequential exposure of cytotoxic agents after a vascular shutdown induced by anti-angiogenesis therapy. This is beneficial as the suppression of tumour growth requires long-term administration of angiogenesis inhibitors and short treatment cycles of chemotherapeutic drug. Both angiogenesis inhibitors and chemotherapeutic drug will be released sequentially and the concentrations of drug can also be maintained due to its controlled release from the inner PLGA shell.

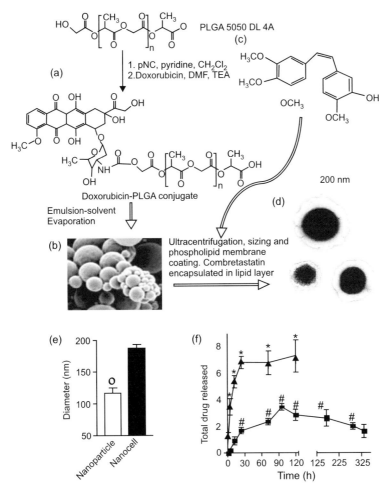

Fig. 11.7 Synthesis and characterization of a combretastatin—doxorubicin nanocell. (*a*) Diagram of conjugation reactions between doxorubicin and PLGA 5050. DMF, dimethylformamide; pNC, p-nitrophenyl-chloroformate; TEA, triethylamine. (*b*) Scanning electron micrograph of heterogeneous nanoparticles. (*c*) Combretastatin is encapsulated in the lipid envelope. (*d*) Transmission electron micrograph of the cross-section of three nanocells shows the dark nuclear nanoparticle within the phospholipid blockcopolymer envelope. (*e*) Dynamic light scattering shows that nanoparticles of defined sizes were used for encapsulation within the envelope. (*f*) Physicochemical release kinetics shows the temporal release of combretastatin (triangles, scale in 102 mg) and doxorubicin (squares, scale in mg). Results are means ± s.e.m. (n = 4). The error is small where not visible. Asterisk, $P \le 0.002$; hash, $P \le 0.001$.

Reprinted from Shiladitya Sengupta 1*, David Eavarone 1*, Ishan Capila 1, Ganlin Zhao 1, Nicki Watson 3, Tanyel Kiziltepe 2 & Ram Sasisekharan 1, Temporal targeting of tumour cells and neovasculature with a nanoscale delivery system, Nature 436, 568–572(2005), by permission from Nature, ©2005 *Nature Publishing Group.*

(*b*) **Liposomes:** Liposomes are composed of phospholipid moieties, mimicking host cells which can prevent its removal from the blood circulation due to RES. Further they can be pegylated so that their half-life in blood circulation increases. Such liposomes can be used to load anti-cancer drugs within it and can be targeted using monoclonal antibodies against cancerous cells. Researchers have successfully exploited nanoliposomes with pegylation to load and target anti-cancer drugs like doxorubicin (Markman, 2006).

(*c*) **Gold nanoparticles:** As already discussed, Gold nanoparticles exhibit surface Plasmon resonance and also possess the property of binding antibodies. Both these properties make gold nanoparticles as paraphernalia for drug delivery. Gold nanorods have two Plasmon absorption peaks, and the position of the second peak can be moved deep into the near infrared region just by controlling the aspect ratio of the nanorod. Such gold nanorods are well-capable enough to absorb infrared radiation and cause localized heating and can consequently lead to thermal destruction of cells. Hence, such nanorods when injected in the body and irradiated with infrared or near infrared light, (This is useful because body tissue is moderately transparent to NIR light) can cause cell destruction due to localized heat. The first account of the use of gold particles in hyperthermal therapy was published in 2003 (Hirsch, *et al.*, 2003).

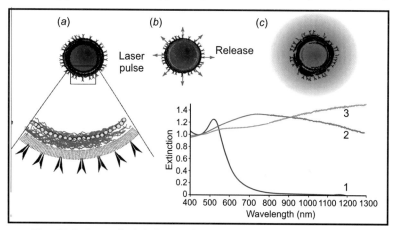

Fig. 11.8 Controlled delivery of a chemical payload. The antibody-functionalized polymer capsule and application of laser light causes the gold nanoparticles in the shell of the capsule to heat up (b), bursting the container and releasing the therapeutic agent in a localized area (c). Representative optical extinction spectra are shown for isolated gold nanoparticles (1), a poly-electrolyte shell containing a single layer of aggregated gold particles (2), and a shell containing three layers of gold particles (3).

Reproduced with permission from Radt, B. (2004) Optically addressable nanostruc-tured capsule. Adv. Mater. 16, 2184–2189
© 2004 *Wiley.*

In this, and follow-up work by Halas and co-workers, gold-silica nanoshells were used to target breast carcinoma cells, actively, using the HER2 antibody (Loo, *et al.,* 2004). In the Halas study, NIR irradiation using a near infrared laser led to a rise in the temperature of the target regions of between 40 to 508C, which selectively destroyed the carcinomas. The survival rate of mice treated in this manner was excellent compared with the controls. The anticancer drugs can also loaded on gold nanoparticles and then after irradiation by a laser can release it in the body.

11.6.2 Nanoscale Drug Delivery and Central Nervous System

Brain is an organ which is seated inside a protective cranium possessing various centres such as pain, vision, odour, fear, logical reasoning, memory and behaviour *etc*. It also comprises of neurons and circuits in a number of discrete anatomical locations in the brain. These systems subserve one of three general functions.

(*a*) Sensory systems represent information about the state of the organism and its environment.

(*b*) Motor systems organize and generate actions.

(*c*) Associational systems link the sensory and motor sides of the nervous system, providing the basis for "higher-order" functions such as perception, attention, cognition, emotions, rational thinking, and other complex brain functions that lie at the core of understanding human beings, their history and their future.

Blood Brain Barrier is a unique membranous permeability barrier that tightly segregates the brain from the circulating blood. Capillaries of the vertebrate brain and spinal cord lack the small pores that allow rapid movement of solutes from circulation into other organs; these capillaries are lined with a layer of special endothelial cells that lack fenestrations and are sealed with tight junctions. This permeability barrier, comprising the brain capillary endothelium, is known as the BBB. Ependymal cells lining the cerebral ventricles and glial cells are of three types:

1. Astrocytes form the structural frame work for the neurons and control their biochemical environment. Astrocytes foot processes or limbs that spread out and abutting one other, encapsulate the capillaries are closely associated with the blood vessels to form the BBB.

2. Oligodendrocytes are responsible for the formation and maintenance of the myelin sheath, which surrounds axons and is essential for the fast transmission of action potentials by saltatory conduction.

3. Microglias are blood derived mononuclear macrophages.

The tight junctions between endothelial cells result in a very high trans-endothelial electrical resistance of 1500–2000 $\Omega.cm^2$ compared to 3–33 $\Omega.cm^2$ of other tissues which reduces the aqueous based Para-cellular diffusion that is observed in other organs. Therefore, only lipid-soluble solutes that can freely diffuse through the capillary endothelial membrane may passively cross the BBB. In capillaries of other parts of the body, such exchange is overshadowed by other non-specific exchanges. Despite the estimated total length of 650km and total surface area of 12 m^2 of capillaries in human brain, this barrier is very efficient and makes the brain practically inaccessible for lipid—insoluble compounds, such as polar molecules and small ions. As a consequence, the therapeutic value of many promising drugs is diminished, and cerebral diseases have proved to be most refractory to therapeutic interventions. Given the prevalence of brain diseases alone, this is a considerable problem.

The BBB is further reinforced by a high concentration of P-glycoprotein (Pgp), active drug-efflux-transporter protein in the luminal membranes of the cerebral capillary endothelium. This efflux transporter actively removes a broad range of drug molecules from the endothelial cell cytoplasm before they cross into the brain parenchyma.

11.6.2.1 Nanotherapy for Neurological Cancers (Glioblastoma Multiforme)

Brain tumours, especially malignant gliomas, belong to the most aggressive human cancers. Despite numerous advances in neurosurgical operative techniques, adjuvant chemotherapy and radiotherapy, the prognosis for patients remains very unfavourable. Even the chemotherapeutic drugs most effective in glioblastoma, such as nitrosoureas, platinum compounds and temozolomide, increase the survival time of patients only marginally. Features responsible for the aggressive character of glioma include rapid proliferation, diffuse growth and invasion into distant brain areas in addition to extensive cerebral edema and high levels of angiogenesis. Glioblastomas develop a distinct neovasculature that is highly permeable to macromolecules and small particles. However, disruption of the blood-brain barrier (BBB) remains a local event, which is evident in the tumour core, but absent at its growing margins. Accordingly, therapeutic drug levels have been found in necrotic tumour areas, while in peritumoral regions the drug levels were markedly lower or non-detectable. Confirming the importance of the BBB, improved survival of patients with malignant gliomas was reported when systemic chemotherapy was applied after hyperosmotic or chemical disruption of the BBB. Therefore, the development of a strategy allowing drug delivery across the BBB is of prime importance and would offer a possibility to use highly active antitumour agents that are usually not employable in the CNS due to effective entrance block by the BBB. The feasibility of this approach to CNS chemotherapy was demonstrated by a number of authors who showed that brain delivery of anticancer drugs, such as the anthracyclines could be achieved using sterically stabilized colloidal carriers such as liposomes or solid lipid nanoparticles. One of the most likely candidates for CNS chemotherapy, doxorubicin, has proven to be effective in glioblastoma cell lines, but lacks the ability to cross the BBB, being a P-glycoprotein substrate. However, a significant increase in survival rate was achieved in patients with malignant gliomas treated with intratumoral injections of doxorubicin, while clinical trials demonstrated that after i.v. injection, doxorubicin did not reach cytotoxic levels in glioma tissue due to delivery problems. Some types of liposomes are capable of crossing the BBB. However, this capacity is dependent on their size and structure. Whereas the large vesicles, which enter the brain only via breaches of the BBB, are retained in glial tumours, small vesicles, which may cross the BBB unrestrictedly, are not specifically enriched in tumours and confer a risk of uncontrolled embolization. In addition, the application of liposome-bound doxorubicin enabled only short-term relief to glioblastoma patients. Therefore, researchers used polysorbate-80 coated solid-lipid butylcyanoacrylate for the delivery of doxorubicin. Polysorbate-80 specifically binds apolipoprotein E of Low density lipoproteins in blood which then bind to LDL receptors present in Brain. Such polysorbate-80 coated nanoparticle-conjugated drug can easily be delivered in the brain. Doxorubicin levels were augmented via such mechanism and apoptosis levels increased in Glioblastoma. (Steiniger, *et al.*, 2004)

11.6.2.2 *Nanomedicine for Neurodegenerative Diseases*

Dementia is a syndrome characterized by failure of recent memory and other intellectual function that is usually insidious in onset but steadily progresses. Alzheimer's disease (AD) is the most common dementia, accounting for 60–80% of cases in the elderly. It afflicts 5–10% of the population over the age of 65, and as much as 45% of the population over 85. The earliest sign is typically an impairment of recent memory function and attention, followed by failure of language skills, visual-spatial orientation, abstract thinking, and judgment. Inevitably, alterations of personality accompany these defects. The tentative diagnosis of Alzheimer's disease is based on these characteristic clinical features, and can only be confirmed by the distinctive cellular pathology evident on post-mortem examination of the brain. The histopathology consists of three principal features (illustrated in the Fig. 11.9):

(1) Collections of intraneuronal cytoskeletal filaments called *neurofibrillary tangles*.

(2) Extracellular deposits of an abnormal protein in a matrix called amyloid in so-called *senile plaques*.

(3) A diffuse loss of neurons.

These changes are most apparent in neocortex, limbic structures (hippocampus, amygdala, and their associated cortices), and selected brainstem nuclei (especially the basal forebrain nuclei). Although, the vast majority of AD cases arise sporadically, the disorder is inherited in an autosomal dominant pattern in a small fraction (less than 1%) of patients. Identification of the mutant gene in a few families with an earlyonset autosomal dominant form of the disease has provided considerable insight into the kinds of processes that go awry in Alzheimer's.

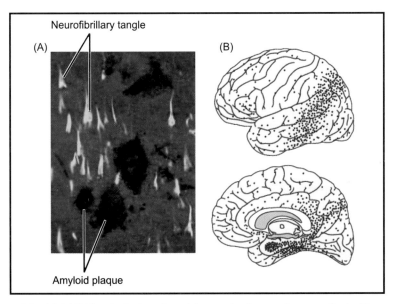

Fig. 11.9 (*A*) Histological section of the cerebral cortex from a patient with Alzheimer's disease, showing characteristic amyloid plaques and neurofibrillary tangles. (*B*) Distribution of pathologic changes (including plaques, tangles, neuronal loss, and gray matter shrinkage) in Alzheimer's disease. Dot density indicates severity of pathology.

A from Roses, 1995, courtesy of Gary W. Van Hoesen; B after Blumenfeld, 2002, based on Brun and Englund, 1981.

The prominence of amyloid deposits in AD further suggested that a mutation of a gene encoding amyloid precursor protein is somehow involved. The gene for amyloid precursor protein (APP) was cloned by D. Goldgaber and colleagues, and found to reside on chromosome 21. The mutant genes underlying two additional autosomal dominant forms of AD have been subsequently identified (*presenilin 1* and *presenilin 2*). Thus, mutation of any one of several genes appears to be sufficient to cause a heritable form of AD.

Like in AD, the symptoms of Parkinson's disease appear gradually but are unique and dependent on the affected brain subregion. They include difficulties in maintaining balance and in ambulation; tremors; inflexibility/stiffness of the limbs and trunk; and bradykinesia (slowness of movement). PD is defined pathologically, by the loss of dopaminergic neurons (dopamine producing neurons) of the substantia nigra pars compacta (SNpc) and subsequent loss of striatal dopamine projecting to the caudate-putamen. Neuronal damage caused by neurotoxic factors initiated from inflammatory responses by immune activated glia are linked to cognitive and motor deterioration, which contribute to the breakdown of the BBB. This allows leukocytes entry into the brain serving to speed a neuroinflammatory cascade. Although the causes of both AD and PD remain unknown, patterns of familial inheritance suggest a possible connection involving abnormal protein processing (Ab for AD and alpha synuclein for PD) and accumulation. After heart disease and cancer within the developed world, stroke is the third largest killer, second most common cause of neurologic disability after AD. There are over five million deaths a year from stroke and over nine million stroke survivors. Between the ages of 45 and 85, 20–25% of men and women, respectively, can expect to have at least one incident of stroke. Atherosclerosis, heart disease, hypertension, diabetes, and life-style habits are risk factors correlated with disease. Unlike AD and PD, many of the risk factors for developing this disease can be modified. The etiology of stroke is brain vascular occlusions (thrombotic stroke) or rupture (hemorrhagic stroke). The neuropathological hallmarks of stroke are necrotic infarcts of variable size with inflammatory gliosis.

Drugs are also becoming available serving to stimulate neuronal repair and differentiation. However, each of these therapeutic modalities shows limitations due to toxicity and BBB penetration. Improving the effectiveness of CNS treatment outcomes through targeted delivery of enzymes and genes into brain cells is just now being developed. This may be achieved through the CNS implantable biocapsules and micro-electro-mechanical system approaches or using nanomedicine approaches. Alzheimer's disease (AD) is marked by a progressive and irreversible damage to memory, thought, and language. It is extremely prevalent and represents the most common form of dementia in geriatric populations over 65 years of age. Current therapies include acetylcholinesterase inhibitors, cholinesterase inhibitors, antioxidants, amyloid-â-targeted drugs, nerve growth factors, γ-secretase inhibitors, and vaccines against amyloid-β. Polymeric nanoparticles may represent a potential means to transport drugs across the BBB. May be designed to mimic LDL and interact with the LDL receptor, consequently triggering uptake by brain endothelial cells.

To facilitate transport through the BBB via the LDL mechanism, the nanoparticles may be further functionalized with apolipoproteins.

11.6.2.3 *Nanocontainers Targeted to Brain*

The therapeutic value of Nanoparticles with its efficacy in crossing the BBB and enhancing the concentrations of the drug in the central nervous system is of paramount importance. The different types of Nanocontainers used for targeting the Brain and delivering the drug are as follows:

(*a*) **Liposomes:** Liposomes are readily taken up by macrophages, microglia and astrocytes in the CNS. PEGylated liposomes accumulate more rapidly in brain when the BBB is compromised such as in experimental autoimmune encephalomyelitis (EAE) (Schmidt, 2003). Interestingly, brain accumulation of liposomes labeled with radioactive isotope, 99mTc is increased during EAE. PEGylated liposomes coupled with monoclonal antibodies to glial fibrillary acidic protein (GFAP), an antigen expressed in astrocytes show altered brain penetrance. Incapable of penetrating a normal BBB, immunoliposomes used to treat glial brain tumours that express GFAP can reach their disease site when the BBB is partially permeabilized (Chekhonin, 2005). The mechanism(s) of drug accumulation in Brain during diseased condition may involve enhanced permeability and retention (EPR) of circulating liposomes at sites of disease-induced BBB compromise. Liposomes have also been conjugated with mannose, transferrin and insulin receptors at the surface of brain capillaries (Pardridge, 1999). In particular, transferrin receptor is necessary to deliver iron across BBB. Expression of this receptor in BBB increases during certain pathologies, for instance after stroke. Hence, transferrin-conjugated liposomes successfully targeted post-ischemic brain endothelium in rats.

(*b*) **Polymeric nanoparticles:** Nanoparticles can be used for drug delivery (Roney, *et al.,* 2005; Liu, *et al.,* 2005). Poly (butylcyanoacrylate) nanoparticles were coated with PEG-containing surfactants, such as Tween 80 and after injection, they localized in the choroid plexus, via ventricles, and, to a lower extent, in the capillary endothelial cells. It has been suggested that increased brain delivery with surfactant-coated poly (butylcyanoacrylate) nanoparticles may be associated with non-specific permeabilization of BBB and toxicity [JC Olivier, *et al., Pharm Res*, 16, 1836, (1999)]. Drugs delivered to CNS in such nanoparticles included, analgesics (Dalargin, Loperamide), anti-cancer agents (Doxorubicin), anti-convulsants (NMDA receptor antagonist, MRZ 2/576), and peptides (Ddalargin and Kytorphin) (Kreuter, *et al.,* 2003 and Steiniger, *et al.,* 2004). In one study, Doxorubicin-laden nanoparticles increased survival in rats with aggressive glioblastoma (Steiniger, *et al.,* 2004).

Carboxylated polystyrene nanospheres (20 nm) were also evaluated for CNS drug delivery (Yang, *et al.*, 2004). After intravenous injection such nanospheres remained in the vasculature under normal conditions. However, they extravasated into brain during cerebral ischemia-induced stress that partially opened the BBB (Kreuter, 2001). Such nanospheres may have potential for CNS delivery of drugs and imaging agents during ischemia, stroke and other conditions that disrupt the BBB.

11.7 Summary

Nanomedicine will be at its helm in near future for therapy of all pathological sufferings of human beings. There are myriad of nanomaterials including polymeric nanoparticles, iron oxide nanoparticles and gold nanoparticles which can be easily synthesized and exploited as drug delivery piggybacks. The pharmacokinetic parameters of these nanoparticles may be altered according to size, shape, and surface functionalization. Such nanoparticles can be effectively engineered for synaphic targeting to specific sites. Along with targeting they can also evade the clearance mechanism of reticuloendothelial system. They can also be used to alter the kinetic profiles of drug release, leading to more sustained release of drugs with a reduced requirement for frequent dosing. Especially interesting applications of Nanomedicine in drug delivery relate to cancer and Neurological disorder. Due to their small size, they can easily target a miscreant cell and unload the drugs within them. They can also breach the Blood Brain Barrier and establish a new frontier for neuropharmacologic agents.

Nanotechnology for Environmental Development

> "Nanotechnology a tool for sustainable development rather than an environmental liability".
>
> —Dr. Vicki Colvin

12.1 Introduction

The microtechnology of the second half of the 20th century has produced a technical revolution that has led to the production of computers and the internet, and eventually lead us into a dynamic emerging era of *nanotechnology*. Any technology when it is viewed through the eyes of an environmentalist, it passes through a scrutinized assay of its applications as well as hazards. Environmental technologies at the nanoscale are no exception to it. It is strongly believed that nanotechnology could substantially enhance the environmental qualities and sustainability through pollution prevention, treatment and remediation. The present day life style is also a source of introducing nanoparticles in the atmosphere. It is demonstrated that a large proportion of nanoparticles in engine exhaust form by condensation of semi-volatile vapours during the dilution of the exhaust gases with colder ambient air.

It cannot be denied that proliferation of nanotechnology can lead to problems like cytotoxicity of nanoparticles, generation of a new class of toxins and environmentally harmful substances. Gustafson, *et al.*, (2006e) have shown reduction of NO_2 to nitrous acid on illuminated titanium dioxide aerosol surfaces.

However, it must be mentioned here that there are some naturally occurring environmental nanoparticles which play a key role in important chemical characteristics and the overall quality of natural and engineered waters. Aquatic nanoparticles have the ability to influence environmental and engineered water chemistry and processes in a much different way than similar materials of larger sizes. Environmental nanoparticles are commonly formed as either weathering by-products of minerals, as biogenic products of microbial activity or as growth nuclei in super-saturated fluids. Environmental nanotechnologies have the potential to contribute to sustainable development and protection of the environment.

Application of environmental technologies demands (*a*) Remediation, (*b*) Prevention, (*c*) Maintenance and (*d*) Enhancement.

12.1.1 Remediation

Various applications of nanotechnologies for environmental remediation have been successful at laboratory scale. However, many of them still require scaling up and their safety as well as efficacy verification. Nanotechnology and environmental remediation includes:

(*i*) Access to clean drinking water using nano particles.

(*ii*) Removal of air borne pollutants.

(*iii*) Cleaning up of industrial pollution.

(*iv*) Solution for ground or soil contamination including ground water.

(*v*) Green chemistry using mesoporous catalyst—Mesoporous and microporous materials are being developed for green chemistry and catalytic applications, including selective catalysis, photocatalysis, membranes and energy applications. These materials are functionalized by depositing catalysts inside their pores. Examples include silica based mesoporous molecular sieves modified with titanium for 4-chlorophenol decomposition, or with manganese for gas-phase oxidation reactions with propene.

The reactivity of nanoparticles is important with toxic contaminants is of chief interest for environmental remediation *e.g.,* abiotic and biotic processes contribute to sequestration or transformation of such compounds. Mayo, *et al.,* (2007) have shown that 12 nm particles of magnetite are 200 times more efficient at removing As(III) and As(V) from water than 20 and 300 nm particles. Magnetite is also important for the degradation of organic contaminants. For example, Vikesland and Rebodo (2007) showed that magnetite reactivity with carbon tetrachloride is not only size-dependent, 9 nm particles were more reactive than 80 nm particles. Moreover, Vikesland, *et al.*, also showed a pronounced effect of nanoparticle aggregation. Nanoparticle aggregation, affects the reactive surface area of the component nanoparticles. Determining the effect of aggregation on other systems is very important in understanding how environmental nanoparticles behave in aquatic systems. Recent studies by Chen, *et al.*, (2007), of the aggregation kinetics of alginate-coated hematite provide some of the first steps towards understanding the reactivity consequences of environmental nanoparticle aggregation.

12.1.2 Prevention

Dendrimers (*see* chapter 1) are tiny branching molecules. *California Institute of Technology* is developing dendrimers that can bind to insidious water pollutants, including carcinogenic perchlorate, thus preventing the contamination. Another effort that is being made by the scientists at *Iowa State University* along with the U.S. Department of Energy and Eaton Corporation is on special nanoboride coatings for hydraulic pumps. Such Nanocoatings will drastically reduce friction, mining pumps could work harder and save power consumption.

12.1.3 Maintenance

Nanopaint is an effort of European scientists to maintain the environmental conditions. Already nanopaints have been developed and are in the market. Self-cleaning coatings for glass, called Nano Titanium Dioxide paint, protects the surfaces from mildew, pollutants and allergens. Better environment can be maintained by decontamination. Decontamination, commonly described as the removal or destruction of toxic substances, is an important aspect of operations in a chemically or biologically contaminated environment. The ultimate decontaminant is a product that will destroy all toxic contaminants without any adverse effects on surfaces or the environment.

12.1.4 Enhancement

Nanotechnologists are focusing their attempts to develop new products and materials; by changing or creating materials at the atomic and molecular level; to enhance the quality of environment. For example, nanosilver particle which is among the most exploited nanoparticles; is being used to kill germs in shoe liners, food-storage containers, air fresheners, washing machines and other about 200 products.

12.2 Nanotechnological Approaches for Better Environment

Nanotechnology can contribute to improve the environment by providing effective monitoring, detection and measurement of pollution, pollution prevention and remediation methods. Various approaches that have been used include:

12.2.1 Photocatalysis

What is photocatalysis? A semiconductor, upon absorption of a photon, acts as a catalyst in producing reactive radicals, mainly hydroxyl radicals. These radicals, in turn, can oxidize organic compounds and completely mineralize them. In photocatalysis, the organic molecules are decomposed to form carbon dioxide, water and mineral acids as final products. Photocatalytic treatment of polluted water is a more efficient method than conventional treatment methods. In conventional method, pollutants are transferred from one medium to another. The oxidative power of the hydroxyl radical is more than twice that of chlorine, thus its potential for oxidizing pollutants that normally are hard to destroy, like halogenated organics, surfactants, herbicides, and pesticides, to carbon dioxide.

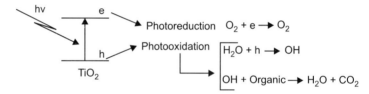

Fig. 12.1 Photocatalytic reactions of TiO_2 nanoparticles.

Titanium dioxide is considered to be the most competently employed semiconductor photocatalytic molecule which can be exploited for the treatment of polluted water. The reason behind using such molecular machine for the water treatment purpose is that TiO_2 is nontoxic, photoreactive and photostable in nature. This photoreactivity leads to artificial generation of photons thus can be used for water treatment purpose. Solar energy has got an important role to play in photocatalysis in the sense that solar light has got a flux of UV light (20 to 30Wm^{-2}) near the surface of the earth. 0.2 to 0.3 mol photons of light from the Sun in the narrow range of 300 to 400nm at the process disposal. These high-energy photons can catastrophically demolish all the pollutants in water, air or on photocatalytically coated materials.

When TiO_2 is illuminated with light of $\lambda < 400$ nm, an electron is promoted from the valence band to the conduction band of the semiconducting oxide to give an electron/hole pair, *i.e.*, the valence band potential is positive enough to generate hydroxyl radicals at the surface and the conduction band potential is negative enough to reduce molecular O_2. The hydroxyl radical is a powerful oxidizing agent and attacks organic pollutants present at or near the surface of the TiO_2 resulting in their complete oxidation to CO_2.

There is burgeoning interest in photocatalytic destruction of pollutants and microorganisms. Electrochemically assisted photocatalysis(EAP) is the most important method used to break down organic molecules in water. Atrazine is an example of organic pollutants which is used as an herbicide and are commonly found in surface and groundwater, which are photocatalytically destroyed. Its persistence is purely due to the existence of s-atriazine ring which limits biodegradation.

Rickerby and Morrison have reported in Report 101 from the work shop on nanotechnologies for environmental remediation by JRC Ispra (207) that a method is being developed for sol-gel synthesis of nanosized TiO_2 particles on polypropylene fibres for the photocatalytic degradation of organic pollutants.

Extensive work of Orlov, *et al.*, (2004, 2006a and 2007) has shown that modification of TiO_2 with nano gold increases the activity of titania for a variety of catalytic processes on the efficiency of decomposition of various organic pollutants. TiO_2 where metal particle sizes < 5 nm give rise to unique physical and chemical properties. However, the effect of Au nanoparticle size on photo-degradation organics in solution is largely unknown. They have designed a large scale chemical reactor and used it to clean up contaminated groundwater at an air force base in Canada.

For photocatalytic degradation of air pollutants, a combination of solar light with photocatalysts *i.e.,* N-doped TiO_2 has been tried (Orlov, *et al.*, 2006 b and c).The advantage of visible light active materials compared to UV light active materials is higher efficiency of energy utilization, as the solar spectrum contains only a small UV component. These materials can also utilize artificial illumination, commonly in use inside the buildings.

Byrne, *et al.*, (2002 and 2007) immobilized the nanoparticles of TiO_2 onto supporting substrates *i.e.*, glass, ITO glass and titanium metal, using dip coating or, in the case of electrically conducting supports, an electrophoretic coating procedure. Immobilized catalyst films were then tested for their efficiency in the photocatalytic degradation of organic contaminants (17β-oestradiol, atrazine, and formic acid as a model pollutant) as well as microbiological pathogens (*E. coli*, *Clostridium perfringens*, and *Cryptosporidium parvuum*). It was found that photocatalysis was effective for the destruction of the parent compounds *i.e.,* 17β-oestradiol, oestriol and oesterone.

PICADA an European industry has practically shown that Titanium dioxide (TiO_2) can degrade air pollutants, especially Nitrogen Oxide (NO), by photocatalytic mechanism.

Rawat, *et al.*, (2006) have synthesized a composite nanoparticle of 10–15 nm; consisting of anatase-titania photocatalytic shell and nickel ferrite magnetic core. They have shown that the photocatalytic shell of titania is responsible for the photocatalytic and anti-microbial activity and nickel ferrite magnetic core is responsible for the magnetic behaviour, studied by superconducting quantum interference device. The anatase TiO_2 coated $NiFe_2O_4$ nanoparticles retains the magnetic characteristics of uncoated nanocrystalline nickel ferrites, super-paramagnetism (absence of hysteresis, remanence and coercivity at 300 K) and non-saturation of magnetic moments at high field. The magnetic measurements result encouraged their application as removable anti-microbial photocatalysts. Moreover, they found that bacterial inactivation with UV light in the presence of titania-coated $NiFe_2O_4$ nanoparticles is faster than the action with UV light alone.

Like TiO_2, Zinc Oxide (ZnO) has photocatalytic characteristics. It is known that ZnO is an n-typed semiconductor with band-gap energy of ~3.2 eV and thus only absorbs UV light with the wavelength $\leq$ 385 nm. However, as far as photocatalytic efficiency and practical applications are concerned, it is desirable that ZnO absorbs not only UV, but also visible light. Zhang, *et al.*, (2005) could synthesize ZnO nanoparticles by a novel dc thermal plasma reactor. This dc plasma affects the nanoparticles morphology and N_2 plasma favours formation of the spherical nanoparticles. Visible light absorption of the ZnO nanoparticles was achieved by doping it with few thousands ppm of nitrogen into the material; resulted into photocatalytic characteristics of the ZnO nanoparticles.

Mesoporous materials for green chemistry and catalytic applications: Mesoporous materials have pores in the range 2–50 nm diameter. They have very high surface areas for adsorption and reactions. Apart from catalysis, they are also used in separation and energy applications. The surfaces of silica based sieves (SBA-15) were modified with manganese and titanium have exhibited significant activity for oxidation of organic pollutants in both air and water Orlov, 2006d)

Photocatalytic killing of microbes: The anatase type TiO_2 absorbs photons in the UV range of the solar spectrum exciting the valence electrons and generating the Electron-Hole Pairs (EHPs). These EHPs then recombine and become adsorbed on or near the surface of TiO_2. These excited electrons and holes have high redox activities and hence react with water and oxygen yielding Reactive Oxygen Species (ROS), such as super oxide anions (O_2^-) and hydroxyl radicals (˙OH). The versatility of the photocatalysis process is complimented by the fact that the different ˙OH radical production possibilities could be adapted for specific treatment requirements (Galvez, *et al.*, 2007). Photocatalytic killing of microbes by TiO_2 have been reported by Matsunaga, *et al.*, (1988), Maness, *et al.*, (1999) Huang, *et al.*, (1999), Cho, *et al.*, (2005), Wong, *et al.*, (2006), Tongpool, *et al.*, (2007) *etc.* Apart from microbes (bacteria) photocatalytic killing by TiO_2 has been conducted intensively on a wide spectrum of organisms including viruses (Cho *et.al.*, 2005; Gerrity *et.al.*, 2008; Lee, *et al.*, 1997), fungi (Mitoraj, *et al.*, 2007), cancer cells (Zhang and Sun, 2004), algal toxins (Srinivasan and Somasundaram, 2003).

Photocatalysis using TiO_2 has also been used to disinfect selective food-borne pathogens such as *Salmonella* sp. and *Listeria monocytogenes*, destroy *Bacillus anthracis* and the spores of *Bacillus subtilis* (Kau, *et al.*, 2009 and Vohra, *et al.*, 2005). The main objectives of all these studies ranged from identifying the factors involved in photocatalytic disinfection, optimizing the conditions for the process, studying the mechanism and kinetics of photocatalytic disinfection on field applications (Cho, *et al.*, 2005 and Sunada, *et al.*, 1998). The major challenge however has been to improve the photocatalytic efficiency of the process (Egerton, *et al.*, 2005).

Reactive oxygen species generated on irradiated TiO_2 surfaces, have been shown to operate in concert to attack polyunsaturated phospholipids in bacteria (Wong, *et al.*, 2006) and to catalyze site-specific DNA damage by generating H_2O_2 (Hirakawa, *et al.*, 2004) which might therefore result in subsequent cell death. TiO_2-mediated photooxidations have emerged as a promising technology for the elimination of microorganisms in many applications, *e.g.*, self-cleaning and self-sterilizing materials (Cho, *et al.*, 2005).

The TiO_2 photocatalytic process is a conceptually feasible technology, however, since the UV region occupies only near 4% of the entire solar spectrum and 45% of the energy belongs to visible light (Yao, *et al.*, 2006), the potential applications of this promising technology are limited. The development of a visible-light responsive photocatalyst is therefore the need of the day.

Recently Desai and Kawsik (2009) have shown photocatalytic activity of nanosized TiO_2 prepared by a modified sol-gel process against some common pathogenic microorganisms, such as *Escherichia coli*, *Pseudomonas aeruginosa*, *Klebsiella pneumoniae* and *Staphylococcus aureus* under visible light illumination.

Photocatalytic killing of *E. coli, S. aureus* and inactivation of bacterial spores of *Cryptosporidium* have been studied by many authors. Photocatalytic disinfection of water containing pathogenic microorganisms is an effective method for providing clean drinking water and works also for chlorine resistant organisms such as *Clostridium perfringens,* which are an indicator of fecal pollution. These results have shown the implications of nanoparticles for producing clean drinking water especially in emergency situations, where water borne diseases constitute a severe threat. Tests on *E. coli* have shown significantly increased disinfection efficiencies using photocatalysis as compared to UVA irradiation alone. Solar photocatalysis will be the main technology breakthrough for water treatment and purification. Findings of Sharon, *et al.*, (2007) has shown that CNT can photocatalytically degrade the membrane protein, thus fatally rupturing the cell wall of microbes.

12.2.2 Nanofiltration using Nanostructured Membranes for Separations

Nanofiltration is one of the most recently developed processes, aiming at a compromise between a high product quality and low energy consumption.

Membrane technologies uses tailor membranes having nanoscale pores for permeability, selectivity, reactivity and low fouling. These nanostructures have narrow pore size distribution, high porosity, high specific surface area and nanostructured asymmetry. This involves the use of a semipermeable membrane that transports one component more readily than the other due to differential transport, separating them into a retentate, enriched in less mobile components, and a permeate, enriched in faster components. Applied pressure drives fluid through the membrane. The pressure required is lower than that for reverse osmosis because nanofiltration membranes have larger free volumes in the polymer. Nanofiltration lies between reverse osmosis and ultrafiltration techniques in terms of the size of the molecules removed. Nanofiltration membranes consist of a multilayer structure with separation occurring in the top layer, which is ultrathin and mechanically supported by a series of asymmetric layers. Membrane fouling depends on the interaction between solute parameters and membrane parameters.

It has potential uses in water softening, removal of natural organic matter (NOM), micro-pollutants and heavy metals, disinfection, desalination, and ion separation (Kim, *et al.*, 2008). For this purpose, they electrostatically immobilized a homogeneous catalyst, iron(III)-tetrasulfophthalocyanine (FeTsPc), was on Amberlite IRA-400 (Amb) anion-exchange resin and evaluated its sorption behaviour and oxidation reactivities using bisphenol-A (BPA), which has received great concern as a strong endocrine disrupting chemical (EDC) due to its adverse effects on human health. FeTsPc-immobilized Amb could completely remove BPA by adsorption and oxidation in the presence of H_2O_2, even though Amb itself could remove BPA to a certain degree by adsorption. Moreover, they improved the stability of FeTsPc by immobilizing it on Amb. This application provides safe discharge and reuse of wastewater, high quality drinking water, groundwater treatment, removal of organic and inorganic pollutants from surface water.

12.2.3 Monitoring Devices of Pollutants

Removal of impurities from water by Nanocrystalline Diamond is being considered as an Environmental Friendly Process; because diamond is non-toxic, biocompatible, erosion resistant, and corrosion resistant against all chemicals. Diamond which is not chemically and thermally stable material but also a good thermal conductor and electrical insulator. Hence, if it is doped with boron it becomes electrical conductor. The conductivity is directly related to boron content. Pleskov, (1987) first floated the idea of using diamond as an electrode material in electrochemistry. Gandini, *et al.*, (2000) reported that Boron Doped Diamond (BDD) has capacity to produce very strong chemical oxidizing agents such as hydroxyl radicals. BDD has the largest electrochemical window of all known electrode materials and are considered as a new versatile electrode material. Diamond-electrodes can be used as anode as cathode as well as bipolar electrodes as reported in Europe-Patent (2000). Homogeneous polycrystalline 0.1–5 mm thick boron doped diamond (BDD) film on a silicon substrate of up to 0.5 m^2 is used as BDD/Si electrodes, it could be made bipolar also; which allow a very high anodic potential to be used to produce very efficient oxidants for water treatment and disinfection. Disinfecting agents are generated from mineral salts present in the water. This method of disinfection is achieved without chlorine, independently of water turbidity, and with low by-product potential. By this method wastewater oxidation is achieved without sludge generation and there is no need of adding any chemical. COD/TOC is reduced by the production of hydroxyl radicals for the destruction and increased biodegradability of organic pollutants (pesticides, phenols, solvents, PCBs).

Moreover, inactivation of viruses and bacteria (*E. coli, Legionella*) by this method is 3–5 times faster than with chlorine dosing. It also destroys algae, fungi and protozoa.

12.2.4 Applying Passive Biotechnology using Microbes as Natural Bioreactors

A study done in Europe has shown that when bacteria were used for precipitation of arsenic present in well-water, due to presence of bacterial exopolymers precipitation was enhanced. Exopolymers are principally carbohydrates that along with iron oxides can co-precipitate arsenic and other pollutants. Moreover, they found that, chemically catalyzed iron oxidation and precipitation is slower than the biological process. Iron precipitating bacteria could co-precipitate both organic and inorganic arsenic pollutants. The kinetics of biological iron oxidation was about 1000 times faster than for non-catalyzed oxidation and precipitation. There is a need to apply the passive biotechnology in environmental remediation, as this process would be a path through green chemistry.

12.2.5 Nanosized Metal Oxides for Environmental Remediation

Various metal oxide nanoparticles are being envisaged in environmental remediation *e.g.*,

1. *Nickel* is a known contaminant of ground water and well *etc. Nickel* even at ppb levels Ni causes painful skin allergies. To minimize nickel concentration, Calcite ($CaCO_3$) is being used in Denmark; which is known to accommodate considerable divalent substitution in its crystal structure. Ni is rapidly adsorbed from solution by calcite and Ni is incorporated as calcite precipitates.

2. *Iron nanoparticles* are being used for cost-effective cleaning up problems of environment, because they have large surface areas and high surface reactivity. Moreover, iron nanoparticles are very effective for the transformation and detoxification of a wide variety of common environmental contaminants, such as chlorinated organic solvents, organo-chlorine pesticides, and PCBs.

   Zhang (2003) has shown that iron nanoparticles (10–100 nm) can be synthesized from common precursors, such as Fe(II) and Fe(III); can remain reactive towards contaminants in soil and water for >4–8 weeks. Fe nanoparticle slurry can flow with groundwater over 20 m distance

3. *Zero-valent iron (nZVI)* is cheap, environmentally innocuous, and effective at reducing chlorinated organics. It is used for remediating aquifers contaminated with trichloroethylene and other halogenated pollutants. Nuxoll, *et al.*, (2003) have incorporated iron nanoparticles into polyvinyl alcohol membranes, forming water-permeable barriers to pollutants like carbon tetrachloride, copper, and chromate. These iron membranes are especially effective at blocking copper, potentially opening a new avenue of metal / radionuclide remediation application. They have proposed a theory to predict how the particle size (and hence, the effective reaction rate of the system) should affect contaminant diffusion through the membranes.

   McDowall (2005) has reported that zero valent iron (ZVI) can actually catalyze oxidative degradation in a Fenton-type reaction.

   Tratnyek and Richard L. Johnson (2006) have reviewed the remediation of contaminated groundwater using nanoparticles containing nZVI and suggested it to have high potential benefits. Since the factors like greater reactivity of nanoparticles which is due to the larger overall surface area, greater density of reactive sites on the particle surfaces, and/or higher intrinsic reactivity of the reactive surface sites, the nZVI particles result in rapid degradation of contaminants that do not react detectably with larger particles of similar material *e.g.*, polychlorinated biphenyls (Lowry and Johnson, 2004); and does not yield undesirable by-products like carbon tetrachloride (Nurmi, *et al.*, 2005).

Schrik, *et al.*, (2004) and Saleh, *et al.*, (2005) have worked out the mobility of iron nanoparticles and have suggested that though it is commonly assumed that nanoparticles are highly mobile in porous media because they are much smaller than the relevant pore spaces, but the fact is that generally, the mobility of nanoparticles in saturated environmental porous media is determined by the product of the number of nanoparticles collision with the porous medium per unit transport distance and the probability that any collision will result in removal of the nanoparticle from the flow system (*i.e.*, the sticking coefficient), that is as described by Tufenkji and Elimelech (2004). Collisions may result from three processes: *Brownian diffusion, interception,* and *gravitational sedimentation.* Modified iron nanoparticles, have been synthesized to further enhance the speed and efficiency of remediation.

4. Zhang, (2003) has shown that Iron nanoparticles can react with CHC and decompose them into relatively harmless substances, according to the (non-stoichiometrically written) reaction scheme

$$Fe + CHC + H+ \rightarrow Fe_{2+} + HC + Cl^- \ (1)$$

where HC is a hydrocarbon.

The strategy for heavy metals is to disperse into the plume iron oxide nanoparticles, on the surface of which the metal ions are strongly adsorbed. The nanoparticles can then be collected using a magnetic field, and thereby eliminated from the vulnerable ecosystem.

5. *Boron Nitride Nanotube:* Hilder, *et al.*, (2009) have proposed boron nitride nanotube based devices for purification, desalination and demineralization of water through their ability to remove salts and heavy metals without significantly affecting the mass flow of water molecules. They have suggested that Boron Nitride nanotubes have better water flow properties than Carbon nanotubes, thus can be a more efficient water purification device. Using molecular dynamic simulation, it has shown that when a (5, 5) boron nitride nanotube is embedded in silicon nitride membrane it should obtain 100% salt rejection at as high as 1M concentration; due to high energy barrier while still allowing water to flow at 10.7 mol/sec or 0.9268 L m-2 h-1. Moreover, ions continue to be rejected under the influence of high

hydrostatic pressure (upto 612 MPa). The interesting observation was that when the boron nitride nanotube radius was increased to 4.14 Å the tube became cation selective and at 5.52 Å the tubes became anion selective.

12.3 The Dark Side of Nanotechnology

Since nanoparticles are as small as a virus or haemoglobin molecule, small irritant dust particles or pollens; they can pose health risks.

Nanotechnology will be impacting manufacturing methods, infrastructure of consumer goods and usage of materials. We have already witnessed nanotechnology associated pollution *e.g.*, gallium arsenide that is used in microchips for computers and cellular phones have already entered the land fills and nanoparticles can enter the food chain too. The large surface area, crystalline structure and high reactivity of the nanoparticles are facilitating the transport of toxic nanomaterials in the environment. Size and chemical composition of nanoparticles are expected to harm the biological system (Gorman, 2002). There are also naturally occurring nanomaterials, which can profoundly influence the quality of the environment. Naturally occurring nanomaterials or 'Mineral dust' are fine particles of crystal origin consisting primarily of silica and silicate minerals. These sub-micron particles have shown their impact on various heterogeneous surfaces on various atmospheric reactions resulting in the formation and destruction of HONO, which is an important atmospheric contaminant affecting atmospheric processes and public health.

12.3.1 Cytotoxicity of Nanoparticles

There is increasing concern about health effects of nanoparticles as they can enter the human system through many pathways *e.g.*, exposure media, uptake pathways, translocation and distribution as well as excretory pathways (Oberdörster, *et al.*, 2005).

Nanoparticles present in the air cannot only be inhaled, but also get deposited on skin like dirt particles do. Inhaled NP will easily get launched into respiratory tract, nasal passage, trachea and finally into the alveoli of lungs. From alveoli it would not be difficult for nano sized particles to enter into blood stream and get lodged into organs like heart, kidney, spleen, bone marrow and muscles; thus causing multiple problems. Moreover, particles deposited on the skin can block sweating. Entry through pulmonary system is suspected of causing detrimental acute and chronic damage to the pulmonary and the cardiovascular system. Based on scientific proofs, air pollution legislations of several countries classify diesel soot and particulate matter (PM) emitted from diesel engines for automotive applications; as carcinogenic. Moreover, several health effects are associated with the ultra-fine particles with diameter below 100 nm (Brown, *et al.*, 2001).

There has been indications that these particles can penetrate the cell membranes, enter the blood and even reach the brain (Oberdörster, *et al.*, 2004) and can induce inheritable mutations (Somers, *et al.*, 2004). Elder and Oberdörster (2006) have found that when rats breathed in nanoparticles, the particles settled in the brain and lungs, which led to significant increase in biomarkers for inflammation and stress response. NP present in water and food will have easy access through ingestion to G I tract and from there getting access to lymph and blood circulatory system.

Drug delivery using NP *i.e.*, through injection can straight way enter into the blood stream, leading their translocation and distribution to all the parts of body.

Hence, like any new technology, risks due to nanotechnology need to be scrutinized and regulated.

12.3.2 Assessing the Cytotoxicity of Nanoparticles

Nanoparticles toxicology is the study of the adverse effects of nanoscale materials on biological systems. Nanotoxicology can be studied both *in vivo* as well as *in vitro*.

12.3.2.1 *In Vitro Toxicological Studies*

They are fast and cheap. It uses immortalized cell lines and gives important information of the mechanism of *in vivo* effect. Hence, Gulumian advocates applying a wide range of *in vitro* tests.

(*a*) Non-cellular tests like free radical generation by particle—Formation of free radicals *in vitro* have been shown by Shvedova, *et al.*, (2003) in response to SWCNT and by Uchino, *et al.*, (2002) in response to TiO_2. More recently working on the testing of silver nanoparticles Hussain, *et al.*, (2005) in BRL 3A; Hsin, *et al.*, (2008) NH_3T_3; Carlson, *et al.*, (2008). Alveolar microphages AshaRani, *et al.*, (2009); Lung fibroblast cells (IMR-90) and Human glioblastoma cells (U25); Arora, *et al.*, (2009). Primary fibroblast and primary liver cells isolated from albino mice found that it has the ability to generate free radicals. Nanoparticles tend to cluster in the granules of the mast cells. Since the granules are essential to the primary function of mast cells, this location of nanoparticles could explain the observed negative electrochemical effect of the addition of nanoparticles (Cornell, 2006).

(*b*) Moreover, silver nanoparticles could induce size dependent apoptosis in U937 cell (macrophages). During *in vitro* testing of SWCNT with cells, Jia, *et al.*, (2005) noted significant impaired AM phagocytosis at very low concentration.

(*c*) *In vitro* cellular system assessment of Physicochemical property of NP also helps in studying the effect of different physicochemical parameters of NP on Cytotoxicity. Cellular or cell based *in vitro* system comprising of either primary cells (*e.g.*, HeLa cells, Phagocytes, Astrocytes, Fibroblasts *etc.*) in mono culture and co-culture (comprising of epithelial cells with microphages along with dentritic cells) or organ culture (heart, lung *etc.*); to study NP effects and translocation (Blank, *et al.*, 2007).

(*d*) Hayens, *et al.*, have studied the nanotoxicology of 24.8 ± 4.1 nm citrate-reduced gold colloids; by exposing the cells to nanoparticles of different size, shape, and surface chemistry. They used Murine Peritoneal Mast Cells obtained from Abdominal Cavity, and measured the release of Serotonin and Histamine as an indication of immune response. Moreover, SEM analysis showed uptake of NP by mast cells which was possibly by caveolae mediated endocytosis. They have hypothesized that each NP displaces 1.41×10^{-17} mL of matrix *i.e.*, it increase halo volume (by fast diffusion) and decrease matrix volume (by slow diffusion).

(*e*) Gojova, *et al.*, (2007) used inflammatory response to NP having different composition *viz* Fe_2O_3, Y_2O_3 and ZnO using Human aortic endothelial cells (HAEC), and measured mRNA and protein level of intra-cellular adhesion molecule-1, interleukin B and monocyte chemotactic protein-1. They observed that Fe_2O_3 did not elicit inflammatory response at any concentration. Whereas both Y_2O_3 and ZnO provoked significant inflammatory response above a threshold concentration of 10 µg/ml. At 50 µg/ml ZnO was toxic and detrimental. Their data concludes that cytotoxic response is composition dependent. Similar conclusions were made by Sayes, *et al.*, (2004) when they studied the dermal toxicity of fullerenes having different functional groups attached on the surface. Dose, shape and size of CNT also had impact on cytotoxicity.

(*f*) Chang, *et al.*, (2006) have analyzed the mitochondrial membrane potential and cytochrome c release in CdSe-core QD treated Human neuroblastoma cells IMR-32 and found loss of mitochondrial membrane potential, mitochondrial release of cytochrome c.

(*g*) Size, shape and surface propertied of carbon nanomaterial have been found to have differential impact on Cytotoxicity *e.g.*, at all the tested concentrations of 1.41, 2.82, 5.65 and 11.3. SWCNT showed maximum Cytotoxicity followed by MWCNT and minimum by fullerene. (Jia, *et al.*, (2005).

So we have seen that NP can produce oxidative stress, affect cellular functions cause apoptosis, affect mitochondria, and are cytotoxic to cells. These effects can be altered by physicochemical properties of NP.

12.3.2.2 *In Vivo Toxicological Studies*

It is rather expensive and slow. It provides critical and little mechanistic information. It is done to find out quantitative relationship between magnitude of dose and adverse health effect. It covers response to dose, the exposure routes, end point measurement, exposure limit, target organ and biopersistence. *In vivo* toxicity of NP may be acute *i.e.*, response would be immediate; sub-acute or chronic that is long term effect that may lead to mutagenesis or carcinogenesis:

(*a*) Dose—definition of dose varies for different assessments *e.g.*, dose could be amount of drug absorbed from an administration (as received by Human Pharmacology); it could be amount of chemical absorbed into the body from an exposure (Human toxicology) or it could be an estimated amount of contaminant taken in by an organism in terms of the body weight (Environmental toxicology). During *in vivo* dose dependent tests; there is a chance of encountering NOAEL (No Observable Adverse Effect); which may be due to the reason that NP will have long term effect. Hence, exposure limit is a major consideration.

(*b*) Exposure route—as mentioned above could be through skin, circulatory system, GI tract or respiratory system.

(*c*) *In vivo* studies done so far have given indications that QD and Nickel oxide can be biopersistent and Fullerene (C_{60}) can damage the DNA. Target organs like Lungs have shown glioma formation in response to CNT; Bile-duct develops hyperplasia in response to silver NP. Kidney also shows accumulation of silver NP and brain accumulated TiO_2. Moreover, anatase NP of TiO_2 can cross the Embryo placental barrier.

(*d*) End points measured for *in vivo* assessment are Target organ, Biopersistence, Translocation-Distribution-Excretion, Immuno-cytotoxicity / inflammatory markers, Genotoxicity and Carcinogenicity.

(*e*) Biopersistence depends on size, surface area, charge loading and surface modification of NP; the injected QDs that maintained their size without binding *in vivo* can be excreted by kidney action and if they bind to proteins they are translocated into liver and excreted with feces, whereas if QDs gets aggregated to larger particles then they are retained in liver for longer period (Chen, *et al.*, 2008).

(*f*) Translocation and distribution—Fletcher, *et al.*, (2002) have shown that NP (30 nm) of manganese oxide inhaled by right nostrils of rats can be taken up, translocated and accumulated into brain. A shocking revelation has been made by Shimizu, *et al.*, (2009) that maternal exposure to anatase TiO_2 can change the expression of gene associated with brain development, cell death, oxidative stress response and mitochondria in brain during prenatal period, and later it changes the expression of genes associated with inflammation and neurotransmitter.

It must be mentioned here that though a NP is deemed as cytotoxic if it disrupts normal cellular function leading to detriment of cell viability and thus is an undesired property for healthy cell. The same cytotoxic property is desirable for killing cancerous cells.

12.4 Summary

Use of nanomaterials in remediation, prevention and enhancement of environmental conditions are being tried. Some of the most used nanoparticles are TiO_2, ZnO, Nickel nanoparticles, Iron nanoparticles, Zero Valent Iron, Boron Nitride nanotubes etc. Silicon and nanodiamond have found its way in creating pollution monitoring devices. However, environmentalists are skeptical also about the use to nanoparticles, as its dark-side has not been fully explored. There is emphasis on making special regulations for use of nanoparticles.

Applications of Nanotechnology in Agriculture, Food and Cosmetics

> "The essence of Nanotechnology is the ability to work at the molecular level, atom by atom, to create large structures with fundamentally new molecular organization. The aim is to exploit these properties by gaining control of structures and devices at atomic, molecular, and supramolecular levels, and to learn to efficiently manufacture and use these devices".
>
> —Statement by the US National Science and Technology Council

13.1 Introduction

By now we know that nanotechnology is manipulating materials and systems at the scale of atoms and molecules. A nanomaterial measures few hundred or less nm and a nanometer is one billionth of one meter. Moreover, the properties of matter (colour, chemical reactivity *etc.*) change at the nanoscale *e.g.*, in bulk form zinc is white and opaque, in nanoform it is transparent.

All previous chapters have dealt with importance of nanoscale activities of living and non-living systems, as well as applications of nanotechnology in different fields. It is because nanotechnology has offered substantially different properties than their macroscopic or bulk counterparts, thus new materials and new technologies having applications in various fields.

Nanotechnology has provided new solutions to problems in food science and cosmetics, offers new approaches to the rational selection of raw materials, or the processing of such materials to enhance the quality of food and cosmetic products. In this chapter, we provide an overview of some current efforts in the area of nanotechnology as it applies to food systems and some morphologically different structures and associated manufacturing technologies that could be used to build functional food systems.

Already there are many major industries involved in nano and food related products *viz.* Kraft, Nestle, Unilever, Pepsi Co., Cargill, Mars, BASF, Syngenta, DuPont, Bayer, Heinz, Hershey, Campina, Friesland Food, Grolsch, ConAgra Foods, General Mills, Danisco *etc*. Their products include: dietary supplements, nutritional additives, colour additives, food processing aids, long-life packaging, anti-bacterial kitchenware, fertilizers and pesticides. Efforts are on to produce following items in future *i.e.,* interactive, personalized foods, edible nano-wrappers, chemical release packaging, extensive nanosurveillance, interactive agrochemicals, nanomanipulation of seeds and synthetic biology. With these efforts, it appears that nano is likely to erode our relationship with real food by enabling junk food to be fat, sugar and carbohydrate reduced and vitamin, protein and fibre-enhanced.

As we have discussed in earlier chapters (Chapter-2), nanotechnology existed in mother nature and functioning of living organisms composed of nanoscale organelles, producing custom made nanosized molecules. Using atoms for complex intricate functions of living system *e.g.,* potassium and sodium (having 0.1 nm ion diameter generate nerve impulses. The vital biomolecules (sugars, amino acids, hormones, and DNA) are in the nanometer range. And above all nature's extensive use of self-assembly principles to create nanoscale structures without the expenditure of large amounts of energy for assembly and creation, nanoscale structures. Taking lesson from the nature; incremental application of nanotechnology in food and cosmetics area can be worked out.

13.2 Food, Agriculture and Nanotechnology

Achievements and discoveries in nanotechnology are beginning to impact the food industry and associated industries. Nanotechnology is being applied in all the phases of the food cycle *i.e.,* from field to table. Even though the food industry is just beginning to explore its applications, nanotechnology has exhibited great potential in areas like: food industry; release systems for pesticides or fertilizers in agriculture; antibacterial or easy-to-clean surfaces in food processing machines; food additives, such as anti-caking in salt, powders and coffee creamers; anti-foaming agents for beer; colour additives for lemonades; encapsulated vitamins for dietary supplements and micelle systems for low-fat foods.

13.2.1 Agro-practices

Heavy use of pesticides is known to affect the soil. Hence, slow and reduced release of pesticides is being tried by using a cloth saturated with nanofibres to slowly release pesticides, thus eliminating the need for additional spraying and reducing chemical leakage into the water supply. Moreover, pest control by early identification of plant diseases using devices like nanosensors would help in protection of crops.

13.2.2 Postharvest and Processing

Food is mainly composed of proteins (globular structures of up to 210 nm in sizes) carbohydrate and lipids or fats (that are linear polymers with less than a nm thickness. Processing of food, for example involves gelatinization of starch based foods so as to influence the nutritional benefits during digestion. Understanding of the nature of nanostructures in foods allows for better selection of raw materials and enhanced food quality through processing. Nanostructures in food are one molecule thick and are examples of 2D nanostructures.

Nanoseives or membranes having nanosized pores are being envisaged for filtration of beer, milk for cheese production.

13.2.2.1 *Food Safety and Packaging*

Packaging industries are diverting themselves towards nanotechnology for food safety. A n edible food film constituting cinnamon or oregano oil or nanoparticles of zinc, calcium are antimicrobial in nature. Nanofibres from Lobster shells or organic corm both are proved to be anti-microbial and biodegradable in nature. The criteria for good packaging material is strength, barrier properties, stability to heat and cold, all of which could be accomplished using nanocomposite materials. Bayer Polymers have produced a nanocomposite film 'Durethan'. It is a film enriched with silicate nanoparticles which reduces the entrance of oxygen and other gases, and preserves moisture, thus preventing food from spoiling. In future, incorporation of silver, magnesium oxide or zinc oxide nanoparticles (which can kill harmful microorganisms) in food or beer packages will save the contamination. Antimicrobial activity can also be imparted through addition of nanosensors to food packages is also anticipated in the future. McHugh has suggested that these nanosensors could be used to detect chemicals, pathogens and toxins in foods. Radio Frequency Identification (RFID) tags could be incorporated into food packages in the future. These do not require line-of-sight for reading like bar-codes and enable registration of hundreds of tags in a second. Use of nanowheels, nanofibres and nanotubes are being tried to improve the qualities of food packages.

13.2.2.2 *Molecular Synthesis to Attain Different Food Manufacturing Techniques*

As discussed in earlier chapter all organisms represent a consolidation of various nanoscale-size objects cell and cell organelles; where atoms and molecules combine to form dynamic structures. In fact, every living organism on earth exists because of the presence and interaction of various nanostructures where molecular synthesis of food molecules, such as carbohydrates, lipids and proteins are synthesized by nanoscale-level mergers between sugars, amino acids, and fatty acids.

13.2.2.3 *Nanoscale Biosensors*

Involving biological molecules, such as sugars or proteins as target-recognition groups could be used as biosensors on foods (Charych, *et al.*, 1996) to detect pathogens and other contaminants. In food industry, nanosensors would provide increased security of manufacturing, processing, and shipping of food products through sensors for pathogen and contaminant detection.

13.2.2.4 *Designing and Formulation of Healthy Food*

Designing novel food, having desired ingredients, such as flavours and antioxidants (Imafidon and Spanier,1994); maximizing or minimizing their concentration are possible to achieve by nanotechnology. As the infusion of novel ingredients into foods gains popularity (Haruyama, 2003), greater exploration of delivery and controlled-release systems for nutraceuticals will occur (Lawrence and Rees, 2000).

13.2.2.5 *Storage of Food*

Nanotechnology may also be useful in encapsulation systems for protection against environmental factors. During storage, there is a need to protect functional ingredients in food (vitamins, antimicrobials, antioxidants and flavourings), taste, texture, and shelf life from chemical or biological degradation, such as oxidation, and controlling the functional ingredient's rate of release under specific environmental conditions. Nanodispersions and nano-encapsulation will be ideal mechanisms for it. Nanostructures like colloids, nanoemulsions, and biopolymeric nanoparticles can be used for encapsulation. Many nanoenhanced barrier keeps oxygen-sensitive foods fresher has also come in the market.

13.2.3 Nanostructures in Association-Colloidal Forms for Delivery of Functional Ingredients

Surfactant micelles, vesicles, bilayers, reverse micelles, and liquid crystals *etc.*, have been found to be ideal nanomaterials for nanodispersions and nano-capsulation for delivery of functional ingredients. *Colloid* is a stable system of a substance containing small particles dispersed throughout in a liquid. Association colloids have been used for many years to deliver polar, non-polar, and amphiphilic functional ingredients (Golding and Sein, 2004; Garti, *et al.*, 2004, 2005; Flanagan and Singh, 2006). Size of colloids range from 5 to 100 nm. The major disadvantage of colloids is that they can spontaneously dissociate if diluted.

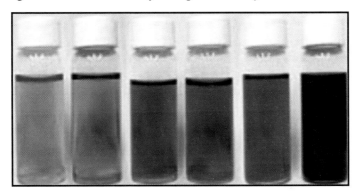

Fig. 13.1 Colloidal solutions of Gold nanoparticles of different sizes showing different colours due to variation in Plasmon Resonance.

13.2.3.1 *Nanoemulsion*

It is a mixture of two or more liquids (such as oil and water) that do not easily combine. In nanoemulsion the diameters of the dispersed droplets are 500 nm or less. Nanoemulsions, can encapsulate functional ingredients within their droplets, which can facilitate a reduction in chemical degradation (McClements and Decker, 2000).

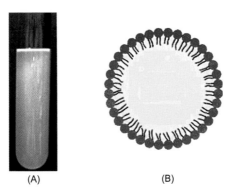

(A)   (B)

Fig. 13.2 (*A*) Nanoemulsion (*B*) Schematic representation of Nanoemulsion.

13.2.3.2 *Biopolymeric Nanoparticles*

Such as food grade proteins or polysaccharides are nm sized polymer molecules. Their nm-sized particles have been produced (Chang and Chen, 2005; Gupta and Gupta, 2005 and Ritzoulis, *et al.*, 2005). These nanoparticles can also be used to encapsulate functional ingredients. However, synthetic biodegradable biopolymeric nanoparticle is polylactic acid (PLA), is being used to encapsulate and deliver drugs, vaccines, and proteins. But the problem with PLA is that it is quickly removed from the bloodstream, and remains lodged in the liver and kidneys. PLA along with polyethylene glycol have shown successful results in this regard (Riley, *et al.*, 1999).

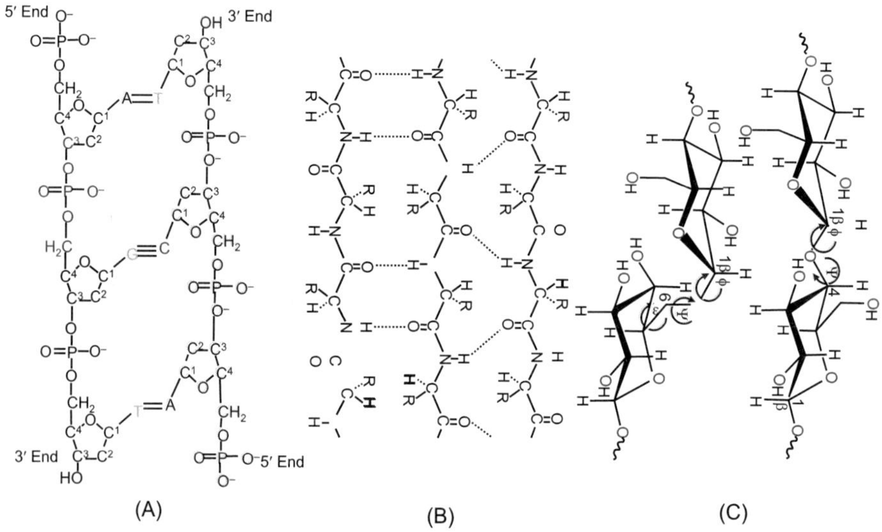

Fig. 13.3 Three main Biopolymers (*A*) DNA, (*B*) Protein and (*C*) Polysaccharide.

Nanolamination: Technique is another viable option for protecting the food from moisture, lipids and gases. Moreover, they can improve the texture and preserve flavour as well as colour of the food. Nanolaminates consist of two or more layers of nanosized (1–100) thin food-grade films which are present on a wide variety of foods: fruits, vegetables, meats, chocolate, candies, baked goods, and French fries (Morillon, 2002; Cagri, *et al.*, 2004; Cha and Chinnan, 2004; Rhim, 2004). Nanolaminates are prepared from edible polysaccharides, proteins, and lipids. Park (1999) has shown that polysaccharide and protein-based nanolaminates are good barriers against oxygen and carbon dioxide, but poor in protecting against moisture. Whereas, lipid-based nanolaminates are good at protecting food from moisture. Trials are on to develop laminates that can protect against all the desired of actors. Coating foods with nanolaminates is done simply by spraying it on the food surface (McClements, *et al.*, 2005).

13.2.4 Food-related Nanoproducts

Nanoparticles of carotenoids and synthetic lycopene added to fruit drinks have shown improved bioavailability. Nanosized micelles of canola oil and nanodrops are already being patented and used as transmucous delivery systems for vitamins, minerals or phytochemicals. Use of copper, gold and silver bhasma has been mentioned in Ayurvedic system of medicine. Now with the advent of nanotechnology, nanosilver or nanogold have also found entry into medicine, cosmetics and neutraceuticals.

13.2.4.1 *Packaging*

Inclusion of nanoparticles in food contact materials can generate novel types of packaging materials and containers. Use of TiO_2 nanoparticles in wraps will have UV absorption characteristics. Such wraps, films or plastic containers where absorption of UV radiation needs to be avoided can be used.

13.2.4.2 *Nanofood Products*

There are more than 105 nanofood products (dietary supplements, nutritional additives, colour additives, food processing aids, long-life packaging, anti-bacterial kitchenware, fertilizers, pesticides and agricultural inputs) that contain manufactured nanoparticles are on sale now.

In future, nanotechnology could enable junk food to be fat, sugar and carbohydrate reduced, and vitamin, protein and fibre-enhanced.

13.2.4.3 *Indirect Applications of Nanotechnology in Food and Cosmetics*

Apart from above mentioned applications of Nanotechnology in food industry, there are some potential indirect applications of Nanotechnology has also been envisaged *e.g.*, increased advances and sophistications in the area of Computing and communications: devices are sure to offer new types of labeling, new ways to store, display and interrogate information on packaging. Thus, allowing access to more information about the source, history and storage of specific cosmetics foods; their nutritional characteristics and their suitability to their genetic makeup and life-style can be improved.

13.3 Cosmetics and Nanotechnology

Though the ancient Greek and Romans did not know about the Nanotechnology, but they used Nanotechnology in cosmetics. According to Michael Berger

(2006), more than 4000 years ago Egyptians knew to synthesize 5 nm diam. Nanocrystals of lead sulfide which have been found to be quite similar to PbS quantum dots synthesized by modern materials science techniques. Walters has stated that "For thousands of years, cosmetics have been used and were made by the judicious combination of naturally available minerals with oils, various creams, or water. Since the Greco-Roman period, organic hair dyes obtained from plants such as henna have been used, but other unusual formulas based on lead compounds, such as the recipes describing several methods to dye hair and wool black, were also common. It is remarkable that these Greco-Roman techniques have been used up to modern times." Moreover, Walter has pointed that more than 2000 years a hair dyeing process was involved, using biomineralization or basic chemistry methods.

The concept that the smaller particles are more readily absorbed into the skin and as such repair damage easier and more efficiently has given impetus to incorporating nanotechnology in cosmetics. Nanotechnology in the cosmetic industry of today involves making products having nanoparticles that can go deeper in the skin's surface to give better results, hence sunscreens with efficient UV protection; long-lasting makeup; anti-ageing creams with an increased intake of vitamins or enzymes; toothpaste; hair care or colouring products and even in contact lenses. Main uses of nanotechnology and nanoparticles in cosmetics are mentioned below:

13.3.1 Anti-aging Products

Currently anti-ageing products are the main cosmetic products in the market that are being made using nanotechnology. Anti-wrinkle cream containing "nanosiomes of Pro-Retinol A" is a very popular product.

13.3.2 Nanotechnology for Sunscreen and UV Protection

Nanoparticles of Titanium dioxide (TiO_2) and zinc oxide (ZnO) are the main compounds used in Sunscreen Creams. They have already received FDA approval.

Nanoparticles of these oxides are more advantageous as they retain the UV filtration and absorption properties, while eliminating the white chalky appearance of traditional sunscreens. Moreover, sunscreen creams containing nanoparticles of ZnO or TiO_2 are transparent and less smelly, less greasy and more absorbable by the skin. Villalobos-Hernàndez, C.C. Müller-Goymann, (2006) have shown that carnauba wax nanoparticles with TiO_2 nanoparticles increase the sun protection factor (SPF). Moreover, Muller, *et al.*, (2002) have proved that lipid nanoparticles have a synergistic effect of the UV scattering when used as vehicles for molecular sunscreens .

13.3.3 Nanoemulsions in Shampoo

Nanoemulsions are dispersed nanosized droplets of one liquid in another. Dingler and Gohla (2002) suggested that emulsion preparation can be manipulated to products *e.g.,* water-like fluids or gels. Nanoemulsions encapsulate active ingredients and are used to carry the active ingredients deeper into the hair shafts. In some products, Nanovectors are used to transport and concentrate active ingredients in the skin. Nanoemulsions are being investigated for many other cosmetic applications. As compared to large size emulsions, nanoemulsions have a number of advantages *e.g.,* (*a*) they can be stabilised to increase the time before creaming occurs, thus increasing the shelf life of products, (*b*) they are transparent or translucent, and have a larger SVR, (*c*) smaller the size of the emulsion, higher the stability and suitability as ingredient carrier Dingler and Goha (2002), (*d*) the components of nanoemulsions are usually GRAS compounds, therefore, they are considered relatively safe. Several cosmetic products, such as sunscreen for hair and skin uses nanoemulsion.

13.3.4 Vesicular Delivery Systems

Vesicular delivery system has explored the use of Lipid nanoparticles in pharmaceutical and cosmetics. It has shown superior advantages for topical purposes over conventional colloidal carriers. Both solid lipid nanoparticles (SLN) and nanostructured lipid carriers (NLC) are being used for cosmetic applications. The protective and impermeable qualities of the skin protect the organism from losing water, minerals and dissolved proteins are well known *i.e.,* Continual exposure of skin to environment, infection by some systemic diseases, vitamin deficiencies and disturbances of endocrine glands; demands efficient supply of nourishment, replenishment of moisture treatment with appropriate drug molecule and hygienic practices. Lipid-based formulations are the most appropriate for topical application of actives because of their protective action, adhesiveness, occlusion and skin hydration qualities. According to Wissing and Müller (2001), approximately 4% of lipid nanoparticles of approximately 200 nm diameter should form theoretically a monolayer film when c. 4 mg of formulation is applied per cm^2 on the skin. Lipids are hydrophobic in character, that's why the mono-layered film of lipid has an occlusive action on the skin retarding the loss of moisture caused by evaporation. Wissing, *et al.*, (2003) have demonstrated different degree of occlusion, depending on the size of the applied particles. Wissing and Muller

(2003) have shown that while ageing, the dermis loses much of its elasticity. The skin stretched by muscular movement then fails to shrink back to its normal smoothness and wrinkles are formed. Due to their hydration properties, lipid nanoparticles enhances the skin, hence can be used to formulate anti-ageing products. Moreover, having a spherical-like shape, lipid nanoparticles impair excellent lubricating action.

Many lipid-based carriers *e.g.*, liposomes [Singh, *et al.*, 2005], oil-inwater (o/w) emulsions [Teichmann, *et al.*, 2005], multiple (w/o/w) emulsions [Gallarate, *et al.*, 1999] and microemulsions [Kreilgaard, *et al.*, 2001 and Park, *et al.*, 2005, 6] are in use. To increase physico-chemical stability of both incorporated actives and the system itself, SLN [Muller and Lucks, 1996] and NLC [Muller, *et al.*, 1998] have been found most suitable.

13.3.4.1 *Liposomes*

They are being used in the cosmetic industry as delivery vehicles. Solid Lipid Nanoparticles (SLN) and Nanostructured Lipid Carriers (NLC) have been found to be better delivery vehicle than liposomes. Moreover, NLCs have been found to be not only delivery vehicle, but could also provide enhanced skin hydration, bioavailability, stability of the agent and controlled occlusion.

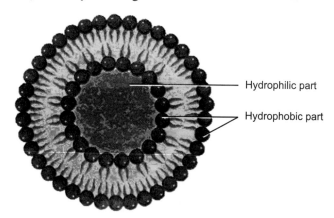

Hydrophilic part

Hydrophobic part

Fig. 13.4 Schematic diagram of a Liposome.

Liposomes are vesicular structures of 15 nm to several μm, having an aqueous core surrounded by a hydrophobic phospholipids bilayer or unilayer. Phospholipids are safe to be used as delivery vehicle. The advantage is that solutes in the core cannot pass through the hydrophobic bilayer; but hydrophobic molecules can be absorbed into the bilayer, enabling the liposome to carry both hydrophilic and hydrophobic molecules. The lipid bilayer of liposomes can fuse with the bilayer of the cell membrane, and release its contents. Moreover, they are easy to prepare. The other advantage is that Liposomes can continuously deliver the ingredients into the cells for a long duration which make them suitable for the delivery of several active ingredients *e.g.*, vitamins A and E and antioxidants (*e.g.*, CoQ_{10}, lycopene and carotenoids) *etc*. Propylene glycol (PG)-coated Liposomes are being used in the treatment of hair loss for delivering Minoxidil sulphate in liposomes (an active ingredient) in products like Phosphatidylcholine (main ingredients of liposomes) has softening and conditioning properties, hence, it is used in many skin care products and shampoos.

13.3.4.1 *Niosomes*

They (developed and patented by L'Oreal in the 1970s and 80s) are another vesicular structure being used as delivery vehicle. Niosome is a non-ionic surfactant based phospholipid vesicles like liposomes and can self-assemble by non-ionic surfactants in aqueous media in bilayer structure when agitated and heated. They can be used to encapsulate aqueous solutes and act as drug and cosmetic carriers. Like liposomes, it has hydrophobic and hydrophilic groups. They are being used for the delivery of anti-inflammatory and anti-infective agents and transdermal drug delivery. Niosomes can increase the stability of entrapped drugs, thus increased bioavailability. Van Hal, *et al.*, reported that Niosome encapsulated estradiol can be delivered through the stratum corneum, which is known to be a highly impermeable protective barrier.

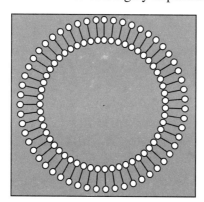

Fig. 13.5 Schematic representation of the Niosome (red rounds are "Hydrophilic head group they face hydrophilic region; and blue lines are "Hydrophobic tail" they are non-polar tails facing each other to form a hydrophobic region; Central light blue is aqueous region where hydrophilic cosmetic ingredients are encapsulated.

The disadvantage is that like liposome, Niosome do not contain GRAS components and are known to be more irritating than liposomes.

Few other synthetic nanopolymers have also shown a potential to be used as nanocarriers that encapsulate and transport Vitamin E deeper into the skin.

Vitamin E is a known antioxidant and has ability to protect skin and hair from damage. But Vitamin E does not easily absorb through the skin's outer layers that is why synthetic nanocarriers are developed that encapsulate and transport Vitamin E deeper into the skin. These novel Vitamin E-toting polymers can combine both fast- and slow-acting components. These can be used for water-based preparations of cosmetics, pharmaceuticals and nutraceuticals.

13.3.4.2 *Transferosomes*

They are a new type of liposomes like substance (200–300 nm), that are more elastic and efficient than liposome. Like liposomes; transferosomes can also self-assemble as lipid droplets with elastic bilayers and quickly penetrate the stratum corneum through intracellular or transcellular routes. This quality is very useful in cosmetics and drug delivery.

The disadvantage is that they are unstable due to their susceptibility to oxidation. However, formulations have been developed to stabilize them.

13.3.4.3 *Cubosomes*

According to Spice, *et al.*, (2003), Cubosome are nanostructured particles of bicontinuous cubic liquid crystalline phase; which is optically clear and very viscous material. Self-assembly of cubosomes is (Spice, *et al.*, 2003) done by mixing liquid crystalline particles of certain surfactants with water and a microstructure at a certain ratio. Cubosomes offer a large surface area, high heat stability and can carry both hydrophilic and hydrophobic molecules (Kesselman, *et al.*, 2007). However, they do not offer controlled release on their own. Trials are on to modify them by using proteins (Anglova, *et al.*, 2005 and Spicer, 2005).

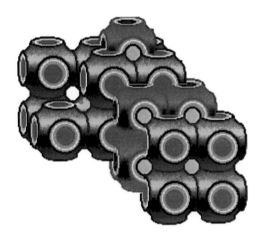

Fig. 13.6 Schematic representation of cubosome, the bio-continuous cubic structures.

13.3.4.4 *Dendrimers*

They are approximately 20 nm in size, unimolecular, monodisperse, micellar, regularly branched symmetrical, robust, covalently fixed, three-dimensional nanostructure having high density of functional end groups at their periphery and an interior core filled with solvent. They are prepared in a step-wise fashion as shown in Fig. 13.7.

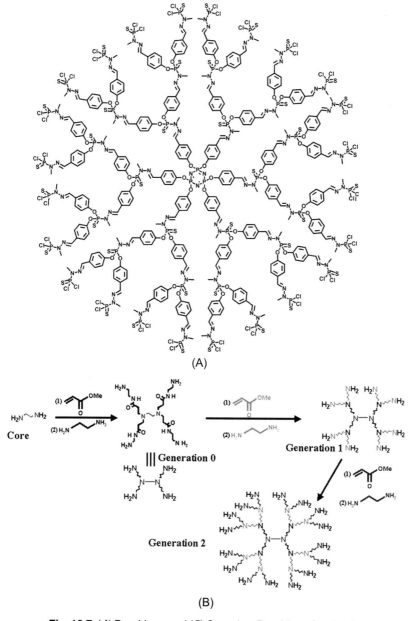

Fig. 13.7 (*A*) Dendrimer and (*B*) Steps in a Dendrimer Synthesis.

13.3.4.5 *Hyperbranched Polymers*

Like dendrimer, Hyperbranched polymers (HBP) also contain many branches, but unlike dendrimers HBP are effectively disorganized and unsymmetrical dendrimers. They are prepared in a single synthetic polymerization step. Hyperbranched polymers or dendrimers formulations are being used in mascara or nail polish.

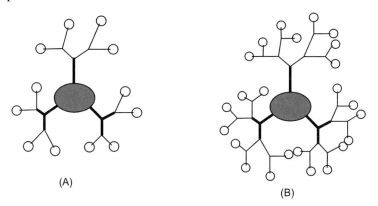

(A)

(B)

Fig. 13.8 Schematic representation of (A) Dendrimer and (B) Hyperbranched polymer both having single central core (○) and to which multiple functional end groups (•) are attached.

13.3.5 Nanocrystals in Cosmetics for Delivery of Poorly Soluble Materials

Nanocrystals are aggregates of many atoms into a "cluster" of 10–400 nm and they exhibit physical and chemical properties somewhere between that of bulk solids and molecules. For utilization of Nanocrystals, its size and surface

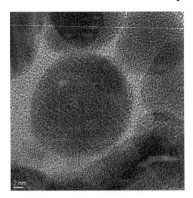

Fig. 13.9 High resolution transmission electron micrograph of a single gold nanocrystal. The alternating light and dark lines are atomic planes. Inset is SAED in which the different shades of grey of the particles come from different orientations of their atomic planes with respect to the incident electron beam.

area is controlled such that its bandgap, charge conductivity, crystalline structure and melting temperature can be altered. To prevent the further formation of larger aggregation crystals needs to be stabilized. Two, poorly soluble, plant glycoside Rutin and Hesperidin have now been made available for dermal use by formulating it as nanocrystal (Patented by Peterson in 2008). Other examples are Resveratrol and Ascorbylpalmitate nanocrystals.

13.3.6 Nanoencapsulation for Controlled Release

Encapsulation is a widely used old technology for drug delivery. With the emergence of nanotechnology a new type of particles are being used for targeted delivery. The nanoparticles can carry drug payloads for localized action as they have a shell and an interior space that can be used to nanoencapsulate active ingredients. Nanocapsules could be of hydrophobic or hydrophilic nature. It can also be functionalized to target specific molecules. Polymers, proteins and different types of biomolecules have been proposed as suitable materials to coat nanocapsules. The release of the payload could be organized by an external trigger (ultrasound or magnetic field *etc.*) or internal factors of the cells *e.g.,* pH, temperature, light exposure *etc.* Hydrophobically modified polyvinyl alcohol 10,000 (PVA) with fatty acids (FAs) have been used to create polymeric nanoparticles for cosmetic applications (Luppi, *et al.,* 2004). They substituted PVA with saturated FAs to make the polymer lipophilic and tested the percutaneous absorption of benzophenone-3 (a UV filter) and found that the PVC nanoparticles can limit the adsorption of BZP, with nanoparticles with a high degree of substitution preventing absorption more efficiently. Recently a hydrogel is developed by Hu, *et al.*, (2008) that responds to temperature and is being used as a facial mask. Their patented technology called Facial Switch™ shrinks the hydrogel if the temperature is increased and releases nutrients. They have also created smart fabrics with anti-bacterial properties that can respond to external stimuli and fabrics that turn from semi-transparent to opaque in response to stimuli. MiCap (http://www.micap.biz/) is exploring the possibilities of using microbial cells and cell walls for the controlled delivery of perfumes.

Use of Nanotechnology in Contact Lenses: Use of nanoparticles in making coloured contact lenses is a big business nowadays.

Human vision can be envisioned to possess extraordinary power, with the discovery of nanotechnology. Terminator and other fictional superheroes have become models for average people who can may now be able to watch movies, read emails or surf internet without a computer in sight but just with their eyes staring straight ahead.

Nanotech Contact Lens Monitors Diabetes by Changing Colour–
Nanoparticles in a hydrogel lens change colour with the glucose level in tears.
Professor Jin Zhang at the University of Western Ontario has developed
anotech contact lens that would change colour as the user's glucose levels
varies. This contact lens is made by embedding nanoparticles into standard
hydrogel. These particles react with glucose in the tears and change colour.
Thus, it could alert diabetics to dangerous sugar levels without the need for
regular blood tests. It is likely that the nanoparticles embedding process is
probably going to find more applications outside of medicine also.

A photochromic contact lens is developed by Singapore's institute of
Bioengineering and Nanotechnology which darkens under sunlight and are able
to change colour to protect eyes from ultraviolet radiation and glare. The dyes
in such nanomaterial can respond very fast to any light changes.

The big cosmetic players in the market today are L'Oreal, Estee Lauder,
Proctor & Gamble, Shiseido and Dupery Cosmetics.

Use of Nanotechnology in Dental Applications: Sun and Chow (2008)
have prepared 10–15 nm sized Calcium Fluoride for Dental Applications.
Calcium fluoride nanoparticle is expected to be a labile F reservoir for more
effective F regimens and as an agent for use in the reduction of dentin
permeability. The nano CaF_2 displayed much higher solubility and reactivity
than its macro counterpart. The CaF_2 ion activity product (IAP) of the solution
in equilibrium with the nano CaF_2 was $(1.52 \pm 0.05) \times 10^{-10}$, which was nearly
four times greater than the Ksp (3.9×10^{-11}) for CaF_2. The reaction of DCPD
with nano CaF_2 resulted in more F-containing apatitic materials compared to
the reaction with macro CaF_2. The F deposition by the nano CaF_2 rinse was
(2.2 ± 0.3) µg/cm^2 (n = 5), which was significantly (p < 0.001) greater than
that (0.31 ± 0.06) µg/cm^2) produced by the NaF solution.

13.4 Is Nanotechnology a Threat to its Application in Food and Cosmetics?

No doubt Nanotechnology has been touted as the next revolution in food,
cosmetics and packaging. However, nanoparticles have the potential to
penetrate deeply into the skin and organs, causing physical as well as
biochemical effects. Experiments using fish have shown that moisturizers
containing carbon fullerenes (buckyballs) can not only penetrate cells and
tissues, but also move through the body and brain, and cause biochemical and
physical damage. It has been found to be toxic to human liver cells also. The
health impacts of nanomaterials in cosmetics and sunscreens remain largely
unknown, but growing evidence indicating that nanomaterials can be toxic for

humans and the environment has lead the Friends of the Earth group to file first-ever legal challenge on the potential health impacts of nanotechnology in a 2006 petition to the FDA, demanding that the agency monitor and regulate nanoparticles in cosmetics.

According to Pascale, *et al.*, (2007), Poly-cationic organic nanoparticles disrupt model biological membranes and living cell membranes at nanomolar concentrations. The degree of disruption is shown to be related to nanoparticle size and charge as well as to the phase, fluid liquid crystalline or gel, of the biological membrane.

Following are the area of concern that needs to be looked into:

- Every new products should be evaluated.
- Anti-microbial nanoparticles used for sterilizing the surfaces, such as cutting boards, refrigerators or utensils should be studied to find out whether they are released on contact with food, do they bioaccumulate and how they impact human health?
- Acceptable levels of intake of nanomaterials and effect of size are other concerns to be looked into.
- Concern should also be there over the long-term fate and disposal and release of these nanoparticles into the environment.
- Toxicity risk of inhaled nanomaterials can cross the blood-brain barrier and gain easy access to tissues and cells.
- Many nanomaterials can photocatalytically may kill the useful bacteria also in the living system (*see* Table 13.1), such adverse effect should also be looked into.
- Some of the established adverse effect of nanoparticles are (*a*) nano silver is toxic to rodent liver, brain and stem cells and may harm beneficial bacteria (*b*) nano zinc oxide is known to be toxic to rat and human cells (*c*) nano silicon oxide even at < 70 nm can cause onset of pathology similar to neurodegenerative disorders and (*d*) nano titanium oxide can damage DNA in human cells, harm algae and water fleas, especially with UV light exposure.

Mathew, *et al.*, (2006) have suggested that Green chemistry metrics need to be incorporated into nanotechnologies at the source. According to them this will:

- Prevent generation of waste.
- The Bottom-up nanotechnology *i.e.*, building atom-by-atom, will be Atom efficient and generates less waste.
- Packaging and transport of the nanoparticles under a liquid will avoid release of fine dust including nanoparticles.

- Designing safer materials *e.g.*, Nanoparticles for drug delivery will help in reducing the side effects.
- Joy (200) has proposed that use of safer solvents and auxiliaries are emerging examples of green syntheses of nanoparticles.
- Green chemistry approach will involve more energy efficient and production of nanoparticles.
- Use of waste biomass for production of carbon nanomaterials *e.g.*, CNT, fullerenes *etc.*
- It will be inherently safer chemistry.
- Real-time analysis for pollution prevention will be possible.

According to Thomas, *et al.*, (2006), though the potential benefits of nanotechnology have been widely reported, little has been done to characterize the safety or identify potential hazards associated with products containing nanoscale materials. Sufficient hazard and exposure data are not yet available to conduct comprehensive risk assessments for products containing nanoscale materials. More studies are needed to evaluate the stability of these matrices in a variety of test systems to fully determine the potential for human exposure to the nanoscale components.

13.5 Summary

It can be concluded that in future, foods produced using nanotechnology will have improved food processing, packaging and safety; enhanced flavour and nutrition. However, there are concerns about the use of nanoparticles in food, cosmetics or contact materials. Hence, it is important to ensure that consumers are able to exercise choice in the use of the products of nanotechnology and that they have the information to assess the benefits and risks of such products. There is a need to address and confirm:

- Will nanotechnology affect the other properties?
- Will it have or generate the potential toxicity?

Nanoimplants: A Fantasy Leading to Reality

> There is something fascinating about science. One gets such wholesale returns of conjecture out of such a trifling investment of fact.
>
> —Mark Twain, Life on the Mississippi, 1883

14.1 Introduction

Science has been fantasized since the concept and application of science was implanted in human understanding. Power of imagination combined with the knowledge has predicted the advancement of science by many. Nanotechnology has been hyped and glamourized. What is the truth behind the hype and glamour of nanotechnology? What can something as small as one billionth of a meter do for you? Fantasies have been unlimited. Some fantasies that are either at the brink of becoming a reality or already a reality are:

- Waterproof clothing using nanowhiskers that make water to bead up.
- Cold-resistant shoes and boots having Nanopores.
- Silver nanoparticles as antimicrobial agent in wound treatment.
- Nanoshells to selectively kill cancer cells by focusing heat from infrared light.
- Light weight strong space ships manufactured using carbon nanotubes.
- Nanosensors to search water and life-supporting chemicals in different planets.
- Low cost nanotech solar cells.
- Platinum nanoparticles to improve fuel cell efficiency.

- Nanowire electrodes will make flat-panel TVs thinner and more flexible.
- Nanobatteries built by viruses are expected to be of the size of human cell.
- Nanocatalysts to turn more toxic emissions from auto engines and industries into harmless gasses.
- Nanofilters to strain lethal viruses from water.
- Onboard nanocomputers will enable self-controlled robotic fighting systems.

Acceptance of any implant in the human body has always been a topic of very careful consideration and scrutinization; same applies to Nanotechnology.

Nanomaterial science approaches atomic level control of material assembly, as their bulk and surface properties are influenced by quantum phenomena that do not govern traditional bulk material behaviour.

Nanoimplants that are being researched on include considerations like (*i*) synthetic material to replace the biological tissues, (*ii*) designing materials for specific medical applications and (*iii*) materials for diagnostics and array technologies.

14.1.1 Synthetic Materials to Replace the Biological Tissues

A material needed (for applications such as direct tissue replacement and tissue engineering) that can be used as replacement has to be compatible as well as similar to naturally occurring biological material. For this purpose, extracellular matrix (ECM) components are being studied as it provides an important model for biomaterials design (Yurchenco, *et al.*, 1994). The ECM is composed of a complex composite of proteins, glycoproteins and proteoglycans. Such ECM-derived macromolecules are collagen and elastin, which has been used for many years in biomaterials applications (Bell, *et al.*, 1979 and Huang, *et al.*, 2000). Now, efforts are on to create artificial analogues of ECM proteins using recombinant DNA technology (Hest and Tirrel, 2001). However, Langer and Tirrel (2004) have suggested that through the design and expression of artificial genes, it is possible to prepare artificial ECM proteins with controlled mechanical properties and with domains chosen to modulate cellular behaviour. They have also mentioned that this approach is better as it avoids several important limitations encountered in the use of natural ECM proteins *e.g.*, (*a*) batch-to-batch (or source-to source) variation in materials isolated from tissues, (*b*) restricted flexibility in the range of accessible materials properties and (*c*) possible disease transmission associated with materials isolated from mammalian sources.

Many workers (Huang, *et al.*, 2000; Lee, *et al.*, 2001 and Nagapudi, *et al.*, 2002) have shown that simple repeating polypeptides related to elastin can be engineered to exhibit mechanical behaviour reminiscent of the intact protein. Elastin polypeptides are cross linked and elastomeric. They have shown that crosslinking can be accomplished via radiative (Nagapudi, *et al.*, 2002) or chemical (MacMillan and Conticello, 2000) means. Huang, *et al.*, (2000) have used electrospinning technique to prepare fibrous forms of engineered elastin (Fig.14. 1). For incorporation, attachment and spreading of cultured cells; elastin cell adhesion Ligands have been used. Molecular self-assembly of peptides or peptide-amphiphiles may also lead to unique biomaterials.

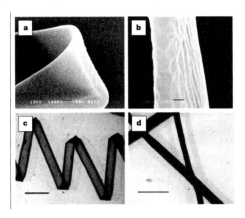

Fig. 14.1 Electrospun fibres of elastin-like artificial proteins made by expression of artificial genes in bacterial cells. a, b, Scanning electron micrographs. c, d, Transmission electron micrographs. Scale bars: a and b, 100 nm; c, 3.3 mm; d, 2.0 mm.

Images Permission Needed: An alternative to synthesizing polymers composed of natural components is the synthesis of biomimetic polymers that has both information content and multifunctional character of natural materials and that can be tailored to degrade at desired times and appropriate mechanical properties *e.g.*, polylactic-glycolic acid (PLGA) that is composed of lactic acid and lysine (Cook, *et al.*,1997). Addition of an amino acid lysine as co-polymer was providing a free amino acid and was helpful in allowing coupling reactions without affecting the overall biocompatibility of the polymer (Barrerra, *et al.*, 1993). Later on Cook, *et al.*, (1997) found that by coupling specific amino acids (such as the tripeptide sequence RGD) to this polymer, cell adhesion can be regulated. PLGA was later on modified by adding PEG to it, to reduce non-specific effects of protein adsorption and colloidal aggregation. Miyata, *et al.*, (1999) have created hydrogel which is a biomimetic reversible system, where corresponding antibody pairs are used to form reversible non-covalent cross-links in a polyacrylamide system. In the presence of excess free antigen, the hydrogel swells, but in its absence, the gel collapses back to a cross-linked network. Swelling does not occur when foreign antigens are added, showing that the system is antigen specific. They have also demonstrated the release of a model protein such as haemoglobin in response to specific antigens.

14.1.2 Designing Materials for Specific Medical Applications

Nanotechnology is being used to create potential new medical devices. Many different types of nanomaterials are being designed for the purpose of use as implantable material for specific medical applications. The last decade has seen the evolution of shape memory material for medical application. Kelch and Lendlein first proposed the shape memory polymers which have one shape at one temperature and another shape at a different temperature. Shape-memory materials have the ability to memorize a permanent shape that can be substantially different from an initial temporary shape. Materials that are liquids at room temperature, but harden in response to a change in temperature or an external stimulus, such as light, are also being studied (Anseth, et al., 1999). New polymers such as phase segregated multi-block copolymers have been synthesized whose starting materials are biocompatible monomers such as e-caprolactone and p-dioxanone. These materials possess two separate phases, each with thermal transition temperatures. The phase having high transition temperature is required for permanent shape while the second phase behave as a molecular switch leading to the fixation of the temporary shape. Shape of the polymers is highly modifiable since we can regulate the temperatures above or below the second phase transition temperature. New polymers have been synthesized with this concept in mind, including phase segregated multiblock copolymers whose starting materials are known biocompatible monomers such as e-caprolactone and p-dioxanone. Generally, these materials have at least two separated phases, each with thermal transition (glass or melting) temperatures. The phase with the higher transition temperature is responsible for the permanent shape, whereas the second phase can act as a molecular switch and enable fixation of a temporary shape. By regulating temperatures or pH above and below the second phase's transition temperature, shape can be shifted from one form to another. For example, smart gel developed by Peppas (1999) that responds (and swell, for example) to temperature or pH, or even specific molecules in the body such as glucose. Such systems may, with further study, be valuable in the treatment of disease: in the case of diabetes, for example, smart gels could provide direct feedback control, allowing more insulin to be delivered in response to excess glucose.

These materials are also being tried for use in gene therapy, especially to replace the viral vectors. To be effective, there are a number of attributes that the material must possess, including the ability to condense DNA to sizes less than 150 nm so that it can be taken up by receptor mediated endocytosis, the ability to be taken up by endosomes in the cell and to allow DNA to be released in active form, and to enable it to travel to the cell's nucleus37. An interesting example of the design of such new materials is provided by the cationic-cyclodextrin polymers developed by Hwang and Davis (2001). Cyclodextrins are relatively non-toxic and do not elicit an immune response.

Nano and micro-fabrication of implantable material is another approach for delivery system and creating sensors.

14.1.3 Materials for Diagnostics and Array Technologies

So far we have discussed the development of materials that display biological information, often in the form of peptide or protein domains, for controlling cell and tissue behaviour. However, difficulties and challenges remain in engineering these materials for use in diagnostics and array technologies. In diagnostics and array technologies, thousands of nucleic acids (Eisen and Brown, 1999) or proteins (Zhu and Snyder, 2003) are presented in a format that allows rapid and highly parallel read-out of information concerning gene expression or protein function. Analogous technologies are being pursued with peptides, carbohydrates (Houseman, *et al.*, 2002a and b) and other small molecules, as well as with cells and tissues (Kononen, *et al.*, 1999). Photo-nanolithographic technique has been used by Fodor, *et al.*, (1991) to prepare an array of 1,024 peptides in just ten sequential operations; and by McGall, *et al.*, (1996) to prepare oligonucleotide arrays. Brown and co-workers introduced an alternative approach to DNA arrays by high-speed robotic spotting of complementary DNAs on treated glass surfaces. Both techniques yield 'DNA chips' characterized by high densities of biological information.

14.2 Desired Properties of Nanomaterials to be used as Implants

Nanomaterials of size 100nm in atleast one dimension, is constituted by metals, ceramics, polymers or composite materials demonstrate novel properties at nanoscale level as compared to bulk. These unique properties in conjunction with recognition capabilities of biomolecules consequently leads to novel tissue substitutes, electronic tools such as biosensors, diagnostic system, and controlled drug delivery systems. Geometrically, nanomaterials can be divided into three major categories *viz.*

  (*i*) Gold, Platinum and titanium nanoparticles possess equiaxed structures. Gold nanoparticles have got their applications in cancer diagnostics and therapy. Platinum nanoparticles possess catalytic properties due to surface-to olume ratio and high surface energy. Titanium nanoparticles are exploited in pigments, transparent UV-scattering sunscreens, orthopaedic coatings *etc.* Dendrimers also possess equiaxed structure having multi-drug delivery system. Quantum dots have got very efficient transport and optical properties thus can be used in Biological sensors.

(*ii*) Carbon nano fibres or carbon nanotubes are also being used as drug carriers.

(*iii*) Lamellar materials like nanohydroxy apatite can be used as orthopaedic implants, bone/cartilage tissue engineering for plethora of problems of bone.

High reactivity of nanoparticles, chances of them clogging the cellular structures, their possible interaction with host organisms; has posed several questions in deciding the use of nanomaterials as implants. Hence, it is imperative that the nanomaterial should be selected only after thorough investigation of their properties and biocompatibility. Critical problems in biocompatibility, mechanical properties, degradation and numerous other areas remain, hence there is a need for new materials including those with improved biocompatibility, stealth properties, responsiveness (smart materials), specificity and other critical properties.

Current analytical tools have helped in elucidating novel properties of nanomaterials for various implant applications will also be covered.

14.2.1 Chemically Inert

Chemically inert material has found special place in controlled drug release implants *e.g.*, poly (ethyleneco-vinyl acetate) (EVAc); which has been used to release protein hormones (Brown, *et al.*, 1986), growth factors (Murray, *et al.*, 1983 and 1993; Edelmann, *et al.*, 1991; Krewson, *et al.*, 1995 and Mahoney & Saltzman, 2001), antibodies (Saltzman and Langer, 1989; Rodomsky, *et al.*, 1992 and Sherwood, *et al.*, 1996), antigens (Pries and Langer, 1979 and Wyatt, *et al.*, 1998) and DNA (Luo, *et al.*, 1999). EVAc matrices remaining inert allow a high degree of control over release and good retention of biological activity.

14.2.2 Biodegradable

Most commonly used biodegradable polymer is poly (lactide-co-glycolide) PLGA which has been used to release growth factors (Camarata, *et al.*, 1992;

Krewson, *et al.*, 1996 and Gombotz, *et al.*, 1993), protein hormones (Mathiowitz, *et al.*, 1988 and Johnson, *et al.*, 1996), antibodies (Sherwood, *et al.*, 1992), antigens (Cleland 1995; Marx, *et al.*, 1993 and Whittum-Hudson, *et al.*, 1996) and DNA (Shea, *et al.*, 1999 and Luo, *et al.*, 1999). Biodegradable materials disappear from the implant site after protein release.

14.2.3 Surface Plasmon Absorbance and Scattering

The metallic nanoparticles are known to exhibit different characteristic colours. Mie was first to explain the origin of this colour theoretically in 1908. the absorption of electromagnetic radiation by metallic nanoparticles originate from the coherent oscillation of the valence band electrons induced by an interaction with the electromagnetic field. These resonances are known as surface Plasmon, which occur only in the case of nanoparticles and not in the case of bulk metallic particles.

Gold nanoparticles are used in Cancer diagnostics and cancer therapy due to their strongly enhanced surface Plasmon absorption and scattering, and are much safer as an implant.

14.2.4 Surface Chemical Properties of the Nanomaterials

The alterations in surface chemistry and/or crystal structures of NP used in bone implants is well known to influence the bone cell functions. Webster, *et al.*, (1999) have shown that the crystal structure of conventional ceramics and crystal structures of their respective nanophase materials show similar characteristics, and only the degree of nanometer surface features was altered. Implanted Titanium NP invariably forms Titania or the titanium oxide.

Nanomaterials due to higher surface areas, higher surface roughness, higher amounts of surface defects (including grain boundaries), altered electron distributions, *etc.*, have unique surface properties and energetic. These properties of nanomaterials affect the interactions with proteins, since all proteins are nanoscale entities. Miller, *et al.*, (2006) examined fibronectin interactions with nanomaterials with various nanoscale surface features under atomic force microscope and demonstrated that increased surface areas and nanoscale surface features on nanomaterials can provide for more available sites for protein adsorption and thus, alter the amount of cellular interactions. It was demonstrated using adsorption of fibronectin on to PLGA surfaces with 500, 200 and 100 nm spherical bumps. The 500 nm bumps showed little to no interconnectivity; 200 nm spherical bumps showed a higher degree of interconnectivity, whereas fibronectin (5 mg/mL) adsorbed to PLGA surfaces with 100 nm spherical bumps showed well-spread fibronectin molecules with the highest degree of interconnectivity.

14.2.5 Properties Associated with Tissue Regeneration

The comprehension of nanomaterials interacting with proteins and cells for the regeneration of various tissues such as bone, cartilage, vascular, bladder and neural systems, needs a close view. The nanomaterials possess unique surface properties and are thermodynamically energetic structures forming stable complexes with Biological nanomachines such as proteins and carbohydrates, thus ultimately conjugate with certain organelles inside the cells.

14.3 Biomaterials used in Health Care Implants

Discovery of synthetic polymers by the end of nineteenth century has been a big boon to their use in health care e.g., polymethyl methacrylate (PMMA), in dentistry, ladies' girdles, in artificial hearts; cellulose acetate for dialysis tubing, Dacron to make vascular grafts; PMMA and stainless steel were used in hip replacements (Peppas and Langer, 1994 and Ratner, et al., 1996). Among the naturally occurring materials collagen is the most widely used biomaterial (Bell, et al., 1979). More modern efforts of developing desired biomolecules as implant material are based on the need to integrate biomaterials design with cell-matrix interactions, cellular signaling processes and developmental as well as systems biology. PMAA—is a synthetic polymer made up of methyl methacrylate.

Methyl methacrylate — Free radical vinyl polymerization → Poly(methyl methacrylate)

Cellulose acetate

14.4 Nano-biomaterials used in Tissue Engineering

Nanomaterial science is governing the atomic level control of material assembly; as their bulk and surface properties are influenced by quantum phenomena that do not govern traditional bulk material behaviour. Based on these properties, nanoparticles (metallic or polymeric molecules) have been developed and studies of their uses in various tissue engineering applications.

14.4.1 Aluminum and Titanium for Adhesion of Osteoblast

Webster (1999 and 2000) was first to report that alumina with grain sizes between 49 and 67 nm and titania with grain sizes between 32 and 56 nm promoted osteoblast adhesion compared to their respective micro-grained materials. Later, Gutwein (2004) also confirmed that nanoceramics, such as alumina, titania, and hydroxyapatite (HA) proliferates osteoblast and enhances long-term functions, such as collagen and alkaline phosphatase, as well as calcium-containing mineral deposition on ceramics with grain or fibre sizes less than 100 nm.

14.4.2 PLGA-Scaffold for Repair of Chondrocytes (Cartilage)

PLGA, is expected to contribute more to rehabilitating damaged bone tissue or catilage due to their controllable biodegradability. While the natural tissue regenerates, PLGA degrade *in vivo* by hydrolysis into non-toxic, natural metabolites the lactic acid, and glycolic acid which enter into normal metabolic pathways of the tricarboxylic acid cycle and eventually gets eliminated from the body in the form of carbon dioxide and water (Athanasiou, 1996).

Li and Webster (2007) have suggested that tissue engineering combined with nanomaterials for cartilage regeneration in a natural way by developing biomaterials that can mimic or closely match the composition, microstructure and properties of natural cartilage. Park, *et al.*, (2005) have prepared nanostructured PLGA by chemically etching PLGA in 1 N NaOH for 10 min. and demonstrated that NaOH-treated PLGA, three-dimensional scaffolds accelerated chondrocyte functions such as adhesion, growth, differentiation, and extracellular matrix synthesis compared to non-treated, traditional PLGA scaffolds. They believed that the material properties that may have enhanced chondrocyte functions include a more hydrophilic surface (due to hydrolytic degradation of PLGA by NaOH), increased surface area, altered porosity (both percent and diameter of individual pores), and a greater degree of nanometer roughness.

14.4.3 PLGA for Regeneration of Vascular Tissues

Miller, *et al.*, (2004) studied the responses of vascular cells (such as endothelial and smooth muscle cells) to PLGA treated with NaOH and found that endothelial and smooth muscle cell densities increased on nanostructured PLGA solely due to nanometer surface features.

Chowdhary, *et al.*, (2006) have also demonstrated increased endothelialization by nanometals prepared by powder metallurgy techniques as compared to conventional metals. Such evidence provides promise for the use of nanometals in vascular stent applications where through the use of conventional materials the formation of an endothelial monolayer is often problematic.

14.4.4 CNT/CNF-Composite in Matrix for Neural Implant

Whenever a neural implant is placed, at the site of implant scar formation occurs, which is due to gliotic response caused by the activity of astrocytes and meningeal cells, and is thought to interfere with the long-term efficacy of neural prostheses, which partitions the implant from the tissues. Neural prostheses provide a means for monitoring and applying electrical signals to neural tissue. Therefore, designing materials which enhance nerve cell interactions and deter astrocyte formation of scar tissue is crucial for neurological applications. Liu and Webster (2007) have suggested that CNT or CNF can transmit and receive electrical signals as well as supporting and enhancing nerve cell neurite/axon extension because of their excellent conductivity and biocompatibility properties. It was confirmed by findings of McKenzie, *et al.*, (2004) that showed that astrocytes adhered and proliferated less on carbon fibres that had the smallest nanometer diameter and the highest surface energy leading to decreased gliotic scar tissue formation and increased neuronal implant efficacy. Khang, *et al.*, (2008) have mentioned that when stem cells are combined with CNT/CNF it reverses neural tissue damage created by stroke *in vivo*. These results are showing promise and highlight the unique electrical and biocompatibility properties CNT/CNF may have with stem cells to treat a wide range of neurological disorder.

14.4.5 Nanoimplant for Bladder Tissue Replacement

Need for bladder replacement is an increasing demand, because more than 90% of bladder cancers are superficial bladder cancers that begin in the urothelium and require bladder tissue replacements. Either a large portions of the bladder or the entire bladder wall needs to be repaired in such cases. The polymer needed for bladder tissue replacement should not only be biocompatible, but should also have the ability to stretch and relax. Though presently most of the data for bladder regeneration with nanomaterials is at the *in vitro* level, but it does offer significant promise for the continued exploration of these materials. The key design parameter for achieving maximal cell responses is the material topography. The topography of natural soft bladder tissues are due to the constituent of extracellular matrix proteins that are of nanometer scale. Therefore, the polymer for bladder tissue replacement should incorporate nano-dimensional surface features. Keeping this in mind Thapa, *et al.*, (2003 a and b) have developed nanostructured PLGA and poly (ether urethane) (PU) formulations with surface feature dimensions ranging between 50 and 100 nm (Figs. 14.2 and 14.3). Results of tests of *in vitro* cytocompatibility with bladder smooth muscle cells showed that bladder smooth muscle cell adhesion and proliferation were enhanced on polymeric surfaces with nano-dimensional, compared to micro-dimensional, features. The bladder smooth muscle cell adhesion and proliferation were greater on two-dimensional nanometer surfaces of PLGA and PU. Similar trends were found by Pattison, *et al.*, (2005) when they used three-dimensional PLGA scaffolds.

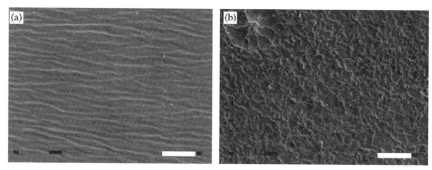

Fig. 14.2 Scanning electron micrographs of chemically treated PLGA surfaces. Representative (conventional) PLGA (feature dimensions 10–15 mm) and (*b*) chemically treated nanostructured bar ¼ 10 microns (*a*) and 1000 nm (*b*).

Adapted and reprinted with permission from Pattison WF, Wurster S, Webster T.J., Biomaterials 26, 2491–500 (2005).

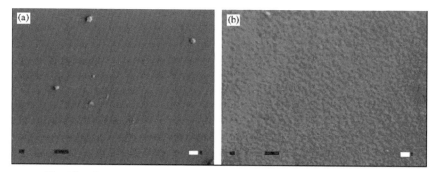

Fig. 14.3 Scanning electron micrographs of chemically treated PU surfaces. Representative scanning electron micrograph images of (*a*) untreated (conventional) PU (feature dimensions 415 mm) and (*b*) chemically treated nanostructured PU (feature dimensions 50–100 nm). Scale bar ¼ 1000 nm.

Adapted and reprinted with permission from Thapa A, Webster T.J., Haberstroh K.M, J. Biomed Material Research 67 A; 1374–83 (2003a).

14.5 Summary

Many attempts for use of nanotechnology, which initially appeared as fantasies; have now become a reality. The invention of such realities is touched upon in this chapter. Both inorganic and organic nanoparticles are used or the purpose. The unique and novel properties of nanoparticle have been found to be a big asset in developing implants for tissue regeneration, healthcare implants, neural implants and implants for bladder replacement, scaffold for cartilage and skin *etc.*, are some of them.

References

A

1. Abrahams JP, Leslie AGW, Lutter R, Walker JE, *Nature*, **370**: 621–628 (1994).

2. Ahmad A, Senapati S, Khan M, Kumar R, and Sastry M; *J. Biomed. Nanotechnol*, 1: 47, (2005).

3. Ahmad A, Senapati S, Khan M, Kumar R, and Sastry M; *Langmuir*, 19: 3550 (2003a).

4. Ahmad A, Senapati S, Khan M, Kumar R, Ramani R, Srinivas B and Sastry M; *Nanotech*, 14: 824 (2003b).

5. Aiking, Govers H and J van't Riet, *Appl. Environ. Microbiol.*, 50: 1262 (1985).

6. Akamatsu K, Kimura M, Shibata Y, Nakano S I, Miyoshi D, Nawafune H and Sugimoto N, *Nano Lett.*, 6, 491 (2006).

7. Akerman ME, Chan WCW, Laakkonen P, Bhatia SN, Ruoslahti E, *Proc. Natl. Acad. Sci. USA*, 99: 12617–21 (2002).

8. Aksay I A, Lange F F, Davis B I, *J. Am. Ceram. Soc.*, 66:190 (1983).

9. Alberti P and Mergny J L, *Proc. Natl. Acad. Sci. U.S.A.*, 100: 1569–1573 (2003).

10. Alivisatos A P, Johnsson K P, Peng XG, Wilson T E, Loweth C J, Bruchez M P and Schultz P G, *Nature*, 382: 609–611 (1996).

11. Alberti P, Bourdoncle A, Sacca B, Lacroix L and Mergny J L, *Org. Biomol. Chem.*, 4: 3383–3391(2006).

12. Alivisatos AP, *Science*, 271, 933 (1996).

13. Alivisatos AP, Gu W and Larabell C, *Annu. Rev. Biomed Eng.*, 7: 55, (2005).

14. Allen M and Willits D, *Adv Mater.*, 14: 1562 (2002).

15. Al Shawi MK, Ketchum CJ, Nakamoto RK, *Biochemistry*, **36**: 12961–12969 (1997).

16. Angelova B, Angelov B, Papahadjopoulos-Sternberg B, Ollivon M and Bourgaux C, *Langmuir*, 21: 4138–4143, 2005.

17. Ankamwar B, Damle C, Ahmad A and Sastry M, *J. Nanosci. Nanotech.*, 5: 1665, (2005).

18. Anseth K, Shastr, V and Langer R, *Nature Biotechnol.*, 17: 156–159 (1999).

19. Aoyama Y, Kanamori T. Nakai T. Sasaki T, Horiuchi S, Sando S, Niidome T, *J. Am. Chem. Soc.*, 125: 3455 (2003).

20. Armaroli N, Balzani V, Collin J- P., Gavina P, Sauvage J P and Ventura B, *J. Am. Chem. Soc.,* 121: 4397(1999).

21. Arnold R, OiChristina TJ and Hoffmann MR, *Appl. Environ. Microbiol.,* 52: 281 (1996).

22. Astumian RD, PNAS 102, 1843–1847 (2005).

23. Arkin M R, Stemp E D A, Homlin R E, Barton J K, Herrman A, Olson E J C, and Barbara P F, *Science*, 273: 475 (1996).

24. Arkin M R, Stemp E D A and Barton J K, *Chem. Biol.*, 4: 389 (1997).

25. Arnold R, OiChristina T J and Hoffmann M R, *Appl. Environ. Microbiol.,* 52: 281 (1996).

26. Arora S, Jain J, Rajwade J M and Paknikar K M, *Toxicology and Applied Pharmacology,* 236(3): 310–318 (2009).

27. Asakura S, *Advances in Biophysics* 1: 99–155 (1970).

28. Asbury C L, Fehr A N and Block S M, *Science* 302: 2130–2134 (2003).

29. Asha Rani P V, *ACS Nano*, 3: 279–290 (2009).

30. Ashton P R, Ballardini R, Balzani V, Constable E C, Credi A, Kocian O, Langford S J, Preece J A, . Prodi L, Schofield E R, Spencer N, Stoddart J F and Wenger S, *Chem. Eur. J.* 4: 2413 (1998).

31. Ashton PR, Balzani V, Kocian O, Prodi L, Spencer N and Stoddart J F, *J. Am. Chem. Soc.,* 120 : 11190 (1998).

32. Atala A, *J. Endourol.*,14 (1): 49–57, (2000).

33. Athanasiou KA, Niederauer GG, Agrawal CM, *Biomaterials*, 17(2):93–102 (1996).

34. Atherton J and Beaumont P C, *J. Phys. Chem.*, 99: 12025 (1995).

35. Atkins P W, 3rd Edition, Oxford University Press, Great Clarendon Street, Oxford (2001).

36. Atwood J L and Steed J W(eds.), Encyclopedia of Supramolecular Chemistry, Dekker, New York, (2004).

37. Aviram A and Ratner M A, *Chem. Phys. Lett.* 29: 277(1974).

38. Ayhan Demirba°, *Energy Sources, Part A: Recovery, Utilization, and Environmental Effects*, 27(14): 1313–1319 (2005).

39. Azzazy H M, Mansour M M, Kazmierczak SC. *Clin Chem*; 52:1238–46 (2006).

40. Azzazy H M, Mansour M M, Kazmierczak SC, *Clin Biochem*; 40:917–27 (2007).

B

41. Babitzke P, Stults J T, Shire S J and. Yanofsky C, *J. Biol. Chem.*, 269: 16597 (1994).

42. Babitzke P and Bear D G, *Proc Nat Acad Sci U S A*, 92: 7916 (1995).

43. Babitzke P, *Mol Microbiol*, 26: 1 (1997).

44. Babitzke P, *Curr. Opin. Microbiol.* 7, 132 (2004).

45. Badjing J D, Balzani V, Credi A, Silvi S and Stoddart J F, Science, 303: 1845 (2004).

46. Bae W, Abdulla R and Mehra R K, *Chemosphere*, 37: 363, (1998).

47. Bai W and Zhang Z, *Mater. Lett.* 63: 764 (2009).

48. Bai H J, Zhang Z M, Guoa Y and Yangc G E, *Colloids and Surfaces B: Biointerfaces* 70: 142 (2009).

49. Ballardini R, Balzani V, Credi A, Gandolfi M T andVenturi M, *Acc. Chem. Res.*, 34: 445 (2001).

50. Ballmoos C., *et al.*, *J. Biol. Chem.* 277, 3504–3510 (2002).

51. Balzani V, edited., Towards a Supramolecular Photochemistry. Assembly of Molecular Components to Obtain Photochemical Molecular Devices. In Supramolecular Photochemistry, Balzani, V. (ed.), Reidel, Dordrecth,, 1(1987).

52. Balzani V, Credi A, Marchioni F and Stoddart J F, *Chem. Commun.*, 1860–1861 (2001).

53. Balzani V, Credi A and Venturi M, *Nanotoday*, 2 (2): 18 (2007).

54. Barak R, Eisenbach M, *Biochemistry*, 31:1821–1826 (1992).

55. Barrera D A, Zylstra E, Lansbury P T and Langer R, J. *Am. Chem. Soc.*, 115, 11010–11011 (1993).

56. Barton J K and Kelly S O (eds Sigel, A. and Sigel, H.), Marcel Dekker, New York, vol. 39, pp. 211(1999).

57. Basavaraja S, Balaji S D, Lagashetty A, Rajasab A H and Venkataraman A, *Mater.Res.Bull* 6: 20 (2007).

58. Baudhuin P, Van der Smissen P, Beavois S, Courtoy, *J.Colloidal Gold: Principles, Methods, Applications,* 2: 1 (1989).

59. Baughman R H, Zakhidov A A and de Heer WA, *Science*; 297: 787 (2002).

60. Baumann H, *J. Biol. Chem.*, 272: 14571–14579(1997).

61. Behkam B and Sitti M, *"Towards Hybrid Swimming Microrobots: Bacteria Assisted Propulsion of Polystyrene Beads",* in Proceedings of the 28th IEEE EMBS Annual International Conference New York City, USA, Aug 30-Sept 3, 2006, pp 2421–2424 (2006).

62. Bell E, Ivarsson B and Merrill C, *Proc. Natl Acad. Sci. USA*, 76:1274–1278 (1979).

63. Belliveau H, Starodub M E, Cotter C, and Trevors J T, *Biotech. Adv.* 5:101 (1987).

64. Ben-Jacob E, Hermon Z and Caspi S, *Phys. Lett.A*, 263: 199 (1999).

65. Beratan D N, Onuchic J N and Hopfield J J, *J. Chem. Phys.*, 86: 4488 (1987).

66. Berg H and Turner L, *Biophys. J.,* 65: 2201–16 (1993).

67. Berg U E, Kreuter J, Spieser P P and Soliva M, *Pharm. Ind.* 48: 75(1986).

68. Berlin Yuri A., Alexander L. Burin, and Mark A. Ratner, *J. Phys. Chem. A*, **2000**, *104* (3): 443–445.

69. Berlin, Y. A.; Burin, A. L.; Ratner, M. A. J. Am. Chem. Soc. 2001,123, 260–268.

70. Beveridge T J, and Murray R G, *J. Bacteriol.,* 141: 876 (1980).

71. Bhalla V, Bajpai R P and Bharadwaj L M, *EMBO reports*, 4: 442, (2003).

72. Bhardwaj S, Maheshwar Sharon, Soga T, Afre R, Sathiyamoorthy D, Dasgupta K, Madhuri Sharon and Jaybhaye S, *International Journal of Hydrogen Energy,* 32: 4238– 4249 (2007).

73. Bhardwaj S, Maheshwar Sharon, Ishihara T, Jayabhaye S, Afre R, Soga T and Madhuri Sharon, *Carbon Lett.* 8(4): 285–291 (2007).

74. Bhardwaj S, Jaybhaye S, Madhuri Sharon, Sathiyamoorthy D, Dasgupta K, Jagdale P, Gupta A, Patil B, Ozha G, Pandey S, Soga T, Afre R, Kalita G and Maheshwar Sharon, *Asian J Exp. Sci.,* 22(2): 89–93, (2008).

75. Bhardwaj S, Maheshawar Sharon, Jaybhaye S, Sathiyamoorthy D, Dasgupta K and Madhuri Sharon, *Asian J. Exp. Sci.*, 22(2): 75–88 (2008).

76. Bhardwaj S, Maheshwar Sharon and Ishihara T, *Current Applied Physics,* 8(1): 71–77(2008).

77. Bhatt R, de Vries P, Tulinsky J, *J. Med. Chem.*, 46, 190 (2003).

78. Bhushan B, *Introduction to Tribology* (Wiley, New York, (2002).

79. Bianco A, Kostarelos K and Prato M, *Curr Opin Chem Biol.* 9(6): 674–9 (2005*)*.

80. Bieri C, Ernst O P, Heyse S, Hofmann K P, Vogel H, *Nat. Biotechnol.*, 17: 1105 (1999).

81. Bifeng Pan, Daxiang Cui, Ping Xu, *Colloid and Surface A*. 2007, 295: 217–222. 21.

82. Bilati U, Allémann E, Doelker E., *Eur J Pharm Sci.*; 24: 67–75 (2005).

83. Birge R R, *Computers* 25: 56 (1992).

84. Bixon M, Giese B, Wessely S, Langenbacher T, Michel-Beyerle M E and Jortner J, *Proc. Natl. Acad. Sci. U. S. A.,* 96: 11713 (1999).

85. Blair D.F., *Annu. Rev. Microbiol.,* 49:489–522 (1995).

86. *Blank F,* Vanhecke D, Ochs M and Gehr P, *Am. J.* Respir. *Cell Mol. Biol.,* 36(6): 669–677 (2007).

87. Block S M, Goldstein L S B and Schnapp B J, *Nature* 348: 348–352 (1990)

88. Bockman R., *Nat. Struct.* 9: 198–202 (2002).

89. Bong D T, Clark T D et al *Angew. Chem. Int. Ed.* 40: 988 (2001).

90. Bootz A, Vogel V, Schubert D, Kreuter J. *Pharm Res,* 20, 409 (2003).

91. Bourret R.B., Stock A.M., *J. Biol. Chem.* 277: 9625–28 (2002).

92. Bowden F P and Tabor D, *The friction and lubrication of solids,* New York: Oxford Univ. Press, (1950).

93. Boyer PD. *FASEB J,* **3**: 2164–2178(1989).

94. Boyer P., *Biochimica et Biophysica Acta* 1140, 215–250 (1993).

95. Boyer, P., *Biochim.Biophys.Acta* 1365, 3–9 (1998).

96. Boyer, P., *Biochim.Biophys.Acta* 1458, 252–262 (2000).

97. Braatz R D, Alkire R and Seebauer C. *Computers & Chemical Engineering,* 30:1634–1656, (2006).

98. Brady ST, *Nature* 317: 73–75 (1985).

99. Braig H R, Guzman H, Tesh R B and O'Neill S L, *Nature,* 367(6462): 453–5 (1994).

100. Bray D., *Proc. Natl. Acad. Sci.* USA 99: 7–9 (2002).

101. Bren A., Eisenbach M., *J. Bacteriol.* 182: 6865–73 (2000).

102. Brower A M, Frochst C, Gatti F C, Leigh D A, Mottier L, Paolucci F, Roffio S and Wurpel G W H, *Science,* 291: 2124–2128 (2001).

103. Brown L, Siemer L, Munoz C, Edelman E and Langer R, *Diabetes* 35: 692–697 (1986).

104. Brown D M, Wilson M R, MacNee W, Stone V, and Donaldson K, *Toxicology and Applied Pharmacology,* 175: 191–199 (2001).

105. Bruchez M, Moronne M, Gin P, Weiss S, Alivisatos AP, *Science;*281: 2013–6 (1998).

106. Brun A M and Harriman A, *J. Am. Chem. Soc.,* 114, 3656 (1992).

107. Brus LE, *Appl. Phys. A,* 53: 465 (1991).

108. Brust M, Walker D, Bethell DJ, Schiffrin R and Whyman, *J. Chem. Soc., Chem. Commun.* 801. 1994.

109. Budhian A, Siegel SJ, Winey KI, *J. Microencapsul.,* 22 (7): 773 (2005).

110. Bueken A and Huang H, *Journal of Hazardous Materials*, 62: 1–33 (1998).

111. Butt T R and Ecker D J, *Microbiol. Rev.* 51: 351(1987).

112. Byrne, J A, Davidson A, Dunlop, P S M, and Eggins B R, *Photochem. Photobiol. A: Chem.,*148: 365–374 (2002).

113. Byrne J A, Dunlop P S M, McMurray T A, Hamilton J W J, Alrousan D and Dale G, *Photocatalytic Water Treatment and Purification in Report from the Workshop onNanotechnologies for Environmental Remediation* JRC Ispra 16–17 April 2007.

C

114. Cagri A, Ustunol Z and Ryser E T, *J. Food Protect.*, 67: 833–848 (2004).

115. Calladine C R, *Nature,* 255: 121–124 (1975).

116. Calvo P, RemunanLopez C, VilaJato JL, Alonso MJ. *J Appl Polym Sci;* 63:125–32 (1997).

117. Camarata P J, Suryanarayanan R and Turner D A, *Neurosurgery,* 30: 313–319 (1992).

118. Capaldi R., and Aggeler R., *Trends Biochem. Sci.* 27, 154–200 (2002).

119. Cappello G, Badoual M, Ott A and Prost j, *Phys. Rev. E* 68: 021907 (2003).

120. Carlson C, Hussain S M, Schrand A M, Braydich-Stolle L K, Hess K L, Jones R L and Schlager J J, *J. Phys. Chem. B.,*112 (43): 13608–13619 (2008).

121. Carr C M and Kim P, *Cell,* 73: 823(1993).

122. Carter F Let al. (eds.) Molecular Electronic Devices, Elsevier, Amsterdam (1988).

123. Carroll D L, Redlich P, Ajayan P M, CharlierJ-C, Blasé X, De Vita A, Car R, *Phys. Rev. Lett.* 78 (2811)363: 364–373 (1997).

124. Castranova A A, Kisin V, Schwegler-Berry E R,, Murray D Gandelsman A R, Maynard V Z, and Baron A, *J Toxicol Environmental Health, Part A,* 666: 1909–1926 (2003).

125. Cavalcanti A and Freitas R A Jr., *IEEE Tr. Nanobioscience*, 4(2): 133–140, (2005).

126. Cavalcanti A, Bijan T H, Hwee S and Liaw C, "Nanorobot Communication Techniques: A Comprehensive Tutorial", *IEEE ICARCV 2006 International Conference on Control, Automation, Robotics and Vision*, (2006).

127. Cha D S and Chinnan M S, Crit. *Rev. Food Sci.. Nutr.* 44:223–237 (2004).

128. Chan WCW and Nie SM, *Science* 201602018 (1998).

129. Chan WC, Maxwell DJ, Gao X, Bailey RE, Han M, Nie S. *Curr Opin Biotechnol*;13:40–6 (2002).

130. Chan WH, Shiao NH, Lu PZ, *Toxicol Letters*, 167: 191–200 (2006).

131. Chang C.-N., Y.-S. Ma, G.-C. Fang, A. C. Chao, M.-C. Tsai, and H.-F. Sung, *Chemosphere*, 56(10): 1011–1017, 2004.

132. Chang Y C and Chen D G H, *Macromol. Biosci.* 5: 254–261 (2005).

133. Charych D, Cheng Q, Reichert A, Uziemko G, Stroh N, Nagy J, Spevak W and Stevens R, *Chem. Biol.* 3: 113 (1996).

134. Chatterjee A K, Sharon Maheshwar and Banerjee R, *J. of Power sources*, 117: 39–44 (2003).

135. Chatterjee A K, Sharon Maheshwar and Banerjee R, *J. of Power sources*, 5251: 1–6 (2003).

136. Chebotareva N A, Kurganov B I, Livanova N B, *Biochemistry (Moscow)* 69: 1239 (2004).

137. Chekhonin VP, Zhirkov YA, Gurina OI, Ryabukhin IA, Lebedev SV, Kashparov IA, Dmitriyeva TB. *Drug Deliv.*, 12: 1 (2005).

138. Chen J and Seeman N C, *Nature* 350: 631–633 (1991).

139. Chen X, Antson A A, *et al.*, *J. Mol. Biol,* 289, 1003 (1999).

140. Chen X and Berg H C *Biophys. J.,* 78:1036–41 (2000).

141. Chen J M and Ho C, *Ann. N.Y. Acad.Sci.*, 1093: 123 (2006).

142. Chen Kai Loon, Steven E, and Menachem E, *Langmuir*, 23 (11): 5920–5928 (2007).

143. Chen X, Murawski A and Patel K, *Toxicol. Appld. Pharmacol*, 230: 364–367, (2008)

144. Chen YF, Rosenzweig Z. *Anal Chem*; 74:5132–8 (2002).

145. Cheney R E, O'Shea M K, Heuser J E, Coelho M V, Wolenski J S, Espreafico E M, Forscher P, Larson R E, and Mooseker M S, *Cell* 75: 13–23 (1993).

146. Cheng Qiaoyuan· Jian Feng· Jianming Chen[2], Xuan Zhu· Fanzhu Li, *Biopharm. Drug Disposition,* 29: 431(2008).

147. Cheung M S, Daizadeh I, Stuchebrukhov A A and Heelis P F, *Biophys J*, 76: 1241(1999).

148. Chih-Ming Ho, *"Fluidics- The Link between Micro and Nano Sciences and Technologies"*, 0-7803-5998-4/01@ 2001 IEEE, pp 375–384 (2001).

149. Chithrani BD, Ghazani AA, Chan WCW, *Nano Lett.,* 6: 662 (2006).

150. Cho M, Chung H, Choi W and Yoon J, *Applied Environ. Microbiol.*, 71: 270–275 (2005).

151. Choi S H and Park TG, *Int J Pharm,* 203: 193–202 (2000).

152. Choudhary A K· Donnelly L F, Racadio J M and Strife J L, *AmerJour Roentgenology* 2007; 188: 1118–1130.

153. Chrusch D D, Podaima B W, and Gordon R, in "Conference Proceedings IEEE Canadian Conference on Electrical and Computer Engineering" (W. Kinsner and A. Sebak, Eds.). IEEE, Winnipeg, Canada (2002).

154. Churchill M E A, Tullius T D, Kallenbach N R and Seeman N C, *Proc. Natl. Acad. Sci. USA* 1985.

155. Cleland J L in *Vaccine Design: the Subunit and Adjuvant Approach* (eds Powell, M. F. & Newman, M. J.) 439–462,Plenum, New York, (1995).

156. Clemens S, Kim E J, Neumann D and Schroeder J I, *EMBO J.* 18, 3325 (1999).

157. Clemens S Planta 212: 475–486 (2001).

158. Clemente-León M Credi A, Martínez-Díaz M-V, Mingotaud C, Stoddart J F, *Adv. Mater.* 18, 1291 (2006).

159. Cobbett C, Goldsbrough P, *Annu Rev Plant Biol* 53: 159–182 (2002).

160. Cobbett C S, *Plant Physiol.*, 123:825–832 (2000).

161. Coblenz A and Wolf K, *FEMS Microbiol. Rev.* 14: 303 (1994).

162. Cohen M.L,. Chou M.Y, Knight W.D, de Heer W.A, *J.Phys.Chem.* 91, 3141 (1987).

163. Cook A D *et al.*, *J. Biomed. Mater. Res.*, 35: 513–523 (1997).

164. Connor EE, Mwamuka J, Gole A, Murphy CJ, Wyatt MD, *Small* 1: 325–327 (2005).

165. Cornell Eva, *NNIN REU Research Accomplishments :* 1–2 (2006).

166. Court D L, Sawitzke J A, *et al.*, *Annu Rev Genet*, 36: 361 (2002).

167. Cram D J, *Angew. Chem., Int. Ed* 27: 1009 (1988).

168. Cramer F, Chaos and Order. The Complex Structure of Living Systems, VSH, Weinheim (1993).

169. Crawford S A, Higgins M J, Mulvaney P and Wetherbee R, *J. Phycol.* **37:** 1 (2001).

170. Cui Y, Wei Q, Park H and Lieber CM, *Science*, 293: 1289, (2001).

171. *Cullis C F, and Hirschler M, Eur. Polym. J.*, 17: 451 (1981).

172. Cumings J and Zettl A, *Science* 289, 602 (2000).

173. Curtis A S G, *IEEE Tr. On Nanobioscience*, 4(2): 201–202 (2005).

D

174. Dahan M, Levi S, Luccardini C, Rostaing P, Riveau B, Triller A, *Science,* 302: 442–445 (2003).

175. Dai H. and Flesher, wrow.medicalreport.com – Medical cases report on *Nanotubes can kill cancer cells, leave healthy tissue intact.* August (2005).

176. Dameron C T, Reese R, Mehra R K, Kortan A R, Carroll P J, Steigerwald M L, Brus L Z and Winge D R, *Nature* 338: 596 (1989).

177. Dandliker P J, Homlin R E and Barton J K, *Science*, 275: 1465, (1997).

178. Davis W B, Hess S, Naydenova I, Haselsberger R, Ogrodnik A, NewtonM D and Michel-Beyerle M-E, *J. Am. Chem. Soc.*, 124: 2422(2002).

179. De Campos AM, Sanchez A, Alonso MJ, *Int. J. Pharm.* 224 (1–2): 159 (2001).

180. Dekker C and Ratner M, *Physics world*, 14: 29 (2001).

181. De Lorenzo AJD, CIBA Foundation Symposium Series; J&A Churchill: London; 151, (1970).

182. Demirok U, Laocharoensuk R, Manesh M, Wang J, *Angew. Chem. Int. Ed.* 47: 9349–9351 (2008).

183. Desai V S and Kowshik M, *Research Journal of Microbiology*, 4 (3): 97–103, (2009).

184. De Pamphilis M L and ADLER J, *Journal of Bacteriology* 105: 376–383(1971c).

185. Derakhshandeh K, Erfan M, Dadashzadeh S, *Eur. J. Pharm. Biopharm.*, 66 (1): 34, (2007).

186. De Vita A, Charlier J-C, Blasé X and Car R, *Appl. Phys. A* 68, 283 (1999).

187. Dhar P, Fischer Th M, Wang Y, Mallouk T. E, Paxton W F and Sen A, *Nano Lett.*,6, 66–72 (2006).

188. Dickerson M B, Naik R R, Sarosi P M, Agarwal G, Stone M O and Sandhage K. H, *J Nanosci Nanotechnol.* 5: 63 (2005).

189. K. E. Drexler K E, *Proc. Nat. Acad. Sci.*, 78: 5275 (1981).

190. Drexler K E, *"Nanosystems: Molecular Machinery, Manufacturing, and Computation."* Wiley, New York, (1992).

191. Dreyfus R, Baury J, Roper M L, Fermigiev M, Stone H A and Bibette J, *Nature,* 437: 862 (2005).

192. Drummond T G, Hill M G and Barton J K, *Nat. Biotechnol.*, 21: 1192 (2003).

193. Drummond T G, Hill M G and Barton J K, *J. Am. Chem. Soc.,*126: 15010 (2004).

194. Dubertret B, Skourides P, Norris DJ, Noireaux V, Brivanlou AH, Libchaber A. *Science*;298:1759–62 (2002).

195. Dubey A, Sharma G, Mavroidis C, Tomassone M S, Nikitczuk K and Yarmushc M L, *J.Computational and Theoretical Nanoscience*, 1: 18 (2004).

196. Duggan D J, Bitttner M, Chen Y, Meltzer P, Trent J M, *Nat. Genet. Suppl.*, 21: 10 (1999).

197. Dujardin E, Peet C, Stubbs G, Culver J N and Mann S, *Nano Lett.*, 3: 413–417 (2003).

198. Duran N, Marcato P, Alves O, D'souza G and Esposito E, *J. Nanobiotechnol.*, 3: 1 (2005).

199. Dyadyusha L, Yin H, Jaiswal S, Brown T, Baumberg JJ, Booy FP, *Chem Commun*: *9*, 3201–3 (2005).

E

200. Ebbesen TW ed., *"Carbon Nanotubes – Preparation and Properties"*, T. W. CRC Press ; ISBN 0-84939-602-6 (1996).

201. Ebbesen T W, *J. Phys.Chem. Solids*, 57: 951(1996).

202. Edelman E R, Mathiowitz E, Langer R and Klagsbrun M, 12: 619–626 (1991).

203. Egerton T A, Samia Kosa A M and Christensen P A, *Phys. Chem. Chem. Phys.,* 8: 398–406 (2005).

204. Einstein A, *Ann. Phys.Lpz.*, 17: 549 (1905).

205. Eisen M B and Brown P O, *Methods Enzymol.*, 303:179–205 (1999).

206. Ekins R and Chu F J, *J. Int. Fed. Clin. Chem.,* 9: 100 (1997).

207. Elder A and Oberdorster G, *Clinics in occupational and environmental medicine*, 5(4): 785–96 (2006).

208. Eley D D and Spivey D I, *Trans. Faraday Soc.,* 58: 411, (1962).

209. Elghanian R, Storhoff J J, Mucic RC, Letsinger RL, Mirkin C. A. *Science*, 277: 1078 (1997).

210. Goresy, A. El., & Donnay, G. 1968, Science, 161, 363.

211. Ellert C, Schmidt M, Schmitt C, Reiners T, Haberland H, *Phys.Rev.Lett.* 75: 1731 (1995).

212. Ellis J R, *Trends Biochem. Sci.,* 26: 597 (2001a).

213. Ellis J R, *Curr. Opin. Struct. Biol.*, 11: 114 (2001b).

214. Elston T C and Oster G, Biophys. Jour,. 73:703–721 (1997).

215. Emili A Q and Cagney G, *Nat. Biotechnol.*, 18: 393 (2000).

F

216. Fabris L, Antonello S, Armelao L, Donkers R L, Polo F, Toniolo C, Maran FJ, *Am. Chem. Soc.*, 128: 326 (2006).

217. Fahlman B D, *Materials Chemistry*; Springer: Mount Pleasant, MI, 1: 282–283 (2007).

218. Falke J.J., Bass RB, Butler SL, Chervitz, SA, Danielson MA.. Annu. Rev. Cell Dev. 13:457–512 (1997).

219. Fan H, Leve EW, Scullin C, Gabaldon J, Tallant D, Bunge S, Boyle T, Wilson MC, Brinker CJ, *Nano Lett* 5: 645–648 (2005).

220. Faraday M, *Phil. Trans.*, 147: 145 (1857).

221. Faraji AH and Wipf P, *Bioorganic & Medicinal Chemistry*, 17: 2950, (2009).

222. Faravelli L and Rossi R, Fuzzy chip controller implementation, *"Proc. 3rd Int. Workshop on Structural Control"*, World Scientific, Singapore, pp. 201–213 (2000).

223. Farajian A A, Ohno K, Esfarjani K, Maruyama Y and Kawazoe YJ, *Chem Phys.*, 111: 2164–8 (1999).

224. Farid R S, Moser C C and Dutton p l, *Curr. Opin. Struct. Biol.* 3: 225(1993).

225. Farokhzad O and Langer R, *ACS Nano (Perspective)*, 3(1): 16, (2009).

226. Fechter L D, Johnson D L, Lynch R A, *Neurotoxicology*, 23: 177–183 (2002).

227. Feringa B L, *Acc. Chem. Res.*, 34 (6): 504–513 (2001).

228. Feringa B L, Koumura N, van Delden R A and Wiel M K J ter, *App. Phys. A,* 75: 301–308 (2002).

229. Feynman R P, *Eng. Sci. Caltech* 23: 22 (1960).

230. Fiebig T, Wan C, Kelley S O, Barton J K and Zewail A H, *Proc.Natl. Acad. Sci. U.S.A.,* 96: 1187 (1999).

231. Fillingame R., et al. *Biochim. Biophys. Acta* 1555, 29–36 (2002).

232. Fink H J and Schönenberger C, *Nature,* 398: 407 (1999).

233. Flanagan J and Singh H, *Crit. Rev. Food Sci. Nutr.* 46: 221–237 (2006).

234. Flitman S S, "Paper Presented at Eighth Foresight Conference on Molecular Nanotechnology," (2000).

235. Fodor S P, Read J L, Pirrung MC, Stryer L, Lu A T, Solas D, *Science,* 251: 767–773 (1991).

236. Fonseca C, Simoes S, Gaspar R, *J. Control. Release*, 83 (2): 273 (2002).

237. Fordham-Skelton A P, Robinson N J and Goldsbrough P B in *Metal Ions in Gene Regulation* (eds S. Silver, W. Walden) Chapman & Hall, London, pp 398 (1998).

238. Forty N, Beveridge T, edited by E. Bacuerien, Willey-VCH, Weinhein; vol. 7 (2000).

239. Fournier-Bidoz S, Arsenault A C, Manners I, Ozin G A, *Chem. Commun.*, 4, 441–443 (2005).

240. Foyer C H, Theodoulou F L, Delrot S, *Trends Plant Sci.,* 6: 486–492 (2001).

241. Francis N R, Sosinsky G E, Thomas D and DeRosier D J. *J.Mol. Biol.* 235: 1261–70 (1994).

242. Freemantle M, *Chem. Eng. News*, 37: (1998).

243. Freitas R A Jr., *Artificial Cells, Blood Design, and Immobility, Biotech.*, 26: 441–430 (1998).

244. Freitas R A Jr., *Nanomedicine*, Volume I: Basic Capabilities, Landes Bioscience, Georgetown, TX, (1999).

245. Freitas R A, Jr., *Foresight Update,* 41: 9 (2000).

246. Freitas R A Jr., *Foresight Update* 44: 11 (2001).

247. Freitas R A Jr. and Phoenix C J, *J. Evol. Technol.* 11 (2002).

248. Freitas R A Jr. Nanomedicine, Volume IIA: Biocompatibility, Landes Bioscience, Georgetown, TX, (2003).

249. Frithjof A.S. Sterrenburg, *J .Nanosci. Nanotech.* 5,100 (2005).

250. Fu T J and Seeman N C, *Biochemistry* 32: 3211–3220 (1993).

251. Fukuda T, Arai F and Dong L, "Assembly of Nanodevices with Carbon Nanotubes through Nanorobotic Manipulations", *Proceedings IEEE*, vol. 91, No. 11, pp. 1803–1818 (2003).

252. Fukui K and Tanaka K, *Angew. Chem.Int. Ed. Engl.,* 37: 158 (1998).

253. Furuta T, Samatey F A, Matsunami H, Imada K, Namba K and Kitao, *Jour. Structural Biol.* 157: 481–490 (2007).

G

254. Gabay T, Jakobs E, Ben-Jacob E and Hanein Y, *Physica A*,250: 611–21(2005).

255. Gabel F C V and Berg H C, *Proc. Natl Acad. Sci. USA*, 100: 8748–8751 (2003).

256. Gallarate M, Carlotti M E, Trotta M and Bovo S, *Int. J. Pharm.* 188: 233–241 (1999)

257. Galvez J B, Ibanez P F and Sixto M R, *J. Solar Energy Eng.*, 129: 12–12 (2007).

258. Gandini D, Mahé E, Michaud P-A, Haenni W, Perret A and Comninellis Ch.. *Journal of Applied Electrochemistry,* 30: 1345–1350 (2000).

259. Gao L,, Leng N, Taihong Wang, Yujun Qin· Zhixin Guo, Dongling Yang· Xiyun Yan, Gardea- Torresdey J L, Parsons J G, Gomez J and Feralta-Videa J, *Nano Lett.* 2: 397 (2002).

260. Gao X, Chan WC and Nie S, *J Biomed Opt* 7: 532–537 (2002).

261. Gao XH, Cui YY, Levenson RM, Chung LWK, Nie SM., *Nat Biotechnol*; 22:969–76 (2004).

262. Garg J, Poudel B, Chiesa M, *J Appl Phys*; 103: 074301 (2008).

263. Garti N, Shevachman M and Shani A, *J. Am. Oil Chem. Soc.* 81: 873–877 (2004).

264. Garti N, Spernath A, Aserin A and Lutz R, *Soft Matter* 1: 206–218 (2005).

265. Gasco M R, *U.S. Patent* 5, 250, 236 (1993).

266. Gauger E and Stark H, *Phy. Rev. E*, 74 (1–10): 021907 (2006).

267. Gerion D, Pinaud F, Williams SC, Parak WJ, Zanchet D, Weiss S, *J Phys Chem B*;105: 8861–71 (2001).

268. Gerrity D, Ryu H, Crittenden J and Abbaszadegan M, *J. Environ. Sci. Health Part A*, 43: 1264–1270 (2008).

269. Giese B, Wessley S, Spormann M, Lindemann U, Meggers E and Michel-Beyerle M E, *Angew. Chem., Int. Ed. Engl.* 38: 996 (1999).

270. Giles J, *Nature*, Mar. 23, (2004)

271. Gojova B, Guo R S, Kota JC, Rutledge I and Kennedy M, *Environmental Health Perspective*, 115 (3): 403–409 (2007).

272. Golding M and Sein A, *Food Hydrocoll.* 18: 451–461 9 (2004).

273. Goldman ER, Medintz IL, Mattoussi H, *Anal Bioanal Chem*; 384: 560–3 (2006).

274. Gollnick P, Babitzke P, Antson A, Yanofsky C *Annu. Rev. Genet.*, 39: 47–68 (2005).

275. Gombotz W R, Pankey S C, Bouchard L S, Ranchalis J and Puolakkainen P, *J. Biomater.Sci. Polym. Edn*, 5: 49–63 (1993).

276. Goodsell DS, Bionanotechnology: Lessons from Nature. *Wiley-Liss, Inc* (2004).

277. Goodsell D.S., *American Scientist*, 88, 230–237 (2000).

278. Goresy El A and Donnay G, Science, 161: 363 (1968).

279. Górzny M, Walton A S, Wnek M, Stockley P G and Evans S D, *Nanotechnology*, 16: 5704 (2008).

280. Govindaraj A, Satishkumar B C, Nath M and Rao C N R., *Chem Mater,* 12 : 202–205 (2001).

281. Gradishar W.J, Tjulandin S, Davidson N, *J. Clin. Oncol.*, 23: 7794 (2005).

282. Gray H B and Winkler J R, *Annu. Rev. Biochem.* 65: 537 (1996).

283. Gresser M. J., Myers J. A., & Boyer P. D., *J. Biol. Chem.* 257, 12030–12038 (1982).

284. Griffiths D J, Introduction to Quantum Mechanics, 2nd edition, Prentice Hall (2004).

285. Gross S P, Vershinin M and Shubeita G T, *Current Biology*, 17, R478–R486: R479, (2007).

286. Gu F, Zhang L, Teply B A, Mann N, Wang A, Radovic-Moreno A F, R Langer, Farokhzad O C, *Proc. Natl. Acad. Sci.* U.S.A., 105: 2586 (2008).

287. Gulbis J M, Kelman Z, Hurwitz J, O'Donnell Mand Kuriyan, *Cell,* 87: 297 (1996).

288. Guo WZ, Li JJ, Wang YA, Peng XG. *Chem Mater;*15: 3125–33 (2003).

289. Gupta A, Ph. D. Thesis, University of Mumbai (2009).

290. Gupta A K, Nair Pradeep R, Demir Akin, Michael R. Ladischt, Steve Broyles, Muhammad A. Alam and Rashid Bashir *PNAS* Sept. 5,103, (2006) 36.

291. Gupta A K and Gupta M, *Biomaterials* 26: 3995–4021 (2005).

292. Gustafsson R J, Orlov A, Griffiths P T, Cox RA and Lambert R M, *Chemical Communications,* 37: 3936–3938 (2006).

293. Gutwein LG, Tepper F, Webster TJ, *Increased osteoblast function onnanofibered alumina.* 26th Annual American Ceramic Society Meeting,Cocoa Beach, FL, (2000).

H

294. Ha S B, Smith A P, Howden R, Dietrich W M, Bugg W, O'Connell M J, Goldsbrough P B and Cobbett C S, *Plant Cell* 11: 1153 (1999).

295. Hackney D D, *Proc. Natl. Acad. Sci. USA* 91: 6865–6869 (1994).

296. Haenni W, Provent Ch, Pupunat H and Rychen Ph; Adamant Technologies SA in Report 101; *Water Purification with Nano-Crystalline Diamondas an Environmental Friendly Process* (2007).

297. Hahm J and Lieber C M, *Nano. Lett.* 4: 51 (2004).

298. Hall J L, *J Exp Bot.,* 53: 1–11 (2002).

299. Hall D B, Homlin R E and Barton J K, *Nature,* 382: 731 (1996).

300. Hall D B and Barton J K, *J. Am. Chem. Soc.*119: 5045 (1997).

301. Hancock W O and Howard J, *Proc. Natl. Acad. Sci. USA* 96: 13147–13152 (1999).

302. Hardy W B, *Proc. Roy.Soc.,* 66: 110 (1900).

303. Hanahan D and Weinberg R, *Cell,* 100: 57 (2000).

304. Harriman A, *Angew. Chem., Int. Ed. Engl.,* 38: 945 (1999).

305. Hartwich G, Caruana D J, Lumley-Woodyear T D, Wu Y, Campbell C N and Heller A, *J. Am. Chem. Soc.,* 121: 10803 (1999).

306. Haruyama T, *Adv. Drug Delivery Rev.* 55: 393–401 (2003).

307. Harada T and Yoshikawa K, *Appl. Phy. Lett.,* 81: 4850–4852 (2002).

308. Hasegawa K, Yamashita I and Namba K, *Biophysical Jour.,*74, 569–575 (1998).

309. Hashimoto H, Yokoyama S, Kusano H, Ikeda Y, Seno M, Takada J, Fujii T, Nakanishi M and Murakami R, *J. Magn.Magn.Mater.* 10: 793 (2006).

310. He S, Guo Z, Zhang Y, Zhang S, Wang J and Gu N, *Materials Letters* 61: 3984 (2007).

311. Heddle J G, Fujiwara I,Yamadaki H, Yoshii S, Nishio K and Addy C, *Small,* 3, 1148–1152 (2007).

312. Heddle J G, Yokoyama T, Yamashita I, Park SY and Tame J R H, *Structure,* 14: 925–933 (2006).

313. Hellinga H W, Richards F M, *J Mol Biol.* 222: 763–85 (1991).

314. Hernandez M E and Newman D.K, *Cell. Mol. Life. Sci.* 58: 1562 (2001).

315. Hess S, Davis W B, Voityuk A A, Rosch N, Michel-Beyerle M E, Ernsting N P, Kovalenko SA and Lustres J L P, *Chem.Phys.Chem.*, 3, 452, (2002).

316. Hess BC, Okhrimenko IG, Davis RC, Stevens BC, Schulzke QA, Wright KC, *Phys Rev Lett*; 86: 3132–5 (2001).

317. Hess S, Gotz M, Davis W B and Michel-Beyerle M E, *J. Am. Chem. Soc.*, 123: 10046 (2001).

318. Hest Van J C; Tirrell D A *Chemical communications (Cambridge, England)* 2001;(19): 1897–904.

319. Higashi N, Kawahara J, Niwa M, *J. Colloid Interface Sci.*, 288: 83, (2005).

320. Higdon J L L, *J. Fluid Mech.*, 90: 685 (1979).

321. Higham D P, Sadler P J and Scawen M D, *J. Gen Microbiol.*, 131: 2539 (1985).

322. Hildebrand M, *J Nanosci Nanotechnol.* 5: 146 (2005).

323. Hilder T A, Daniel Gordon and Shin-Ho Chung, *Small,* 5 (19): 2183–2190 (2009).

324. Hill T L, Prog. *Biophys. Mol. Biol.*, 28: 267 (1974).

325. Hirahara K, Suenaga K, Badow S, Kato H and Okazaki T, *Phys Rev Lett*, 85: 5384–7 (2000).

326. Hirakawa K, Mori M, Yoshida M, Oikawa S and Kawanishi S, *Free Radic. Res.*, 38: 439–447 (2004).

327. Hirsch LR, Stafford RJ, Bankson JA, Sershen SR, Rivera B,. *Proc. Natl. Acad. Sci. U. S. A.* 100, 13549 (2003).

328. Hoffman J, *J. Metals.* 40: 40 (1998).

329. Hoffmann F, Cinatl J, Kabickova Jr.H, Cinatl J, Kreuter J, Stieneker F, *Int.J. Pharm.*, 157: 189 (1997).

330. Holmes J D, Richardson D J, Saed S, Evans-Gowing R, Russell D.A and Sodeau J R, *Microbiology* 143, 2521 (1997).

331. Homuth M, Valentine-Weiganz P, Rohde M and Gerlach :, *Infect Immun.* 66, 710 (1998).

332. Houseman B T, Huh J H, Kron S J and MrksichM, *Nature Biotechnol.*, 20: 270–274 (2002a).

333. Houseman B T and Mrksich M, *Chem. Biol.*, 9: 443–454 (2002b).

334. Howard J, *Annu. Rev. Physiol.* 58: 703–729 (1996).

335. Hong Y, Blackman N, Kopp N, Sen A and Velegol D, *Phys. Sci. Rev.* 99: 178103–178107 (2007).

336. Hsan-Yin Hsu, *Photochemical Synthesis of Gold Nanoparticles with Interesting Shapes Electrical Engineering*, Purdue University*2004 NNIN REU Research Accomplishments*.

337. Hsiu-Mei C, Qingsu X, Tao Chen and Bai H,*Toxicol Lett*;178: 77–82, 87 (*2008*).

338. Hsiu-Fen Lin, Shih-Chieh Liao and Sung-Wei Hung The dc thermal plasma synthesis of ZnO nanoparticles for visible-light photocatalyst.

339. Hu H, Ni Y, Montana V, Haddon R C and Parpura V, *Nano Lett.,* 4, 507–511 (2004).

340. Huang Z, Maness P C, Blake D M, Wolfrum E J, Smolinski S L and Jacoby W A. *J. Photochem. Photobiol. A. Chem.,* 130: 163–170 (1999).

341. Huang L, McMillan RA, Apkarian RP,Pourdeyhimi B, Vincent P, *Macromolecules*, 33: 2989–2997 (2000).

342. Hummelen J C, Bellavia-Lund C and Wudl F, Heterofullerenes, in *Fullerenes and Related Structures*, Springer-Verlag, GmbH, p. 93 (1999).

343. Hussain SM, Hess K L, Gearhart J M, Geiss K D and Schlager J J, *Toxicol. In Vitro;*19: 975–983 (2005).

344. Husseiny M I, AbdEl-Aziz M, Badr Y and Mahmoud M A, *Spectrochimica Acta Part A* 67: 1003 (2007).

345. Hutchins S R, Davidson M S, Brierley I A and Brierley C.L., *Annual Reviews of Microbiology*, 40: 311 (1986).

346. Huxley A F, *Prog. Biophys.* 7, 255 (1957).

347. Huxley A F and Simmons R M, *Nature,*233: 533 (1971).

348. Huxley H E, *Science*, 164: 1365 (1969).

349. Hwang S J and Davis M E,*Current Opinion in Molecular Therapeutics,* 3: 183–191 (2001).

350. Hynes A J, Stoker R B, Pounds A J, McKay T, Bradshaw J D, Nicovich J M and Wine P H, *Journal of Physical Chemistry*, 99 (46): 16967–16975 (1995).

I

351. Ibarra J, Koski A and Warren R F, *Orthop. Clin. North Am.*, 31: 411 (2000).

352. Ijima S. *Nature* 354, 56–58 (1991).

353. Ilari A, Latella M C, Ceci P, Ribacchi F, Su M, Giangiacomo L and Chiancone, E. *Biochemistry,* 44, 5579–5587 (2005).

354. Imafidon G I and Spanier A M, *Trends Food Sci. Technol.* 5: 315–321 (1994).

355. Immoos C E, Lee S J and Grinstaff M W, *J. Am. Chem. Soc.,* 126: 10814 (2004).

356. Ishiyama K, Arai K I, Sendoh M and Yamazaki A, *J. Micromechatronics* 2: 77 (2003).

357. Ismagilov R F, Schwartz A, Bowden N, Whitesides G M, *Angew. Chem., Int. Ed.* 41, 652–65 (2002).

358. Israelachvilli J.N, Intermolecular and Surface Forces, *Academic Press,* (1992).

359. Iversen G, Friis E P, Kharkats Y I, Kuznetsov A M and Ulstrup J, *J. Biol. Inorg. Chem.*, 3: 229 (1998).

J

360. Jae-Hyuk Kim, Sejoong Kim, Chung-Hak Lee, Heock-Hoi Kwon and Sangho Lee, A novel nanofiltration hybrid system to control organic micro-pollutants: application of dual functional adsorbent/catalyst.

361. Jagendorf AT and Uribe E, *Proc Natl Acad Sci U S.*, 55(1): 170–177 (1966).

362. Jain R, *Cancer Res.,* 47: 3039 (1987).

363. Jain KK. *Expert Rev Mol Diagn*;3: 153 (2003)

364. Jaiswal JK, Simon SM. *Trends Cell Biol*;14: 497–504 (2004).

365. Jaybhaye S, Maheshwar Sharon, Madhuri Sharon and Singh L N, *Int. J. of Synthesis and Reactivity in Inorganic, Metal-Organic, and Nano-Metal Chemistry,* 36 :2, 37–42 (2006).

366. Jaybhaye S, Sharon Madhuri, Sharon Maheshwar, Sathiyamoorthy D and Dasgupta K, Int.J. *Synthesis and Reactivity in Inorganic, Metal-Organic, and Nano-Metal Chemistry,* 37(6): 473–476 (2007).

367. Jenison R, La H, Haeberli A, Ostroff R and Polisky B, *Clin. Chem.,* 47: 1894 (2001).

368. Jenning V and Gohla S, *J. Microencapsul.* 18: 149–158 (2001).

369. Jia G, WangH, Lei Y, Wang X, Rongjuan P. Tao Y, Zhao Y and Guo X *Environ Science & Technol,* 39 (5): 1378–1383 (2005).

370. Jiale H, Qingbiao Li, Sun D, Yinghua Lu, Yuanbo Su, Xin Y, Wang H, Wang Y, Wenyao S, Ning He, Jinqing H and Chen C, *Nanotechnology* **18** 105104 (2007).

371. Jiang W, Kim B YS, Rutka JT, Chan WC W, *Nat. Nanotechnol.,* 3: 145 (2008).

372. Jin R, Cao Y, Mirkin C A, Kelly K L, Schatz G, Zheng C, *J. Science*, 294: 1901 (2001).

373. Jin R, Cao Y C, Hao E, Metraux G S, Schatz, G C, Mirkin, C A, *Nature*, 425: 487 (2003).

374. Johnson, O. L., Cleland, J., Lee, H. J., Charnis, M., Duenas, E., Jaworowicz, Shepar D, Shahzamani A, Jones AJH, Putney A, *Nature Med*, 2: 795–799 (1996).

375. Joho M, Yamanaka C and Murayama T, *Microbios.*, 45: 169, (1986).

376. Jortner J, Bixon M, Langenbacher T and Michel-Beyerle M, *Proc. Natl. Acad. Sci. U. S. A.*, 95: 12759 (1998).

377. Joy B, Why the future doesn't need us, 2000, http://www.wired.com/wired/archive/8.04/joy_pr.html.

378. Julicher F, Ajdari A and Prost J, *Rev. Mod. Phy.*,69: 1269–1281 (1997).

K

379. Kagawa W, Kurumizaka H, *et al.*, *Mol Cell*, 10, 359 (2002).

380. Kalita G, Adhikari S, Aryal H R, Umeno M, Afre R, Soga T, and Maheshwar Sharon, *Appl. Phys. Letts*, 92 (06): 3508 (2008).

381. Kalita G, Adhikari S, Hare Ram Aryal, Umeno M, Afre R, Soga T, Maheshwar Sharon, *Appl. Phys. Letts.* 92(06) 3508 (2008).

382. Kalita G, Adhikari S, Aryal H R, Afre R, Soga T, Maheshwar Sharon, Koichi W and Umeno M, *J. Phys. D: Appl. Phys.*, 42: *115104 (5pp)* (2009).

383. Kanamaru S, Leiman P G, et al. *Nature*, 415, 553 (2002).

384. Kaseda, K., Higuchi, H., and Hirose, K., *Nat. Cell Biol.* 5: 1079–1082 (2003).

385. Kathiresan K, S. Manivannan, M.A. Nabeel, B. Dhivya, *Colloids and Surfaces B: Biointerfaces* 71, 133 (2009).

386. Kau, J.H., D.S. Sun, H.H. Huang, M.S. Wong, H.C. Lin and H.H. Chang, *PLOS* One, 4: 1–8, (2009.)

387. Kaushik M, W. Vogel, J. Urban, S. Kulkarni, K. M. Paknikar, *.Adv. Mater.*14, 815, (2002).

388. Kavan, L. and Kastner, *J.*, *Carbon,*,32: 1533 (1994).

389. Kawaguchi, K., and Ishiwata, S., *Science*, 291: 667–669 (2001).

390. Kay A and Goldberg A, *IEEE Computer.*, 10 (3) (1977).

391. Kayalar C, Rosing J and Boyer PD, *J Biol Chem*, **252**: 2486–2491 (1977).

392. Kelch S and Lendlein A, *Angew. Chem. Int. Edn Engl.*, 41, 2034–2057 (2002).

393. Kelly T R, De Silva H and Silva R A, "Unidirectional Rotary Motion in a Molecular System", Nature (London), vol. 401, Issue 6749, pp 150–152, (1999).

394. Kelley S O, Jackson N M, Hill M G and Barton J K, *Angew.Chem., Int. Ed.*, 38: 941 (1999).

395. Kelley S O, Barton J K, Jackson N M, McPherson L D, Potter A B, Spain E M, Allen M J and Hill M G, *Langmuir,* 14: 6781 (1998).

396. Kelley SO, Homlin R E, Stemp E D A and Barton J K, *J. Am. Chem. Soc.*, 119: 9861 (1997).

397. Kern J M and Sauvage J P, *Angew. Chem. Int. Ed,.* 43: 2392 (2004).

398. Kesselman E, Efrat R, GartiN and Danino D, Formation of cubosomes as vehicles of biologically active substances,

http://materials.technion.ac.il/ism/Docs/2007/Life-Abstracts/Poster/ E_Kesselman.pdf.

399. Khairnar V, Maheshwar Sharon, Jaybhaye s and Michael Neumann-Spallart, *Int. J. Synthesis and Reactivity in Inorganic, Metal-Organic, and Nano-Metal Chemistry*, 36:2, 171–173 (2006).

400. Khairnar V, Jaybhaye S, Chi- Chang Hu, Afre R, Soga T, Madhuri Sharon and Maheshwar Sharon, *Carbon Letters*, 9(3) (2008).

401. Khan S, Khan I H and Reese T S, *Journal of Bacteriology* 173: 2888–2896 (1991).

402. Khan I H, Reese T S and Khan S, *Proc. Natl. Acad. Sci.USA,* 89: 5956–60 (1992).

403. Khang DK, Kim JY, Lee YE, Webster TJ., *Int J Nanomed,* in press.

404. Kille P, Winge D R, Harwood J L and Kay J A, *FEBS Lett,.* 295: 171 (1991).

405. Kim S and Bawendi MG. *J Am Chem Soc*; 125: 14652–3 (2003).

406. Kim S.Y and Lee YM, *Biomaterials* 22 (13): 1697 (2001).

407. Kim, Y-H and Carraway ER, *Environmental Sc. and Technol*,34(10): 2014–2017 (2000).

408. Kim M J and Powers T R, *Phy. Rev. E*, 69: 061910 (2004).

409. Kim *et al.*, – (2008) US patent Applications: Appartus and Method for Improving Fourier Transform Ion Cyclotron Resonance Mass Spectrometer Signal May, 2008–20080099672.

410. Kimura A and Kakitani T, *Chem. Phys. Lett.*, 298: 241(1998).

411. Kinosita K Jr, Yasuda R, Noji H, Ishiwata S and Yoshida M, *Cell,* 93: 21–24 (1998).

412. Kinosita K Jr, Ali M Y, Adachi K, Shiroguchi K and Itoh H, *Adv. Exp. Med.Biol.,*565: 205–219 (2005).

413. Kitao A, Yonekura K, Maki-Yonekura S, Samatey F A, Imada K, Namba K and Go N, *Proceedings of the National Academy of Sciences USA,* 103: 4894–4899 (2006).

414. Klaus T, Joerger R, Olson E and Granqvist G, *Proc. Nat. Acad. Sci.* 96: 13611 (1999).

415. Kline T R, Paxton W F, Mallouk T E and Sen A, *Angew. Chem., Int. Ed.,* 44: 744–746 (2005).

416. Knez M, Bittner A M, *et al., Nano Lett.,* 3: 1079 (2003).

417. Knezevic V, Leethanakul C, Bichsel V E, Worth J M, Prabhu V V, Gutkind J S, Liotta L A, Munson P J, Petricoin E F, Krizman D B, *Proteomics,* 1: 1271 (2001).

418. Kohane D. S., Lipp M., McKinney R. C., Anthony D. C, Louis, Lotan N, Langer R, *J. Biomed. Mater. Res.,* 59: 450 (2002).

419. Kohane D S, Tse J Y, Yeo Y, Padera R, Shubina M, Langer R, *J. Biomed. Mater. Res.,* 77: 351 (2006).

420. Kojima S and Blair D F, *Biochemistry,* 40: 13041–13050 (2001).

421. Konishi Y, Tsukiyama T, Ohno K, Saitoh N, Nomura T and Nagamine S, *Hydrometallurgy* 81: 24 (2006).

422. Konishi Y, Ohno K, Saitoh N, Nomura T, Nagamine S, Hishida H, Takahashi Y and Uruga T, *Journal of Biotechnology,* 128: 648 (2007).

423. Kononen J0, Bubendorf L, Kallionimeni A, Bärlund M, Schraml P, LeightonS, Torhorst J, Mihatsch MJ, Sauter G & Olli-P Kallionimeni, *Nature Med.,* 4: 844–847 (1998).

424. Kopf H, Joshi RK, Soliva M and Speiser PP, *Pharm. Ind.,* 38: 281 (1976).

425. Kopriva S and Rennenberg H, *J Exp Bot,* 55: 1831–1842 (2004).

426. Koutyukhova, *J. Phys. Chem. C,* 112: 6049–6056 (2008).

427. Kowshik M, Deshmukh N, Vogel W, Urban J, Kulkarni S K and Paknikar KM,*Biotechnology and Bioengineering,* 583 (2001).

428. Koumura N, Zijlstra R W, van Delden R A, Harada N and Feringa B L, J, *Nature,* 40: 1152–155 (1999).

429. Kozielski F, Sack S, Marx A, Thormählen M, Schönbrunn E, Biou V, Thompson A, Mandelkow E-M and Mandelkow E, *Cell,* 91: 985–994 (1997).

430. Krämer U, *New Phytol.,* 158: 4–6 (2003).

431. Krämer U, and Chardonnens A N, *Appl Microbiol Biotechnol* 55: 661–672 (2001).

432. Kratschmer W, Lamb L D, Fostiropoulos K and Huffman D R, *Nature,* 347, 354 (1990).

433. Kreilgaard M, *Adv. Drug Deliv. Rev.*, 54(Suppl. 1): S77–S98 (2002).

434. Kreuter J and Kreuter J (Ed.), *Colloidal Drug Delivery Systems*, Marcel Dekker, (1994).

435. Kreuter J. *Adv Drug Deliv Rev*, 47: 65 (2001).

436. Krewson C E, Klarman M and Saltzman W M, *Brain Res.,* 680: 196–206 (1995).

437. Krewson C E, Dause R B, Mak M W and Saltzman W M, *J. Biomater. Sci.,* 8: 103–117 (1996).

438. Kröger N, *Science,* 298 : 584 (2002).

439. Kröger N, Deutzmann R and Sumper M, *Science,* **286:** 1129 (1999).

440. Kröger N, Deutzmann R, Bergsdorf C and Sumper M, *Proc. Natl. Acad. Sci,.* **97**: 14133 (2000).

441. Kröger N and Wetherbee R, *Protist,* **151:** 263 (2000).

442. Kröger N, Deutzmann R and Sumper M, *J. Biol. Chem.,* **276:** 26066 (2001).

443. Kroto H W, Heath J R, O'Brien S C, Curl R F and Smalley R E, *Nature,* 318: 216 (1985).

444. Kruse K, Joanny J F, Julicher F, Prost J and SekimotoK, Phy. *Rev. Lett,* 92(1–10): 078101 (2004).

445. Kshirsagar D. E, Puri V, Maheshwar Sharon and Madhuri Sharon, *Carbon Science,* 7(4): 245–248, (2006).

446. Kukkadapu R, Zachara J, Fredrickson J, Kennedy D, Dohnalkova A and Mccready D, *Am. Minerol.,* 90: 510 (2005).

447. Kumar M, Zhao X and Ando Y*, Int. Symposium on Nanocarbons*, Nagano, Japan, Extended Abstract: 244–245(2001),

448. Kumar A, Abott N L, Kim E, Biebuyck H A and Whitesides G M, *Acc. Chem. Res.,* 28: 219 (1995).

449. Kumari A, Yadav S, Yadav SC, *Colloids and Surfaces B: Biointerfaces* 75: 1, (2010).

450. Kuo-Hsiung Tseng, Jen-Chuen Huang, Chih-Yu Liao, Der-Chi Tien, Tsing-Tshih Tsung, *Journal of Alloys and Compounds*, 472 (1–2): 446–450, (2009).

L

451. LaBean T, Yan H, Kopatsch J, Liu F, Winfree E, Reif J H and Seeman N C, *J. Am. Chem. Soc.*,122: 1848–1860 (2000).

452. Lakhno V D and Sultanov V B, *Biophysics*, 48: 5, 741(2003).

453. Langer K, Stieneker F, Lanbrecht G, Mutschler E and Kreuter J, *Int. J. Pharm.*, 158: 211. (1997).

454. Langer K, Coester C, Weber C, Von Briesen H and Kreuter J, *Eur. J. Phar. Biopharm.*, 49: 303 (2000).

455. Langer K, Coester C, Weber C, H. Von Briesen and Kreuter J., *Int. J. Pharm.*, 196:147 (2000).

456. Langer R, Tirrell DA, *Nature*, 428 (6982): 487–92, 2004.

457. Lanza RP, Langer RS and Vacanti J, *"Principles of Tissue Engineering,"* 2nd ed.Academic Press, San Diego, 2000.

458. Laocharoensuk R, Burdick J and Wang J, *ACS Nano*, 2: 1069–1075 (2008).

459. Larson DR, Zipfel WR, Williams RM, Clark SW, Bruchez MP, Wise FW, Webb WW, *Science*,300:,1434–1436 (2003).

460. Lawrence M J and Rees G D, *Adv. Drug Delivery Rev.*, 45: 89–121 2000.

461. Lee J, Macosko C W and Urry D W, *Macromolecules,* 34: 5968–5974 (2001).

462. Lee H Y, Tanaka H, Otsuka Y, Yoo K H, Lee J O and Kawai T, *Appl. Phys. Lett.,* 80: 1670 (2002).

463. Lee S, Otaki N M and Ohgaki S, *Water Sci. Technol.,* 35: 101–106 (1997).

464. Lehn J.-M, *Angew. Chem., Int. Ed.* 27: 8 (1988).

465. Lehninger A, Nelson D L, Cox M M, 4th edition W.H. Freeman and Company, New York (2005).

466. Lengke M and Southam G, *Geochimica et Cosmochimica Acta* 70: 3646 (2006).

467. Lengke M F, Fleet M E and Southam G, *Langmuir*, 23: 2694 (2007).

468. Li X, Yang X, Qi J and Seeman N C, *J. Am. Chem. Soc.,* 118: 6131–6140 (1996).

469. Li Y, Brown J H, Reshetnikova L, Blazsek A, Farkas L, Nyitray L and Cohen C, *Nature,* 424: 341–345 (2003).

470. Li Y, Ma Q, Wang X, Su X. *Luminescence*; 22: 60–6 (2007).

471. Li C, *Adv Drug Deliv Rev*, 54: 695 (2002).

472. Lighthill J, *SIAM Rev.*, 18: 161 (1976).

473. Lil CY, Hsu C K, Lin F H and Stobinski L, *European Cells and Materials* 13,(3 Suppl.): 24, ISSN 1473–2262 (2007).

474. Lilley D M J and Clegg R M, *Annu. Rev. Biophys. Biomol. Struct.*, 22, 299–328 (1993).

475. Lin T, Chen Z, *et al.*, *Virology*, 265: 20 (1999).

476. Lin Z, Su X, Mu Y, Jin Q. *J Nanosci Nanotechnol*; 4: 641–5 (2004).

477. Liu H, Schmidt J J, Bachand G D, Rizk S S, Looger L L, *Nat Mater.* 1: 173–177 (2002).

478. Liu X, Chen C, *Curr Opin Drug Discovery Dev*; 8: 505 (2005).

479. Liu G, Garrett MR, Men P, Zhu X, Perry G, Smith MA, *Biochim Biophys Acta,* 1741: 246–252 (2005).

480. Liu WT, *J Biosci* Bioeng; 102: 1–7 (2006).

481. Liu H and Webster T J, *Biomaterials*, 28 : 354–369 (2007).

482. Lloyd J R, Yong P and Macaskie L E, *Appl. Environ. Microbiol.,* 64: 4607 (1998).

Loo C. Lin A, Hirsch L, Lee M, Jennifer Barton J, Halas N, West J, Drezek R, *Technol. Cancer Res. Treat.* 3: 33 (2004).

483. Lorenzo, A.V., Bresnan, M. J., and Barlow, C. F. Archives ofNeurology (Chic.), 30: 387–393 (1974).

484. Lovely D and Phillips E, *Appl. Environ. Micrbiol.*, 53: 2636 (1987).

485. Lovrinovic M and Niemeyer CM, *Angew. Chem., Int. Ed.,* 44: 3179 (2005).

486. Lowndes D H, Geohegan D B, Puretzky A S, Norton D P and Rouleau C M, *Science*, 16: 273 (5277): 898–903 (1996).

487. Lowry G V and Johnson K M, *Environ. Sci. Technol.*, 38: 5208 (2004).

488. Lu Z, *Patterson KH, Wang A P, Marquez L, Atkinson RT, Baggerly EN, Ramoth KA, Rosen LR, Liu DG, Hellstrom J,Clin. Cancer Res.* 10: 7677 (2004).

489. Lubbe A.S, Bergemann C, Huhnt W, Fricke T, Riess H, Brock JW, Huhn D., *Cancer Res*, 56: 4694 (1996).

490. Luby-Phelps K, *Int. Rev. Cytol.*, 192: 189 (2000).

491. Luo C, Nobusawa E, Nakajima K, *J. Gen Virol.*; 80 (Pt 11): 2969–76, (1999).

M

492. MacBeath G and Schreiber S L, *Science,* 289: 1760 (2000).

493. MacBeath G, *Nat. Genet.*, 32: 526 (2002).

494. Madhuri Sharon, Soga T, Afre R, Sharon Maheshwar., Gupta Arvind, *Abstract National Seminar on Nanotechnology—The Road Ahead*, Lucknow (2005e).

495. Madhuri Sharon, Datta S, Shah S, Maheshwar Sharon, Soga T and Afre R; *Carbon Letters*, 8 (3): 184–190 (2007).

496. Mahoney M J and Saltzman W M, *Nature Biotechnol.* 19: 934–939 (2001).

497. Mainardes RM, Gremiao MP, Brunetti IL, da Fonseca LM, *J. Pharm. Sci.* 98 (1): 257, (2009).

498. Malam Y, Loizidou M, Seifalian AM, *Trends in Pharmacological Sciences* 30 (11): 592, (2009).

499. Maness P C, Smolinski S, Blake D.M, Huang Z, Wolfrum E J and Jacoby WA,*Applied Environ. Microbiol.*, 65: 4094–4098 (1999).

500. Mann S, *Nature,* 365: 499 (1993).

501. Manna L, Scher EC, Li LS, Alivisatos AP, *J Am Chem Soc*;124:7136–45 (2002).

502. Mao C, Sun W and Seeman N C, *Nature* 386: 137–138 (1997).

503. Mao C, Sun W, Shen Z and Seeman N C, *Nature* 397: 144–146 (1999).

504. Marcus R A, Nobel Lecture (1992).

505. Marcus R A, *J. Chem. Phys.*, 24: 966 (1956).

506. Marcus R A and Sutin N, *Biochim.Biophys. Acta*, 811: 265 (1985).

507. Marcus R A, *Adv. Chem. Phys.*, 106: 1 (1999).

508. Markman M, *Expert Opin Pharmacother*, 7: 1469 (2006).

509. Martin C R and Kohli P, *Nat Rev Drug Discov.*, 2: 29–37(2003).

510. Marquardt C L, Williams R T and Nagel D J, *Mater. Res.Soc.Proc.* 38: 32 (1985).

511. Marx P A, Compans RW, Gettie A, Staas JK, Gilley RM, Mulligan MJ, Yamshchikov GV, Chen D, Eldridge JH, *Science,* 260: 1323–1327 (1993).

512. Mata Y N, Torres E, Blázquez M L, Ballester A, González F, Mu~noz J A, *Journal of Hazardous Materials*, 166: 612 (2009).

513. Mathews C K and van Holde K, 2nd Edition. Benjamin-Cummings Publishing Co., Menlo Park, CA. (1995).

514. Matthew A A, Cameron W E and Colin L R, *Green Chem.*, 8: 417–432 (2006).

515. Mathieu J B, Martel S, Yahia L, Soulez G and Beaudoin G, in "Proceedings of the Canadian Conference on Electrical and Computer Engineering (CCECE)." Montréal, Canada, (2003).

516. Mathiowitz E, Saltzman W M, Domb A, Dor P and Langer R, *J. Appl. Polym. Sci.,* 35, 755–774 (1988).

517. Matsunaga T, Tomoda R, Nakajima T, Nakamura N and Komine Y, *Applied Environ.Microbiol.*, 54: 1330-1333 (1988).

518. Mattson M P, Haddon R C and Rao A M, *J. Mol. Neurosci.*, 14: 175–182 (2000).

519. Mattoussi H, Mauro JM, Goldman ER, Anderson GP, Sundar VC, Mikulec FV, *J Am Chem Soc*; 122: 12142–50 (2000).

520. May M, Vernoux T, Leaver C, Van Montagu M and Inze D, *J Exp Bot.,* 49: 649–667 (1998).

521. Mayo J T, Yavuz C, Yean S, Cong L, Shipley H, Yu W, Falkner J, Kan A, Tomson M and Colvin V L, *Sci. Technol. Adv. Mater.,* 8 (1-2): 71–75 (2007).

522. Mazzola L T and Fodor S P, *Biophys. J.,* 68: 1653 (1995).

523. McCarter L L, *Journal of Bacteriology*, 176:5988–5998 (1994a).

524. McCarter L L, *Journal of Bacteriology,* 176: 4219–4225 (1994b).

525. McClements D J and Decker E A, *J. Food Sci,.* 65: 1270–1282 (2000).

526. McClements D J, Decker E A and Weiss J, Inventors; University of Massachussetts, assignee. 2005. UMA 05-27: Novel procedure for creating nanolaminated edible films and coatings, U.S. patent application. (2005).

527. McKenzie J L, Waid M C, Shi R, Webster T J, *Biomaterials,* 25 (7–8): 1309–17 (2004).

528. McConnell H M, *J. Chem. Phys.* 35: 508 (1961).

529. McDowall Lyndal, Degradation of Toxic Chemicals by Zero-Valent Metal Nanoparticles -A Literature Review (2005) 1 - 40DSTO-GD-0446; Published by Human Protection & Performance Division DSTO Defence Science and Technology Organisation 506 Lorimer St Fishermans Bend, Victoria 3207 Australia (2005).

530. McGall G, Labadie J, Brock P, Wallraff G, Nguyen T, Hinsberg W *Proc. Natl Acad. Sci. USA* 93: 13555–13560 (1996).

531. McMillan R A and Conticello V P, *Macromolecules*, 33: 4809–4821 (2000).

532. McMillan R A, Paavola C D, Howard J, Chan S L, Zaluzec N and Trent J D, *Nat. Mater.* 1 247–52 (2002).

533. McNeil SE., *J Leukoc Biol*; 78: 585–94 (2005).

534. Meagher R B, *Curr Opin.Plant Biol* 3: 153–162 (2000).

535. Medintz IL, Uyeda HT, Goldman ER, Mattoussi H. *Nat Mater*; 4: 435–46 (2005).

536. Meggers E, Kusch D, Spichty M, Wille U and Giese B, *Angew. Chem., Int. Ed. Engl.,* 37: 460 (1998).

537. Meggers E, Michel-Beyerle M E and Giese B, *J. Am. Chem. Soc.* 120: 12950 (1998).

538. Mehra R K and Mulchandani P, *Biochem. J.* 307: 697 (1995).

539. Mehra R K, Mulchandani P and Hunter T C, *Biophys. Res. Commun.* 200: 1193 (1994).

540. Mehta A, *J. Cell Sci.* 114: 1981–1998 (2001).

541. Mei H, Wang K, Peffer N, Weatherly G, Cohen D S, Miller M, Pielak G, Durham B and Millett F, *Biochemistry*, 38: 6846 (1999).

542. Merten D F, Bowie J D, Kirks D R and Grossman H, *Radiology*, 142(2): 361–5 (1982).

543. Metzger R M and Panetta C A, *New J. Chem.* 15: 209 (1991).

544. Meyer J D and Manning M. *Pharm Res.*, 15: 188–193 (1998).

545. Michalet X, Pinaud FF, Bentolila LA, Tsay JM, Doose S, Li JJ, *Science*; 307: 538–44 (2005).

546. Mie G, *Ann.Phys. Lpz.*, 25: 377 (1908).

547. Miller DC, Thapa A, Haberstroh KM, *Biomaterials*, 25: 53–61 (2004).

548. Miller DC. Nanostructured polymers for vascular grafts, PhD thesis,Purdue University, West Lafayette, IN, 2006.

549. Miller R A, Presley A D and Francis M B, *J. Am. Chem. Soc.*, 129: 3104 (2007).

550. Minamina T and Namba K, *J. Mol. Microbiol. & Biotechnol.* 7: 5–17(2004).

551. Minamino T, Imae Y, Oosawa F, Kobayashi Y and Oosawa K, *J. Bacteriol.* 185: 1190–1194 (2003).

552. Mirkin C A, LetsingerR L, Mucic R C and Storhoff JJ, *Nature* 382: 607–609 (1996).

553. Mitchell P, *Nature* 144 (1961).

554. Mitchell P, *Biol. Rev. Cambridge Phil Soc.*, 41: 445–502 (1966).

555. Mitchell P., *Science* 206, 1148–1159 (1979).

556. Mitchell J C and Yurke B, in DNA Based Computers VII, No. 2340 in LNCS, (eds N. Jonoska and N. C. Seeman (Springer Verlag, Heidelberg, (2002).

557. Mitoraj D, Janczyk A, Strus M, Kisch H, Stochel G, Heczko P B and Macyk W, *Photochem. Photobiol. Sci.*, 6: 642–648 (2007).

558. Mitra R S, *Appl. Environ. Microbiol.* 47: 1012 (1984).

559. Miyata T, Asami N and Uragami T, *Nature*, 399, 766–768 (1999).

560. Miyazaki S, Yamaguchi H, Takada M, Hou WM, Takeichi Y, Yasubuchi H. *Acta Pharm Nordica*; 2: 401–6 (1990).

561. Miyoshi D and Sugimoto N, *Biochimie*, 90: 1040 (2008).

562. Mobian P, Jean-Marc K, Jean-Pierre Sauvage DOI: 10.1002/anie. 200352522 *Angewandte Chemie International Edition,* (2004).

563. Moghimi SM, Hunter AC, Murray J.C, *Pharmacol Rev*, 53: 283 (2001).

564. Mokhtari N, Daneshpajouh S, Seyedbagheri S, Atashdehghan R, Abdi K, Sarkar S, Minaian S, Shahverdi, H R and Shahverdi A R, *Mater. Res. Bull.* 44: 1415 (2009).

565. Moloni K, Lal A and Lagally M, *Proc. SPIE;* 4098 : 76 (2000).

566. Monthioux M, Mith B W, Burteaux B, Ciaye A, Fischer JE and Luzzide, *Carbon*, 39: 1251–52 (2001).

567. Morillon V, Debeaufort F, Blond G, Capelle M and Voilley A, *Crit. Rev. Food Sci. Nutr.*, 42: 67–89 (2002).

568. Moser C C, Keske J M, Warncke K, Farid R S and Dutton P L, *Nature*, 355, 796, (1992).

569. Moser C C, Page C C, Chen X, Dutton P L, *J. Biol. Inorg. Chem.*, 2: 393 (1997).

570. Mukherjee S, Senapati S, Mandal D, Ahmad A, Khan M I, Kumar R and Sastry,M, *Chem Biochem.*3: 461 (2002).

571. Mukherjee P, AhmadA, Mandal D, Senapati S, Sainkar S.R, Khan M I, Ramani R, Parischa R, Kumar P V A, Alam M, Sastry M and Kumar R *Angew Chem. Int. Ed.* 40: 3585 (2001).

572. Mu L and Feng SS, *J. Control. Release,* 86 (1): 33 (2003).

573. Müller R H and Lucks *J S, European Patent No*. 0605497 (1996).

574. Müller R H, Radtke M, Wissing S A, Adv. *Drug Deliv. Rev. Suppl.* 54, S131–S151 (2002).

575. Müller R H, Dingler A, *Wirkstoffe. PZ Wiss* 49: 11–15 (1998).

576. Muller R H and Lucks J S, Arzneistofftra"ger aus festen Lipidteilchen - Feste Lipid Nanospha"ren (SLN). In European Patent 0605497: Germany, (1996).

577. Murray J, Brown L, Langer R and Klagsburn M, *In Vitro* 19: 743–748 (1983).

578. Murphy C J, Arkin M R, Jenkins Y, Ghatlia N D, Bossman S H, Turro N J and Barton J K, *Science*, 262: 1025 (1993).

579. Murphy CJ and Coffer JL, *Appl.Spetcrosc.,* 56 (1): 16 A- 27A (2002).

N

580. Nagapudi K, Brinkman WT, Leisen JE, Huang L, McMillan RA, Apkarian RP, *Macromolecules*, 35, 1730–1737 (2002).

581. Nair B and Pradeep T, *Cryst. Growth. Des.* 293: (2002).

582. Navaratnam S and Parsons B J, *J. Chem. Soc. Faraday Trans.*, 94: 2577 (1998).

583. Nakai T, Kanamori T, Sando S, Aoyama Y, *J. Am. Chem. Soc.*, 125: 8465 (2003).

584. Nevin K P and Lovley D R, *Environ. Sci. Technol.* 12: 2472 (2000).

585. Newman and Kolter R, *Nature* 405: 94 (2000).

586. Nguyen C V, So C, Stevens R M D, Li Y, Delzeit L, Sarrazin P and Meyyappan M, *J. Phys. Chem. B*, 108 (9) : 2816 (2004).

587. Niemeyer C M, *Angew. Chem.Int.* 36: 585-587 (1997).

588. Niemeyer C M and Blohm D, *Angew. Chem., Int. Ed.,* 38: 2865 (1999).

589. Nicewarner-Peña Sheila R., Anthony J. Carado, Kristen E. Shale, and Christine D. KeatingScience 294, 137–141 (2001).

590. Ning YM, He K, Dagher R, *Oncology* 21, 1503 (2007).

591. Noctor G, Arisi A, Jouanin L, Kunert K, Rennenberg H and Foyer C J, *J Exp Bot.,* 49: 623–647 (1998).

592. Noji H, Yasuda R, Yoshida M and Kinosita Jr K,. *Nature,* **386**: 299–302 (1997).

593. Nosonovsky M and Bhushan B, *ASME J. Tribol.,* 127: 37-46 (2005).

594. Nurmi J T, Bandstra J. Z and Tratnyek, P G, *Environ. Sci. Technol.* 39: 1221–1230 (2005).

595. Nuxoll E E, Shimotori T, Arnold W A and Cussler E L, Iron Nanoparticles in Reactive Environmental Barriers. Presentation at the AIChE Annual Meeting, November 20 (2003).

O

596. Oberdörster G, Sharp Z, Atudorei V, Elder A, Gelein R, Kreyling W and Cox C, *Inhalation Toxicology*, 16: 437–445 (2004).

597. *OberdörsterG, Oberdörster E and Oberdörster J, Environmental Health Perspectives,*113(7): 823–839 (2005).

598. Ogawara K, Un K, Minato K, Tanaka K, Higaki K, Kimura T, *Int. J. Pharm.* 359, 234 (2008).

599. Ogawara K, Un K, Tanaka K, Higaki K, Kimura T, *J. Control. Release* 133, 4 (2009).

600. Olafson R W, McCubbin W D and Kay C M, *Biochem J.* 251: 691 (1988).

601. Olivier JC, Fenart L, Chauvet R, Pariat C, Cecchelli R, Couet W.. *Pharm Res*, 16: 1836 (1999).

602. Ortiz D F, Ruscitti T, McCue K F and Ow D W, *J. Biol. Chem.*, 270: 4721 (1995).

603. Ortiz-Lombardía M, Verma C S, *Journal of Physics: Conference Series,* 34: 7 (2006).

604. Orlov A, Jefferson D, Tikhov M and Lambert R. *Catalysis Communications*, 8: 821–824, (2007)

605. Orlov A, Chan M, Jefferson D A, Zhou D, Lynch R J and Lambert R M., *Environmental Technology*, 27(7): 747–752, (2006a).

606. Orlov A, Jefferson D, Macleod N, Lambert R, *Catalysis Letters*, 92(1–2) (2004).

607. Orlov A, García F, Tikhov M, Wright D, Lambert R M, *Chem. Communications*, 40: 4236–4238 (2006b).

608. Orlov A, Tikhov M, Lambert R, *Comptes Rendus Chimie*, 9 (5-6), 2794–2799 (2006c).

609. Orlov A, Zhai Q Z, Klinowski J', Jour. of Mat. Sc., 41, 2187–2193 (2006d).

610. Oster G, and Wang H., *J.Bioenrg.Biomembr.* 332: 495–496 (2000).

611. Oster G and Wang H, *TRENDS in Cell Biology*, 13(3): (2003).

612. Ozkan M. *Drug Discov Today*; 9:1065–71 (2004).

613. Ozin G A, Manners I, Fournier-Bidoz S, Arsenault A, *Adv. Mater.*, 17: 3011–3018 (2005).

P

614. Page C C, Moser C C, Chen X, Dutton P L, *Nature*, 402, 47(1999).

615. Pal B, Sharon M and Kamat D, Chem. *and Environ. Res.*, 5: 51–56 (1996).

616. Panayam J and Labhasetwar V, *Curr. Drug Deliv.* 1: 235, (2004)

617. Pantarotto D, Singh R, McCarthy D, Erhardt M, Jean-Paul B, Maurizio, Prato,Kostas Kostarelos, Alberto Bianco, Angewandte Chemie 43 (39): 5242–5246

618. Pantarotto D, Briand J P, Prato M and Bianco A, *Chem. Commun.,* 615: 16–17 (2004).

619. Pardridge WM, *Adv Drug Deliv Rev*; 36: 299 (1999).

620. Park H J, *Trends Food Sci. Technol.*, 10: 254–260 (1999).

621. Park GE, Pattison MA, Park K, *Biomaterials,*26: 3075–82 (2005).

622. Park S J, Taton T A and Mirkin C A, *Science*, 295: 1503 (2002).

623. Park E S, Cui Y, Yun B J, Ko I J and Chi S C, *Arch. Pharm. Res.* 28: 243–248 (2005).

624. Parsegian V A, Rand R P and D.C. Rau, *Proc. Natl. Acad. Sci. U.S.A.* 97: 3987 (2000).

625. Pascale R L, Seungpyo H, Almut M, Baker J R Jr., Bradford G, Orr M M and Banaszak H, *Acc Chem Res.*, 40(5): 335–342 (2007).

626. Parihar S, Sharon Maheshwar Sharon Madhuri, *Synthesis and Reactivity in Inorganic, Metal-Organic, and Nano-Metal Chemistry*, 36(1): 107–113 (2006).

627. Passy S I, Yu X, Li Z, Radding C M, Masson J.-Y, West, S C and Egelman E H, *Proc Nat Acad Sci U S A*, 96, 1068410684–10688 (1999).

628. Pathak S, Choi SK, Arnheim N, Thompson ME. *J Am Chem Soc*; 123:4103 (2001).

629. Patolsky F and Lieber C M, *Mater. Today* 8 (4): 20 (2005).

630. Patolsky F, Zheng G and Lieber C M, *Anal. Chem.* 78: 4261 (2006a).

631. Patolsky F, Zheng G and Lieber C M, *Nanomedicine* 1: 51 (2006b).

632. Patolsky F, Zheng G, Hayden O, Lakadamyali M, Zhuang X W and Lieber, C M *Proc. Natl. Acad. Sci. USA* 101:14017 (2004).

633. Pattison MA, Wurster S, Webster TJ, *Biomaterials*, 26:2491–500 (2005).

634. Patungwasa W and Hodak J, *Materials Physics and Chemistry*. 108: 45–54 (2008).

635. Paxton W F, Kistler K C, Olmeda C C, Sen A, St Angelo S K, Cao Y Y, Mallouk T E, Lammert P E, Crespi V H, *J. Am. Chem. Soc.*, 126: 13424–13431 (2004).

636. Paxton W F, Sen A, Mallouk T E, *Chem.-Eur. J.*, 11: 6462–6470 (2005).

637. Paxton W F, Sundararajan S, Mallouk T E, Sen A, *Angew. Chem., Int. Ed.*, 45: 5420–5429 (2006).

638. Pyatenko, M Yamaguchi, M Suzuki - *J. Phys. Chem. B*, (2005).

639. Pazirandeh L A C, Mauro J M, Campbell J R and Gaber B P, *Appl. Microbiol. Biotechnol.* 43: 1112 (1995).

640. Pedersen C J, *Angew. Chem., Int. Ed.* 27: 1021 (1988).

641. Pellegrino T, Manna L, Kudera S, Liedl T, Koktysh D, Rogach AL, *Nano Lett* ; 4:703–7 (2004).

642. Peppas N A and Langer R, *Science*, 263: 1715–1720 (1994).

643. Peppas N A, *Curr. Opin. Colloid Interf. Sci.*, 2: 531–537 (1997).

644. Perez O D and Nolan G P, *Nat. Biotechnol.*, 20: 155–62 (2002).

645. Petersen R, Nanocrystals for use in topical cosmetic formulations and method of production thereof. Abbott GmbH & Co., US Patent 60/866233 (2008).

646. Philip D0, *Spectrochimica Acta Part A* 73: 374 (2009).

647. Phillips R and Quake S, *Physics Today*, 38 (2006).

648. Pickett-Heaps J, Schmid AM M and Edgar LA, in *Progress in Phycological Research*, (Eds F. E. Round and D. J. Chapman), Biopress, Bristol, UK vol. 7, pp 1–169 (1990).

649. Pietryga J., Schaller R., Werder D., Stewart M., V. Klimov, Hollingsworth J.,*J.Am.Chem.Soc.* 126 (38, 11752 (2004).

650. Pike A, Horrocks B, Connolly B and Houlton A, *Aust. J. Chem.*, 55: 191 (2002a).

651. Pike A R, Lie L H, Eagling RA, Ryder L C, Patole S N, Connolly B A, Horrocks A and Houlton A, *Angew. Chem., Int. Ed.* 41: 615(2002b).

652. Pleskov Y V, Sakharova A Y, Krotova M D, Bouilov L L, Spitsyn B V, *Journal of Electroanalytical Chemistry,* 344, 401–404 (1987).

653. Pooley F D, *Nature (London),* 296: 642–643 (1982)

654. Porath D, Bezryadin A, De Vries S, Dekker C, *Nature*, 403: 635 (2000).

655. Porier G E, *Chem. Rev.,* 97: 1117 (1997).

656. Poteete A R. *FEMS Microbiol Lett*, 201: 9 (2001).

657. Powers M, Matsuura J, Brassel J, Manning MC, Shefter E, *Biopolymers.*, 33: 927–932 (1993).

658. Pries I and Langer R, *J. Immunol. Methods,* 28: 193–197 (1979).

659. Purcell E M, *Am.Journal of Physics*, 45(1): 3-11 (1977).

Q

660. Qu LH, Peng XG, *J.Am.Chem.Soc*, 124 (9): 2049 (2002).

661. Qiu L, Jing N, Jin Y*Int. J. Pharm.* 361: 56 (2008).

662. Qiu H, Dewan J C and Seeman N C,. *J. Mol. Biol.,* 267: 881–898 (1997).

663. Quintanar –Guerrero D, Allemann E, Fessi H, Doelker E., *Pharm. Res.*, 14: 119–127 (1997).

R

664. Radomsky M L, Whaley K J, Cone R A and Saltzman W M, *Biol. Reprod,* 47: 133–140 (1992).

665. Rannarda S. and Owen A, *Nano Today* 4: 382 (2009).

666. Rauser W E, *Annu. Rev. Biochem.* 59: 61 (1990).

667. Rauser W, *Plant Physiol.* 109: 1141 (1995).

668. Rauser W E, *Cell Biochem Biophys* 31: 19–48 (1999)

669. Ratner, B., Hoffman, A., Schoen, F. & Lemons, J. (eds) *Biomaterials Science: an Introduction to Materials inMedicine* (Academic, San Diego (1996).

670. Ratner M, *Nature,* 397: 480 (1999).

671. Rawat J, Rana S, Srivastava R and R. Misra D K, :10.1016/j.msec. 2006.05.021.

672. Rea PA, Vatamaniuk OK, Rigden D J, *Plant Physiol,* 136: 2463–2474 (2004).

673. Reed MA, Randall JN, Aggarwal RJ, Matyi RJ, Moore TM, Wetsel AE *Phys Rev Lett* 60 (6): 535–537(1988).

674. Regan J J, Ramirez B E, Winkler J R, Gray H B and Malmström B G, *J Bioenerg. Biomembr.*30: 35 (1998).

675. Rege K, Raravikar N T, Kim D-Y, Schadler L S, Ajayan P M and Dordick J S, *Nano Lett,.* 3: 829–832 (2003).

676. Reis CP, Ronald J. Neufeld, Ribeiro AJ, Francisco V, *Nanomedicine* 2 (1): 8–21 (2006).

677. Ren H., and Allison W., *Biochim.Biophys.Acta* 1458, 221–233 (2000).

678. Renger T and Marcus R A, *J. Phys. Chem. A,* 107: 8404 (2003).

679. Requicha A A G, "Nanorobots, NEMS and Nanoassembly", *Proceedings IEEE,* Vol. 91, no. 11, pp 1922–1933 (2003).

680. Rhim J W, *Food Sci. Biotech.*13: 528–535 (2004).

681. Ricci F, Lai R Y, Heeger A J, Plaxco K W and Sumner J J, *Langmuir,* 23: 6827 (2007).

682. Rickerby D G and Morrison M, *Science and Technology of Advanced Materials,*8: 19–24 (2007)

683. Riddin T L, Govender Y, Gericke M and Whiteley C G, *Enzyme and Microbial Technology,* 43(4-5), 362–368 (2008).

684. Rief M, Rock R S, Mehta A D, Mooseker M S, Cheney R E and Spudich J A,*Proc. Nat.Acad. Sci. USA,* 97: 9482–9486 (2000).

685. Rieux A, Virginie F, Garinot M, Schneider Y, Preat V., *J. Control. Release* 116: 1 (2006).

686. Riley JR, Reynolds DR, Smith AD, Edwards AS, Osborne JL, Williams IH and McCartney HA, *Nature,* 400 (6740): 1(1999).

687. Ritzoulis C, Scoutaris N, Papademetriou K, Stavroulias S and Panayiotou C, *Food Hydrocolloids,* 19: 575-581M (2005).

688. Roca B, Messer and Warshel A, *FEBS Lett.* 581: 2065 (2007).

689. Roitberg A E, Holden M J, Mayhew M P, Kurnikov I V, Beratan D N and Vilker V L, *J. Am. Chem. Soc.,* 120: 8927 (1998).

690. Romero M, Figueroa R and Madden C, *Med. Dev. Diag. Ind. Mag.* (2000), http://www.devicelink.com/mddi/archive/00/10/004.html

691. Romeyer F M, Jacobs F A and Brousseau R, *Appl. Environ. Microbiol.,* 56: 274 (1990).

692. Roney C, Kulkarni P, Arora V, Antich P, Bonte F. *J Control Release*, 108: 193 (2005).

693. Rorrer G L, Chang C, Liu S, Jeffrys C, Jiao J and Hedberg J A, *J.Nanosci.Nanotech.*5: 41 (2005).

694. Rosi NL, Mirkin CA., *Chem Rev*; 105: 1547–62 (2005).

695. Rossetti R and Brus L, *J Phys Chem*; 86: 4470–2 (1982).

696. Rothmund M. published online: 7 DEC 2005. DOI: 10.1002/bjs.1800810203

697. Roukes M, *Scientific American,* **285:** 48 (2001).

S

698. Sabbatini P, Aghajanian C, Dizon D, *J.Clin. Oncol.* 22: 4523 (2004).

699. Saito Y, Hamaguchi K, Hata K, Uchida K, Tasaka Y, Ikazaki F, Yumura M, Kasuya A and Nishina Y, *Nature*, 389 : 554 (1997).

700. Sakamoto T, Amitani I, Yokota E and Ando T, *Biochem. Biophys. Res. Commun.* 272: 586–590 (2000).

701. Salata O, *J Nanobiotech*; 2: 3 (2004).

702. Saleh N, Sarbu T, Sirk, K, Lowry G, Matyjaszewski V, Tilton K, *Nano Lett.*, 5: 2489–2494 (2005).

703. Salt D E, Smith R D and Raskin I, *Annu Rev Plant Physiol Plant Mol Biol,* 49: 643–668 (1998)

704. Saltzman W M and Langer R, *Biophys. J.,* 55: 163–171 (1989).

705. Saltzman W M, *Pharm. Res.,* 16: 1300–1308 (1999).

706. Sano K and Shiba K, *J. Am. Chem. Soc.* 125: 14234 (2003).

707. Sapra S and Sarma DD, *Phys.Rev.B,* 69 (12): 125304 (2004).

708. Satishkumar B C, Govindaraj A, Mofokeng J, Subbanna G N and Rao C N R, *J. Phys. B,* 29: 4925–4934 (1996).

709. Sathishkumar M, Sneha K, Won S W, Cho C W, Kim S and Yun Y S, *Colloids and Surfaces B: Biointerfaces,* 66 (2): *2008,* (2008).

710. Sau T K, Murphy C J, *J. Am. Chem. Soc,* 126*:* 8648 (2004).

711. Sauter N K, Bednarski M D, Wurzburg B A, Hanson J E, Whitesides G M, Skehel J J and Wiley D C, *Biochemistry*, 28: 8388 (1989).

712. Sauvage J-P, and Dietrich-Buchecker C, (eds.), Molecular Catenanes Rotaxanes and Knots, Wiley-VCH, Weinheim, (1999).

713. Sayes C M, Fortner J D, Guo W, Lyon D., Adina M and Boyd Kevin, *Nano Lett.,* 4 (10): 1881–1887 (2004).

714. Schabes-Retchkiman P S, Canizal G, Herrera-Becerra R, Zorrilla C, Liu H B, Ascencio J A, *Optical materials,* 29: 95 (2006).

715. Schena M, Shalon D, Davis R W, Brown P O, *Science* 270: 467 (1995).

716. Schiemann O, Turro N J and Barton J K, *J. Phys. Chem. B*, 104: 7214 (2000).

717. Schliwa, M. (ed.), *Molecular Motors*, Wiley-VCH, Weinheim, (2003).

718. Schmid A M M and Schulz D, *Protoplasma,* **100**: 267 (1979).

719. Schmidt J, Metselaar JM, Wauben MH, Toyka KV, Storm G & Gold R *Brain*, 126, 1895 (2003).

720. Schrick B, Blough J L, Jones A D, Mallouk T E, *Chem. Mater.* 16: 2187 (2004).

721. Schrick B, Blough J, Jones A and Mallouk T E, *Chem. Mater*, 14 (12): 5140–5147 (2002).

722. Schroeder I, Johnson E and De Vries S, *FEMS Microbiol. Rev.*27: 427 (2003).

723. Schrödinger E, Cambridge University Press (1992).

724. Schulze H, *J. Prakt.Chem.*, 25: 431 (1882).

725. Schweitzer B, Roberts S, Grimwade B, Shao W, Wang M, Fu Q, Shu Q,Laroche I, Zhou Z, Tchernev V T, Christiansen J, Velleca M and Kingsmore S F, *Nat. Biotechnol.*, 20: 359 (2002).

726. Sengupta S, Eavarone D, Capila I, Zhao G, Watson N, Kiziltepe T, Ram Sasisekharan, *Nature,* 436: 568, (2005).

727. Sherwood J K, Zeitlin L, Whaley K J, Cone R A and Saltzman W M, *Nature Biotechnol.,* 14: 468–471 (1996).

728. Seeman N C, *Nature*, 421: 427–431 (2003).

729. Seeman N C, *Nano Lett.* 1: 22–26 (2001).

730. Senda S, Tanemura M, Sakai Y, Ichikawa Y, Kita S, Otsuka T, Haga A and Okuyama F, *Rev. Sci. Ins.,* 75 (5): 1366 (2004).

731. Senior E., *J. Bioenerg. Biomembr.* 24: 479–484 (1992).

732. Sevrioukova I F, Li H, Zhang H, Peterson J A and Poulos T L, *Proc. Natl. Acad. Sci. USA*, 96: 1863 (1999).

733. Shaligram S N, Bule M, Bhambure R, Singhal R S, Singh S, Szakacs G, Pandey G A, *Process Biochemistry,* 44: 939 (2009).

734. Sharon Maheshwar, Mukhopadhyay K, Mukhopadhyay I and Krishna KM, *Carbon*, 33 (3) (1995).

735. Sharon Maheshwar, Sundarakoteeswaran N, Kichambre PD, Kumar M, Ando Y, Xinluo Z, *Diamond and Related Materials*, 8, 485–489 (1999).

736. Sharon Maheshwar and Pal B, *Bull. Electrochem.*, 12(3–4): 219 (1996).

737. Sharon M, Krishna K M, Soga T, Mukhopadhyay K and Umeno M, *Solar Energy Materials and Solar Cells*, 48 (1–4): 25 (1997a).

738. Sharon Maheshwar, Mukhopadhyay I, Mukhopadhyay K, *Sol. Energy Mater. Sol.Cells*, 45: 35 (1997b).

739. Sharon Maheshwar, Mukhopadhyay K, Mukhopadhyay I, Soga T and Umeno M, *Carbon*, 35:863–864 (1997c).

740. Sharon M, Jain S, Kichambre P D and Kumar M, *Mater. Chem. Phys.*, 8: 331 (1998).

741. Sharon Maheshwar, Kumar M, Kichambre P D, Avery N R and Black K J, *Mol. Cryst. Liq. Cryst.*, 340: 523 (2000).

742. Sharon Maheshwar, Pradhan D, Ando Y and Xinluo Z; *Current Appd Physics*, 2 : 445 (2002).

743. Sharon Madhuri and Gupta A., *Abstract 4th National Symposium & Conference on Solid State Chemistry and Allied Areas*, (ISCAS). Goa (2005a).

744. Sharon Maheshwar, Pal B and Kamat D.V. *Jour. Biomedical Nanotechnology* 1(3), 365–368, (2005b).

745. Sharon Maheshwar, Pal B and Kamat D V, *J. Biomed. Technol.* 1: 365 (2005c).

746. Sharon Maheshwar, Pradhan D, Zacharia R and Puri V, *J. Nanosci. Nanotechnol.*, 5(12): 2117 (2005d).

747. Sharon Maheshwar, Soga T, Afre R, Sathiyamoorthy D, Dasgupta K, Bhardwaj S, Madhuri Sharon and Jaybhaye S, *International Jour of Hydrogen Energy*, 32: 4238–4249, (2007a).

748. Sharon Maheshwar, Soga T, Afre R., Sathiyamoorthy D, Dasgupta K, Bhardwaj S, Madhuri Sharon and Jaybhay S, *Int. J. Hydrogen Energy*, 32: 4238 (2007c).

749. Sharon Maheshwar, Khairnar V, Jaybhaye S, Chi-Chang Hu, Afre R, Soga T and Madhuri Sharon,*Carbon Science*, 9(3), 188–194 (2008a).

750. Sharon Maheshwar, Rusop M, Soga T and Afre R, *Carbon Science*, 9(1): 17–22 (2008b).

751. Sharp L L, Zhou J and Blair D F, *Biochemistry*, 34: 9166–9171 (1995).

752. Nicewarner-Peña SR, Griffith F, Reiss B D,LHe L, Peña D J, Walton ID, Cromer R, Keating C D, Natan M J; *Science*,5 (294): 5540, 137–141 (2001).

753. Sherman W B and Seeman N C, *Nano Lett.* 4: 1203 (2004).

754. Shirai Y, Osgood A J, Zhao Y, Kelly K Fand Tour J M, *Nano Lett.*, 5: 2330–2334 (2005).

755. Shea L D, Smiley E, Bonadio J and Mooney D J, *Nature Biotechnol.*, 17: 551–554 (1999).

756. Shenoy DB and Amiji MM, *Int. J. Pharm.* 293 (1–2): 261 (2005).

757. Sherwood J K, Dause R B and Saltzman W M, *Bio/Technology,* 10: 1446–1449 (1992).

758. Sherwood J K, Zeitlin L, Whaley K J, Cone R A and Saltzman W M, *Nature Biotechnol.,* 14: 468–471 (1996).

759. Shvedova, AA, Castranova V, Kisin ER, Schwegler-Berry D,Murray AR, Gandelsman VZ Maynard A, Baron P, *J. Toxicol. Environ.Health* A., 66: 1909–1926 (2003).

760. Singh A, Chan J, Chern JJ, Choi KW, Genetics 171(1): 169 (2005).

761. Shiv Shankar S, Rai A, Ahmad A and Sastry M, *Applied Nanoscience* 1: 69 (2004).

762. Shiv Shankar S, Rai A, Ankamwar B, Sinha A, Ahmad A and Sastry M, *Nature Materials* 3: 482 (2004).

763. Shrestha S, Yeung C, Nunnerley C, Tsang S. A, *Phys*; 136: 191–8 (2006).

764. Shuichi N, Yusuke V. Morimoto, Nobunori Kami-ike, Tohru M, Keiichi N, *J. Mol. Biol.,* 393 (2): 300–307 (2009).

765. Silver S, *Plasmid,* 27: 1 (1992).

766. Simmel F C and Dittmer W U, *Small* 1, 284–299 (2005).

767. Simmel F C and Yurke B, *Phys. Rev. E,* 63: 041913 (2001).

768. Singaravelu G, Arockiamary J S, Kumar G V and Govindaraju K, *Collooids and Surfaces B: Biointerfaces,* 57: 97 (2007).

769. Singleton M R, Wentzell L M Liu Y, West S and Wigley D, *Proc Nat Acad Sci U S A,* 99: 13492–13497 (2002).

770. Sirdeshmukh R, Teker K, Panchapekesan B, *Proceedings of the 2004, International Conference on MEMS, NANO and Smart Systems* (2004).

771. Sitti M, "Micro- and Nano-Scale Robotics", *Proceedings of the 2004 American Control Conference*, Massachusetts, June 30 -July 2, 2004, pp 1–8 (2004).

772. Skehel J J, Bayley P M, Brown E B, Martin S R, Waterfield M D, White J M, Wilson I A and Wiley D C, *Proc Nat Acad Sci U S A*, 79: 968 (1982).

773. Sloan J, Hammer J, Zwiefta-Sibley M, Green M L H, *Chem Commun.,*: 347–8 (1998).

774. Somers C M, McCarry B E, Malek F, and Quinn J S, *Science*, 304: 1008–1010 (2004).

775. Smith AM, Dave S, Nie S, True L, Gao X., *Expert Rev Mol Diagn,*; 6: 231–44 (2006).

776. Smith H M and Turner A F, *Appl. Opt.*, 4: 147 (1965).

777. Somers C M, McCarry B E, Malek F and Quinn J S, *Science,* 304 (5673): 1008–1010 (2004).

778. Son S J, Bai X and Sang Bok Lee *J. Control. Release*, 114: 143–152 (2006).

779. Song C, Liu S, *International Joul of Biol. Macromolecules* 36: 116–119 (2005).

780. Soong R K, Bachand G D, Neves H P, Olkhovets A G, Craighead H G and Montemagno C D, *Science*, 290 (5496): 1555–1558 (2000).

781. Soppimath K S, Aminabhavi T M, Kulkarni A and Rudzinski, *J. Controlled Rel.*, 70: 1(2001).

782. G. Southam, and T. Beveridge, *Geochim Cosmochin.Acta* 60, 4367 (1996).

783. Sowa Y, Rowe A D, Leake M C, Yakushi T, Homma M, Ishijima A and Berry RM, *Nature*, 437: 916–919 (2005).

784. Speir J A, Munshi S G. Wang, T. S. Baker and J. E. Johnson *Structure*, 3: 63–78 (1995).

785. Srinivasan, C. and N. Somasundaram, *Curr. Sci.*, 85: 1431–1438 (2003).

786. Stasiak A Z, Larquet E, *Stasiak A*, Muller S, Engel A, Van Dyck E, West S C and E.H. Egelman.. *Curr Biol,*.10: 337 (2000).

787. Steiniger SC, Kreuter J, Khalanski AS, *Int J Cancer,* 109: 759 (2004).

788. Stemp E D A, Arkin M R and Barton J K, *J. Am. Chem. Soc.* 119: 2921 (1997).

789. Stevenson C L, *Curr Pharm Biotechnol.*, 1: 165–182 (2000).

790. Stillman M J, Shaw C F, and Suzuki K T, eds.VCH Publishers, New York, (1992).

791. Stillman T J, Upadhyay M, Norte V A, Sedelnikova S E, Carradus M and Tzokov S, *Mol Microbiol,* 57, 1101–1112 (2005).

792. Stroscio J A and Eigler D M, *Science*, 254(5036): 1319–1326 (1991).

793. Sugie H, Tanemura M, Filip V, Iwata K, Takahashi K and Okuyama F, *Appl. Phys. Lett.;* 78 (17) : 2578 (2001).

794. Sun L and Laurence C. C., *Dent Mater*, 24(1): 111–116 (2008).

795. Sun Y G and Xia T N, *Science*, 298: 2176 (2002).

796. Sunada K, Kikuchi Y, Hashimoto K and Fujishima A, *Environ. Sci. Technol.*, 32: 726–728 (1998).

797. Sumper M, *Science* **295**: 2430 (2002).

798. Spicer P, *Curr. Opin. Colloid Int. Sci.*, 10: 274–279 (2005).

799. Spicer P, *Chemical Engg. Res. and Design,* 83: 1283–1286 (2005).

800. Spudich JA, *Nature*, 348: 284 (1990).

801. Srinivasan C and Somasundaram N, *Current Science.*, 85: 10, 1431–1438 (2003).

802. Stock D, Leslie AGW and Walker JE, *Science,* **286**: 1700–1705 (1999).

803. Stowell M B H and Rees D C, *Adv. Prot. Chem.*, 46: 279 (1995).

804. Svoboda K, Schmidt C F, Schnapp B J and Block S M, *Nature,* 365: 721–727 (1993).

T

805. Tawada K, and Sekimoto K, *J. Theoret.Biol.* 150, 193–200 (1991).

806. Teichmann A, Jacobi U, Weigmann H J, Sterry W and Lademann J, *Skin Pharmacol. Physiol.*, 18: 75–80 (2005).

807. Thapa A, Webster TJ, Haberstroh KM, *J Biomed Mater Res*, 67A: 1374–83 (2003a).

808. Thapa A, Miller DC, Webster TJ, *Biomaterials*, 24: 2915–26 (2003b).

809. Timea K, Zoltan K and Imre K, *Langmuir*, 20: 1656–1661 (2004).

810. Togashi F, Yamaguchi S, Kihara M, Aizawa S-I and Macnab R M, *J. Bacteriol.*, 179: 2994–3003 (1997).

811. Tongpool R, Inphonlek S, Ninlapat J, Na Ubol P and Setwong K, *Technique,* 6: 141–148 (2007).

812. Torres-Martinez C L, Nguyen L, Kho R, Bae W, Bozhilov K, Klimov V and Mehra R K, *Nanotechnol.*, 10 (1999).

813. Tratnyek P G and Johnson R L, *Nano Today* 1(2): 44 –m 49 (2006).

814. Thomas T, Karluss T, Nakissa S, Nora S, Patricia A and Robert B, *Toxicological Sciences,* 91(1): 14–19 (2006).

815. Trevors J T, *J. Basic Microbiol.* 26: 499 (1986).

816. Tsukamoto R, Muraoka M, Seki M, Tabata H and Yamashita I, *Chem. Mater.,* 19: 2389–2391 (2007).

817. Tufenkji N and Elimelech M, *Environ. Sci. Technol.*, 38: 529 (2004).

818. Turick C E, Tisa L S and Caccavo F, *Appl. Environ. Microbiol.* 68: 2436 (2002).

819. Turkevich, P. C. Stevenson, J. Hillier, *Discuss. Faraday. Soc.*, 11: 55–75 (1951).

U

820. Uchino H. Tokunaga, M. Ando and H. Utsumi, *Toxicol In Vitro*, 16: 629–635 (2002).

821. Uemura S, Kawaguchi K, Yajima J, Edamatsu M, Toyoshima Y Y and Ishiwata S, *Proc. Natl. Acad. Sci. USA,* 99: 5977–5981 (2002).

822. Ugarte D, Stockli T, Bonard J M, Chatelian A and de Heer WA, *Appl. Phys. A*, 67: 101 (1998).

823. Ueno T, Oosawa K and Aizawa S, *J. Mol. Biol.,* 227: 672–77 (1992).

V

824. Vale R D, Reese T S and Sheetz M P, *Cell* 42: 39–50 (1985).

825. van de Poll W H, Vrieling E G and Gieskes W W, *J. Phycol.,* **35**: 1044 (1999).

826. Van den Heuvel M G L, Dekker C, *Science*, 317: 333–336 (2000).

827. Van Hest J C M and Tirrell D A, *Chem. Commun.*, 19: 1897–1904 (2001).

828. Van Holde K E, Johnson W C and Ho P S, Prentice-Hall, Inc. Simon & Schuster/ A Viacom company Upper Saddle River, New Jersey (1998).

829. Van Way S M, Hosking E R, Braun T F, Manson M D, *J. Mol. Biol.,* 297: 7–24 (2000).

830. Vatamaniuk O K, Mari S, Lu Y P and Rea P A, *Proc. Natl. Acad. Sci.,* 96: 7110 (1999).

831. Vighneswaran N, Ashtaputre N M, Varadrajan P V, Nachane R P, Paralikar K M and Balasubramanya R H, J. *Mater. Lett.*, 7: 42 (2006).

832. Vikesland P J and Rebodos R L, *Environ. Sci. Technol.,* 41: 5277–5283 (2007).

833. Villalobos-Hernàndez J R, Müller-Goymann C C, *Jour Pharmaceutics,* 322, 161–170 (2006).

834. Voet D and Voet J G, 2nd Ed. John Wiley & Sons, New York (1995).

835. Vohra A, Goswami D Y, Deshpande D A and Block S S, *J. Ind. Microbiol. Biotechnol*, 32: 364–370 (2005).

836. Vora J, Bapat N and Broujerdi M, *Drug Deliv. Ind. Pharm.*, 19: 759 (1993).

837. Voityuk A A, Michel-Beyerle M E and Rosch N, *Chem. Phys. Lett.,*342: 231 (2001).

838. Volk M, Aumeier G, Langenbacher T, Feick R, Ogrodnik A, Michel-Beyerle M E, *J. Phys. Chem.* B, 102: 735 (1998).

839. Vonck J., Krug von Nidda T., Meier T., Matthey U., Mills D.J, *J. Mol. Biol.* 321, 307–316 (2002).

840. Von Smoluchowski M, *Phys.Z.,*17: 557–585 (1916).

841. Vu TQ, Maddipati R, Blute TA, Nehilla BJ, Nusblat L, Desai TA, *Nano Lett* 5: 603–607 (2005).

W

842. Wan C, Fiebig T, Kelley S O, Treadway C, Barton J K and Zewail A H, *Proc. Natl. Acad. Sci. U. S. A.*, 96: 1014 (1999).

843. Wang W U, Chen C, Lin K H, Fang Y and Lieber C M, *Proc. Natl. Acad. Sci.USA,* 102: 3208 (2005).

844. Wang B, Feng W-Y, Wang T-C, Jia G, Wang M, Shi J-W, Zhang F, Zhao Y-L and Chai Z-F, *Toxicology Letters*, 161: 115–123 (2006).

845. Wang J, *Electroanalysis*, 17: 7-14 k 7–14.

846. Wang J, Kawde A N and Jan M R, *Biosens Bioelectron,* 20(5): 995–1000 (2004).

847. Wang X, Zhuang J, Peng Q, Li Y. *Nature*; 437: 121 (2005).

848. Wang Y, Hernandez R M, Bartlett D J, Bingham J M, Kline T R, Sen A and Mallouk T E, *Langmuir*, 22: 10451–10456 (2006).

849. Wang Y, He X, Wang K, Zhang X and Tan W, *Colloids and Surfaces B: Biointerfaces* (2009).

850. Wang YA, Li JJ, Chen HY, Peng XG. *J Am Chem Soc*;124: 2293–8 (2002).

851. Warshaw D M, Kennedy G G, Work S S, Krementsova E B, Beck S and Trybus K M, *Biophys. J.* 88:L30-L32 (2005).

852. Watanabe M, Mishima Y, Yamashita S.-Y, Park J and Tame RH, *Protein Sci.*, 17: 518–526 (2008).

853. Watson J H P, CresseyB A, Roberts A P, Ellwood D C, Charnock J M and Soper A K, *J. Magn. Magn. Mater.*, 214: 13 (2000).

854. Weber J. and Senior A. E., *Biochim. Biophys. Acta.* 1319: 19–58 (1997).

855. Weber J and Senior A E, *FEBSLetters,* 545: 63 (2003).

856. Webster TJ, Siegel RW, Bizios R., *Biomaterials,*20 (13): 1221–7(1999).

857. Webster TJ, Ergun C, Doremus RH, *Biomaterials*, 21(17): 1803–10 (2000).

858. Webster T J, Waid M C, McKenzie J L, Price R L and Ejiofor J U, *Nanotech*, 15 (1): 48–54 (2004).

859. Wehrle, F, Kaim G, Dimroth,*J. Mol. Biol.* 322, 369–381 (2002).

860. Weiss S., *Science;*283: 1676–83 (1999)

861. Wei-xian Zhang *Journal of Nanoparticle Research* 5: 323–332 (2003).

862. Welch M, Oosawa K, Aizawa S-I, Eisenbach M, *Proc Natl Acad Sci,USA,*90: 8787–8791 (1993).

863. Welte M A, *Current Biology*, 14: R525–R537 (2004).

864. Weng JF, Ren JC. *Curr Med Chem*;13: 897–909 (2006).

865. West JL,.Halas NJ, *Annu Rev Biomed Eng*, 5 : 285 (2003).

866. Wetherbee R, *Science,* 298: 547 (2002).

867. Whittum-Hudson J A, Prendergast R A, Saltzman W M, An L L and MacDonald A B,.*Nature Med.,* 2: 1116–1121 (1996).

868. Williams P, Keshavarz-Moore E and Dunhill P, *J. Biotech.,* 48: 259 (1999).

869. Wilson D S and. Szostak J W, *Annu. Rev. Biochem.* 68: 611 (1999).

870. Winfree E, Liu F, Wenzler L A and Seeman N C, *Nature,* 394: 539–544 (1998).

871. Winkelmann D A., Bourdieu L, Ott A, Kinose F and Libchaber A,*Biophys. J.* 68: 2444 (1995).

872. Winter P M,Caruthers S D, Kassner A, Harris T D, Chinen L K, Allen J S, Lacy E K, Zhang H Y, Robertson J D, Wickline S A, *Cancer Res.* 63: 5838 (2003).

873. Wissanu P and Hodak J, *Materials Physics and Chemistry,* 108: 45–54 (2008).

874. Wissing S A and Müller R H, Int. Jour of Pharmaceutics, 254: 65–68 (2003).

875. Wissing S A and Müller R H, *Int. J. Cosm. Sci.,* 23: 233–243 (2001).

876. Witte L Green, Mishra T K and Silver S, *Antimicrob Agents Chemother.,* 29(4): 663 (1986).

877. Wong M S, Chu W C, Sun D S, Huang H S and Chen J S, *Applied.Environ. Microbiol.,* 72: 6111–6116 (2006).

878. Wu X., Liu H., Liu J., Haley K.N., Treadway J.A., Larson J.P., Ge N., Peale F. and. Bruchez M.P, *Nat Biotechnol,* 21: 41–46 (2003).

879. Wyatt T L, Whaley K J, Cone R A and Saltzman W M, *J. Control. Release,* 50: 93–102 (1998).

Y

880. Yan H, Zhang X, Shen Z and Seeman N C, *Nature* 415: 62–65(2002).

881. Yang R, Chen G, Dresselhaus MS, *Phys. Rev. B* 72, 125418 (2005).

882. Yang GY, Gong C, Qin Z, Liu XH, Lorris Betz A. *Anal Chem*; 76: 4465 (2004).

883. Yamaguchi S, Aizawa S-I, Kihara M, Isomura M, Jones C J and Macnab R*M, J. Bacteriol.,* 166: 187–193 (1986).

884. Yamashita I, *Thin Solid Films,* 393: 12 (2001).

885. Yamashita I, Hayashi J and Hara M *Chem Lett.,* 33: 1158 –1159 (2004).

886. Yamashita I, Kirimura H, Okuda M, Nishio K, Sano K, Shiba K, Hayashi T, Hara M and Y. Mishima. *Small,* 2: 1148–1152 (2006).

887. Yao HT, Kuo C S, Huang C H, Li Y Y, Chou P W, Cheng C L and Wong M S, *Nanotechnology,* 17: 2490–2497 (2006).

888. Ye J-S, Wen Y, Zhang W D, Cui H-F, Gan L M, Xu G Q and Sheu F-S,*J. Electroanal. Chem.* 562, 241–246 (2004).

889. Yeu G Z, Qui Q, Gao B, Cheng Y, Zhang J, Shimoda H, Chang S, Lu J P and Zhou O, *Appld. Phys. Lett.*; 81 (2) : 355 (2002).

890. Yildiz A, Forkey J N, McKinney S A, Ha T, Goldman Y E and Selvin P R, *Science* 300 :2061–2065 (2003).

891. Yildiz A, Tomishige M, Vale R D and Selvin P R, *Science*, 303:676-678 (2004).

892. Yin P, Yan H., Daniell X, Turberfield A, Reif, *Angew. Chem., Int. Ed.*, 43: 4906 (2004).

893. Yonekura K, Maki S, Morgan D G, Derosier D J,Vonderviszt F, Imada K and Namba K, *Science*, 290: 2148–2152 (2000).

894. Yoshida M, Eiro Mi and Toru H, *Nature Reviews- Molecular Cell Biology*, 2: 673 (2001).

895. Yoshizawa K, Mishima Y, Park S Y, Heddle J G and Tame, J. R, *J Biochem, (Tokyo)*, 142: 707–713 (2007).

896. Yong P, Rowson N A, Farr J P G, Harris IR and Macaskie L E, *Biotechnology. Bioengg.*, 80: 369 (2002).

897. Yu, H., Y.J. Kaufman, M. Chin, G. Feingold, L.A. Remer, T.L. Anderson, Y. Balkanski, N.Bellouin, O. Boucher, S. Christopher, P. DeCola, R. Kahn, D. Koch, N. Loeb, M.S. Reddy,M. Schulz, T. Takemura, and M. Zhou, *Atmos. Chem. Phys.*, 6, 613–666 (2006).

898. Yurchenco P D, Birk D E and Mecham R P (eds) *Extracellular Matrix Assembly and Structure*, Academic, San Diego (1994).

899. Yurke B, Turberfield A J, Mills A P, Simmel Jr F C, and Neumann J L, *Nature*, 406: 605 (2000).

Z

900. Zenk M H, *Gene*, 179: 21 (1998).

901. Zhang Lin-Li, Guo Chang-Xin, Zhao Jun-Jing and Hu Jun-Tao, *Chinese Phys Lett*, 22 (5) (2005).

902. Zhang A P and Sun Y P, *World J. Gastroenterol.*, 10: 3191–3193 (2004).

903. Zhang W.-X., *J. Nanoparticle Res.*, 5: 323–332 (2003).

904. Zhang Y and Seeman N C, *J. Am. Chem. Soc.*, 114: 2656–2663 (1992).

905. Zhang Y and Seeman N C, *J. Am. Chem. Soc.*,116: 1661–1669 (1994).

906. Zhang L, Zhao G-C, Wei X-W and Yang Z-S,*Chem. Lett.* 33: 86–87 (2004).

907. Zheng G, Patolsky F, Cui Y, Wang W U and Lieber C M, *Nat. Biotechnol.*, 23: 1294 (2005).

908. Zhong X.H., Feng Y.Y., Knoll W., Han M.Y., *J.Am.Chem.Soc*, 125 (44), 13559 (2003).

909. Zhou J, Lloyd S A and Blair D F, *Proc. Natl Acad. Sci. USA*, 95: 6436–6441 (1998).

910. Zhu H and Snyder M, *Curr. Opin. Chem. Biol.,* 5: 40 (2001).

911. Zhu H and Snyder M,. Curr. Opin. Chem. Biol. 7, 55–63 (2003).

912. Zimmerman S B and Minton A P, *Annu. Rev. Biophys. Biomol. Struct.* 22: 27 (1993).

913. Zimmerman S B and Trach S O, *J. Mol. Biol.,* 222: 599 (1991).

914. Zusman L D and Beratan D N, *J. Chem. Phys.*, 110: 10468 (1999).

Index

M

Macrophage 226

Magnetic resonance imaging (MRI) 69, 71

Marcus theory 118, 119, 120, 121, 126

Microwave absorption 191, 192

Microbivores 225

N

Nanobomb 206

Nanocarriers 258

Nanocomposites 210

Nanocones 267

Nanocrystals, 238, 239, 240

Nanofibers 158

Nanoparticles 258

    Distribution and clearance 273

    Toxicity 275

Nanoparticle engineering 258

Nanoprobes 196

Nanorobots 196

Nanosensor 19, 53

Nanoshell 206

Nanostructures 255

Nanotechnology 255

Nanotubes 305

Nanotweezers 196

Nanovectors 319

Nanowires 12

Neurite extension 210

Neurodegenerative disease 258

Neuronal cells 210

Neurotransmitter 19

O

Oligonucleotide 113

Organelle 195

Osteoblast 337

Oxidation 10

Oxidative stress 87

P

Parkinson's disease 208

Particle size 173

Phagocytosis 261

Pharmaceutical 270

Pinocytosis 277

Phonon 14

Photobleaching 269

Photocatalysis 296

Photonic nanocrystals 239, 240

Photons 244

Photovoltaic solar cells 192

p:n junction 192

Polyethylene 261

Polymer 261

Polymeric Nanoparticles 261

    PLGA 261

    N-(2-hydroxypropyl)-methacrylamide copolymer (HPMA) 284

    Polystyrene-maleic anhydride copolymer 284

    Polyethylene glycol (PEG) 284

    Poly-L-glutamic acid (PGA) 284

Polymerization 324

Protein Cages 138

Protein ring 140

Protein tubes 141

❑❑❑